METHODS IN MOLECULAR BIOLOGY

Series Editor
John M. Walker
School of Life and Medical Sciences
University of Hertfordshire
Hatfield, Hertfordshire, UK

For further volumes:
http://www.springer.com/series/7651

For over 35 years, biological scientists have come to rely on the research protocols and methodologies in the critically acclaimed Methods in Molecular Biology series. The series was the first to introduce the step-by-step protocols approach that has become the standard in all biomedical protocol publishing. Each protocol is provided in readily-reproducible step-by-step fashion, opening with an introductory overview, a list of the materials and reagents needed to complete the experiment, and followed by a detailed procedure that is supported with a helpful notes section offering tips and tricks of the trade, as well as troubleshooting advice. These hallmark features were introduced by series editor Dr. John Walker and constitute the key ingredient in each and every volume of the Methods in Molecular Biology series. Tested and trusted, comprehensive and reliable, all protocols from the series are indexed in PubMed.

Manuscript Instructions https://drive.google.com/file/d/1K4Az5vpAFVzYadJ3jKx2-QSo7Dqa_SNA/view?usp=sharing

Bioinformatics for Mass Spectrometry

Edited by

Frédérique Lisacek

Proteome Informatics Group, SIB–Swiss Institute of Bioinformatics, Geneva, Geneva, Switzerland

Humana Press

Editor
Frédérique Lisacek
Proteome Informatics Group
SIB–Swiss Institute of Bioinformatics
Geneva, Geneva, Switzerland

ISSN 1064-3745 ISSN 1940-6029 (electronic)
Methods in Molecular Biology
ISBN 978-1-0716-5518-4 ISBN 978-1-0716-5519-1 (eBook)
https://doi.org/10.1007/978-1-0716-5519-1

This Humana imprint is published by the registered company Springer Science+Business Media, LLC, part of Springer Nature.
The registered company address is: 1 New York Plaza, New York, NY 10004, U.S.A.

If disposing of this product, please recycle the paper.

Preface

Over the past decades, mass spectrometry (MS) has become the analytical method of choice for identifying non-nucleic acid molecules, especially in large-scale studies relying on -omics technologies. Modern biology that encompasses systems biology promotes the most comprehensive screening of molecular actors, and MS is key to identification and quantification in proteomics, metabolomics, lipidomics, glycomics, and interactomics, to name only the most prevalent. But, comprehensiveness comes at the cost of increasing volumes of experimental data that no human mind can process. This spurred the development of numerous strategies, each relying on software tools designed to automate the analysis of mass spectra. When shotgun proteomics was put to the test in 2009 [1], "Mount Bioinformatics" was shown to be highest to climb, and the situation has not fundamentally changed as MS techniques are gradually refined and regularly set new challenges in automation.

The extensive range of technologies underlying MS generates a broad variety of experimental workflows, in stark contrast with nucleic acid sequencing. This diversity is subsequently reflected in a plethora of software solutions for managing, storing, processing, and interpreting MS data. Note that the resulting automated pipelines often entail instrument dependency. From a data perspective, the pressing need for standardization prompted several international initiatives. For example, 15 years ago, the ProteomeXchange project [2] set the basis of coordinated proteomics data submission and collection through communicating repositories in various geographical locations. Nowadays, each publication reporting proteomics-related results includes a unique MS dataset identifier (PXDxxxxxx accession number), and this is exemplified in the chapters of this book.

This volume highlights the usage of bioinformatics at stake in processing and interpreting MS data in various -omics. It is primarily devoted to didactic presentations of bioinformatics resources for MS, rather than to the detailed descriptions typically published in specialized issues (e.g., [3]). Nevertheless, it is important to acknowledge recent trends shaping the field. Most, yet not all, tools for MS-based -omics are released as open source, and most, yet not all, published data comply with the FAIR (Findable, Accessible, Interoperable, and Reusable) principles widely promoted in bioinformatics. These factors have important implications for the quality, sustainability, and long-term impact of resources as clearly reported in the case of proteomics over the past three decades [4].

This book is divided into four parts mirroring the purposes of bioinformatics usage in MS-based studies and emerging trends requiring further customization or innovation. The focus of the first part is MS raw data management, especially related to handling data from diverse instrument sources. This is illustrated in metabolomics in Chapter 1 that demonstrates the relevance of custom-built spectral libraries, and in proteomics in Chapter 2 that emphasizes the added value of preprocessing and metadata standard compliance to enable data reanalysis. Standard use and integration are also key to Chapter 3, which describes a robust package to store, manage, and visualize large-scale metabolomics or proteomics data.

The second part of this book presents a range of software tools for the analysis and interpretation of MS data, reflecting ongoing technological advances (e.g., improved mass accuracy), as well as the evolution of experimental strategies. A major development over the past decade has been the diversification of data acquisition modes in tandem MS—embedded stages of mass spectrometry. Recall that following ionization, the first step in a mass spectrometer, produced ions are fragmented and analyzed in successive stages. The key distinction between acquisition strategies lies in how precursor ions are selected. In data-dependent acquisition (DDA) mode, ions are chosen based on their abundance above a defined threshold. In contrast, the data-independent acquisition (DIA) mode proceeds systematically and applies no selection. These techniques are widely used in the analysis of complex proteomes. However, DIA generates a substantially larger number of spectra per sample, elevating data management challenges and usually necessitating preprocessing stages, as explained in Chapter 4, which details sophisticated protocols for filtering, normalizing, and correcting MS data to ensure reliable results. Maximizing the quality of identification and raising confidence in spectral matches is a strong focus in Chapter 5 that introduces a package enabling the exploration of identification results (irrespective of the identification software), as well as in Chapter 6 that refines identification with a strategy of leveraging peptide uniqueness.

Another central objective of proteomics is the assessment of protein differential expression, and in that respect, handling quantification has challenged bioinformaticians for some time. Again, a wealth of experimental approaches has been proposed to capture abundance. Methods are commonly partitioned into *labeled* (protein/peptide labeling to distinguish conditions) and *label-free*—the latter being the most frequently implemented. Chapter 7 describes a tool for the comparative and quantitative analysis of proteins applicable to DDA label-free data. Chapter 8 puts forward a workflow for the analysis of DDA-labeled phosphoproteins to monitor the subtle changes underlying cell signaling. This theme is also under scrutiny in Chapter 9, which demonstrates the usage of a tool for interpreting DDA-labeled phosphoproteomics data with the goal of estimating kinase activity in cancer-related pathways. Another partition of quantitative proteomics splits methods into *untargeted*, i.e., discovery-oriented (DDA and DIA mentioned earlier) and *targeted*, designed to accurately monitor the expression of a selection of proteins. The latter is exemplified in Chapter 10 that introduces a dedicated bioinformatics pipeline applicable to protein production in bioengineering.

The chapters in the third part cover a nonexhaustive collection of topics that are not yet mainstream, but likely to gain importance in the near future. These use cases all face the challenge of automatically identifying molecules produced by a non-template-driven process, such as peptidic natural products (most are non-ribosomal and synthesized by non-ribosomal synthases) in Chapter 11 and glycopeptides (glycans are synthesized by glycosyltransferases) in Chapter 12. As a result, the search space is boundless, which significantly raises the level of uncertainty of identification.

To conclude, the fourth part unsurprisingly illustrates the use of artificial intelligence (AI) methods in support of MS data analysis in proteomics and metabolomics. In recent years, the sheer volume of accumulated spectra has become a rich source for data mining, pattern extraction, and therefore feeds AI-based prediction methods, as reviewed in Chapter 13 for both proteomics and metabolomics. Chapter 14 demonstrates the power of this approach in metabolomics through protocols relying on spectral models for automatic annotation and small molecule discovery. Chapter 15 proposes a comparable scheme in proteomics focused on predicting peptide properties from spectra to refine and speed up identification. Finally, Chapter 16 highlights an immunopeptidomics application that integrates an AI-generated mapping between sequences of peaks in tandem mass spectra and sequences of amino acids (also known as de novo sequencing) along with transcriptomics data.

This volume is intended for readers who wish to understand the bioinformatics challenges of processing MS data in a broad range of application fields. This may attract graduate students or researchers tackling molecule identification in complex samples, as well as bioinformaticians motivated to integrate new or established tools in pipelines. The diversity of themes covered in this book also makes it relevant to an extended readership in disciplines encompassing biotechnological, pharmaceutical, biological, and medical sciences, as well as computational and engineering sciences.

This preface is written in the shadow of the recent loss of Amos Bairoch[1] and Peter Roepstorff,[2] two visionaries who laid the ground for bioinformatics in proteomics. Their passing leaves a tangible void in the field, yet their impact on protein identification from MS data remains indelible. Many of us who carry forward their scientific legacy will deeply miss their forward thinking and generosity.

My warmest thanks go to all authors and coauthors for their dedication and the excellence of their contribution to this volume.

Geneva, Switzerland *Frédérique Lisacek*

References

1. Aebersold R (2009) A stress test for mass spectrometry–based proteomics. Nat Methods 6:411–412. https://doi.org/10.1038/nmeth.f.255
2. Deutsch EW, Bandeira N, Perez-Riverol Y et al (2026) The ProteomeXchange consortium in 2026: making proteomics data FAIR. Nucleic Acids Res 54:D459–D469. https://doi.org/10.1093/nar/gkaf1146
3. Hoopmann MR, Schwämmle V, Palmblad M (2023) 2023 Special issue on software tools and resources: accelerating research with new and evolving open source software. J Proteome Res 22:285–286. https://doi.org/10.1021/acs.jproteome.3c00033
4. Perez-Riverol Y, Bittremieux W, Noble WS et al (2025) Open-source and FAIR research software for proteomics. J Proteome Res 24:2222–2234. https://doi.org/10.1021/acs.jproteome.4c01079

[1] Created and expanded the Swiss-Prot database (that became UniProt, the reference protein sequence resource) and cofounded the SIB Swiss Institute of Bioinformatics.

[2] Created and shaped the Protein Research group in Odense (Denmark) into an internationally praised proteomics hub.

Contents

Contributors

PETER R. BAKER • *Department of Pharmaceutical Chemistry, University of California San Francisco, San Francisco, CA, USA*
KRISTOFFER BASSE • *Department of Analytical Biochemistry, Faculty of Science and Engineering, University of Groningen, Groningen, Netherlands*
AIVETT BILBAO • *Pacific Northwest National Laboratory, Richland, WA, USA; US Department of Energy Agile BioFoundry, Emeryville, CA, USA*
ANTON BUSHUIEV • *Czech Institute of Informatics, Robotics and Cybernetics, Czech Technical University, Prague, Czech Republic*
ROMAN BUSHUIEV • *Institute of Organic Chemistry and Biochemistry of the Czech Academy of Sciences, Prague, Czech Republic; Czech Institute of Informatics, Robotics and Cybernetics, Czech Technical University, Prague, Czech Republic*
ROBERT J. CHALKLEY • *Department of Pharmaceutical Chemistry, University of California San Francisco, San Francisco, CA, USA*
SEUNGHYUK CHOI • *School of Software, College of Computer Science, Kookmin University, Seoul, Republic of Korea*
TINE CLAEYS • *VIB-UGent Center for Medical Biotechnology, VIB, Ghent, Belgium; Department of Biomolecular Medicine, Ghent University, Ghent, Belgium*
TITO DAMIANI • *Institute of Organic Chemistry and Biochemistry of the Czech Academy of Sciences, Prague, Czech Republic*
GUILLAUME DEFLANDRE • *de Duve Institute, UCLouvain, Brussels, Belgium*
ANTJE DITTMANN • *FGCZ—Functional Genomics Center Zurich, Zurich, Switzerland*
YUQIAN GAO • *Pacific Northwest National Laboratory, Richland, WA, USA; US Department of Energy Agile BioFoundry, Emeryville, CA, USA*
LAURENT GATTO • *de Duve Institute, UCLouvain, Brussels, Belgium*
SEBASTIEN GIBB • *University Medicine Greifswald, Greifswald, Germany*
MARILYN DE GRAEVE • *Eurac Research, Bolzano, Italy*
JONAS GROSSMANN • *FGCZ—Functional Genomics Center Zurich, Zurich, Switzerland; SIB—Swiss Institute of Bioinformatics, Zurich, Switzerland*
TÉO HEBRA • *Institute of Organic Chemistry and Biochemistry of the Czech Academy of Sciences, Prague, Czech Republic*
GÉRARD HOPFGARTNER • *University of Geneva Life Sciences Mass Spectrometry, Geneva, Switzerland*
PETER HORVATOVICH • *Department of Analytical Biochemistry, Faculty of Science and Engineering, University of Groningen, Groningen, Netherlands*
FILIP JOZEFOV • *Institute of Organic Chemistry and Biochemistry of the Czech Academy of Sciences, Prague, Czech Republic*
EVANGELOS KONTOPODIS • *Proteomics Research Unit, Biomedical Research Foundation of the Academy of Athens (BRFAA), Athens, Greece*
ANNE-CLAIRE KRÖGER • *Institut Curie, PSL University, Paris, France; Institut National de la Santé et de la Recherche Médicale (INSERM), U1331, Paris, France; Mines Paris, PSL University, CBIO—Centre for Computational Biology, Paris, France*
LAURA KUNZ • *FGCZ—Functional Genomics Center Zurich, Zurich, Switzerland*

LEV LEVITSKY • *Department of Biochemistry and Molecular Biology, University of Southern Denmark, Odense M, Denmark*
STÉPHANE LIVA • *Institut Curie, PSL University, Paris, France; Institut National de la Santé et de la Recherche Médicale (INSERM), U1331, Paris, France; Mines Paris, PSL University, CBIO—Centre for Computational Biology, Paris, France*
PHILIPPINE LOUAIL • *Eurac Research, Bolzano, Italy*
LOREDANA MARTIGNETTI • *Institut Curie, PSL University, Paris, France; Institut National de la Santé et de la Recherche Médicale (INSERM), U1331, Paris, France; Mines Paris, PSL University, CBIO—Centre for Computational Biology, Paris, France*
CHRISTOPH B. MESSNER • *Precision Proteomics Center, Swiss Institute of Allergy and Asthma Research (SIAF), University of Zurich, Davos, Switzerland; Swiss Institute of Bioinformatics (SIB), Lausanne, Switzerland; The LOOP Zurich, Zurich, Switzerland*
EUNOK PAEK • *Department of Computer Science, Hanyang University, Seoul, Republic of Korea; Department of Artificial Intelligence, Hanyang University, Seoul, Republic of Korea*
CHRISTIAN PANSE • *FGCZ—Functional Genomics Center Zurich, Zurich, Switzerland; SIB—Swiss Institute of Bioinformatics, Zurich, Switzerland*
VASILEIOS PIERROS • *Proteomics Research Unit, Biomedical Research Foundation of the Academy of Athens (BRFAA), Athens, Greece; Section of Cell Biology and Biophysics, Department of Biology, School of Science, National and Kapodistrian University of Athens (NKUA), Athens, Greece*
TOMÁŠ PLUSKAL • *Institute of Organic Chemistry and Biochemistry of the Czech Academy of Sciences, Prague, Czech Republic*
PATRICK POULLET • *Institut Curie, PSL University, Paris, France; Institut National de la Santé et de la Recherche Médicale (INSERM), U1331, Paris, France; Mines Paris, PSL University, CBIO—Centre for Computational Biology, Paris, France*
JOHANNES RAINER • *Eurac Research, Bolzano, Italy*
MARKUS RALSER • *Department of Biochemistry, Charité Universitätsmedizin Berlin, Berlin, Germany; The Centre for Human Genetics, Nuffield Department of Medicine, University of Oxford, Oxford, UK; Max Planck Institute for Molecular Genetics, Berlin, Germany*
EMMA RICART ALTIMIRAS • *Broad Institute of MIT and Harvard, Cambridge, MA, USA; Krantz Family Center for Cancer Research, Massachusetts General Hospital Cancer Center, Boston, MA, USA; Department of Biomedical Informatics, Harvard Medical School, Boston, MA, USA*
MICHAEL RICHARD • *Institut Curie, PSL University, Paris, France; Institut National de la Santé et de la Recherche Médicale (INSERM), U1331, Paris, France; Mines Paris, PSL University, CBIO—Centre for Computational Biology, Paris, France*
QUENTIN ROUCHON • *University of Geneva Life Sciences Mass Spectrometry, Geneva, Switzerland; Swiss Institute of Bioinformatics Proteome Informatics, Geneva, Switzerland*
FAHAD SAEED • *Knight Foundation School of Computing and Information Sciences, Florida International University (FIU), Miami, FL, USA*
BORIS SCHNIDER • *Swiss Institute of Bioinformatics Proteome Informatics, Geneva, Switzerland*
VEIT SCHWÄMMLE • *Department of Biochemistry and Molecular Biology, University of Southern Denmark, Odense M, Denmark*
BILAL SHABBIR • *Knight Foundation School of Computing and Information Sciences, Florida International University (FIU), Miami, FL, USA*

Josef Šivic • *Czech Institute of Informatics, Robotics and Cybernetics, Czech Technical University, Prague, Czech Republic*
Dimitrios J. Stravopodis • *Section of Cell Biology and Biophysics, Department of Biology, School of Science, National and Kapodistrian University of Athens (NKUA), Athens, Greece*
Frank Suits • *IBM Research, Melbourne, VIC, Australia*
Xindong Sun • *Precision Proteomics Center, Swiss Institute of Allergy and Asthma Research (SIAF), University of Zurich, Davos, Switzerland; Swiss Institute of Bioinformatics (SIB), Lausanne, Switzerland*
Beata Szeitz • *Division of Oncology, Department of Internal Medicine and Oncology, Semmelweis University, Budapest, Hungary; Department of Biochemistry & Molecular Pharmacology, New York University Grossman School of Medicine, New York, NY, USA; Institute for Systems Genetics, New York University Grossman School of Medicine, New York, NY, USA*
Usman Tariq • *Knight Foundation School of Computing and Information Sciences, Florida International University (FIU), Miami, FL, USA*
George Th, Tsangaris • *Proteomics Research Unit, Biomedical Research Foundation of the Academy of Athens (BRFAA), Athens, Greece*
Robert Winkler • *Advanced Genomics Unit (UGA), Center for Research and Advanced Studies (Cinvestav) Irapuato, Irapuato, Gto., Mexico*
Witold Wolski • *FGCZ—Functional Genomics Center Zurich, Zurich, Switzerland; SIB—Swiss Institute of Bioinformatics, Zurich, Switzerland*
Shanji Zhang • *Department of Computer Science, Hanyang University, Seoul, Republic of Korea*

Part I

Mass Spectrometry Data Management

Chapter 1

MSDMS: An Open-Source Web Application to Design and Manage Custom Tandem Mass Spectrometry Libraries

Boris Schnider, Quentin Rouchon and Gérard Hopfgartner

Abstract

Mass Spectrometry Data Management System (MSDMS) is a comprehensive software solution for managing small molecule mass spectrometry (MS or MS/MS) libraries in MassBank format. Starting from large public libraries, MSDMS enables the creation of one or more custom libraries without having to run reference compounds or manage unknowns. Custom libraries improve the quality and speed of identification by LC–MS/MS screening of exogenous and endogenous compounds in complex biological samples. With a container-based architecture and Docker Compose configuration, MSDMS can be deployed as a website or as a local software application. Its use is explained in a stepwise manner and illustrated with human urine sample screening.

Keywords Mass spectrometry, MS/MS, Libraries, MassBank, Metabolites, Drug of abuse

1 Introduction

Separation sciences, such as gas chromatography (GC), liquid chromatography (LC), or supercritical fluid chromatography (SFC) coupled to mass spectrometry (MS), have become the methods of choice for the qualitative and quantitative analysis of metabolites, lipids, pharmaceuticals, contaminants, or pesticides in complex biological samples or water. MS requires that molecules are ionized and electron ionization is the most widely used technique with GC/MS, while atmospheric pressure ionization is the technique of choice with LC/MS and SFC/MS. Electron ionization (EI) is a hard ionization technique where the radical precursor undergoes extensive but reproducible fragmentation at 70 eV and standardized libraries can be generated on most instruments [1]. Atmospheric pressure ionization (API) includes ElectroSpray Ionization (ESI), Atmospheric Pressure Chemical Ionization (APCI) and Atmospheric Pressure Photoionization

Boris Schnider and Quentin Rouchon both authors contributed equally to this work.

Frédérique Lisacek (ed.), *Bioinformatics for Mass Spectrometry*, Methods in Molecular Biology, vol. 3071,
https://doi.org/10.1007/978-1-0716-5519-1_1,

Ionization (APPI), which are soft ionization techniques generating mostly protonated or deprotonated ions as well as adduct [2–4]. For precursor ions formed by API, structural information can be obtained by dissociation of positively or negatively charged precursor ions by interactions with neutral gases (CID), electrons (EAD), or photons (UVPD) [5, 6]. The technique is generally referred to as Tandem Mass Spectrometry and produces so-called MS/MS spectra [7]. As each MS/MS spectrum represents a specific fingerprint of a molecule, the generation of libraries is of large interest for compound characterization in untargeted screening and for quantitative analysis. Several analytical fields rely on mass spectrometry for the general screening of low molecular weight compounds, including quality control, forensic, food, environmental, and pharmaceutical sciences. Over several decades public and commercial MS spectral libraries have been compiled containing hundreds of thousands of analytes and millions of spectra [8]. Those libraries have different data formats and can be used with commercial or open-source software. The ratio between the number of compounds and the number of spectra is strongly dependent on the ionization technique. For example, in the NIST 2023 Mass Spectra Library [9], the EI library contains 347,100 compounds and 394,054 spectra (ratio 0.88), while the MS/MS library with collision-induced dissociation contains 51,501 compounds and 2.4 M spectra (ratio 0.02). The ratio difference between EI and API libraries comes from the fact that EI ionization has been normalized at 70 eV and that most commercial instruments generate the same spectrum for a given molecule. The situation is completely different with tandem mass spectrometry and API, where the fragments present depend on the type and geometry of the collision cell, the collision gas (e.g., He, N_2, Ar), the collision energy, and the nature of the precursor ion (e.g., $[M+H]^+$, $[M+Na]^+$) [10]. Besides commercial solutions, open-source databases and tools are available for library search, such as the Global Natural Product Social Molecular Networking (GNPS) [11] (https://gnps.ucsd.edu/), MassBank of North America (MoNA) (https://mona.fiehnlab.ucdavis.edu) and MassBank Europe (https://massbank.eu/MassBank/) [12, 13], and spectrum and metafile databases can be downloaded in different file formats such as JSON, MSP, SDF, or MassBank (MB) and used directly for general screening or integrated into local platforms. The size and quality (spectra curation, metadata, and spectra redundancy) of these databases vary significantly. In addition, the important number of spectra in libraries acquired on different MS systems leads to an increasing number of redundancies and false positives, diminishing confidence in the results. Most of the database files are in text format. While many open-source tools are available to create and curate databases, mostly commercial solutions allow MS spectrum database management. For supporting their projects, many users favor the use of smaller custom libraries with specific chromatography and MS conditions for metabolomics or exposomics [14–16].

We developed MSDMS as an open-source data-management package that runs locally and allows users to generate custom MS or MS/MS spectrum libraries from public or private sources using metadata and spectral information in the MB format [12]. MSDMS is a lightweight, open-source, and user-friendly solution for designing and managing custom mass spectrometry (MS) libraries. Its software architecture is organized into three layers, as illustrated from left to right in Fig. 1: (1) a user interface (UI) for browsing, filtering, and visualizing MB libraries; (2) an application layer composed of an Application Programming Interface (API) coupled with a reverse proxy, which coordinates communication between the UI and the underlying data resources; and (3) a data layer that combines a local file volume for storing MB records with a PostgreSQL database optimized for fast and structured spectrum queries. Each containerized component (shown in blue in Fig. 1) is deployed using Docker, and the full system is

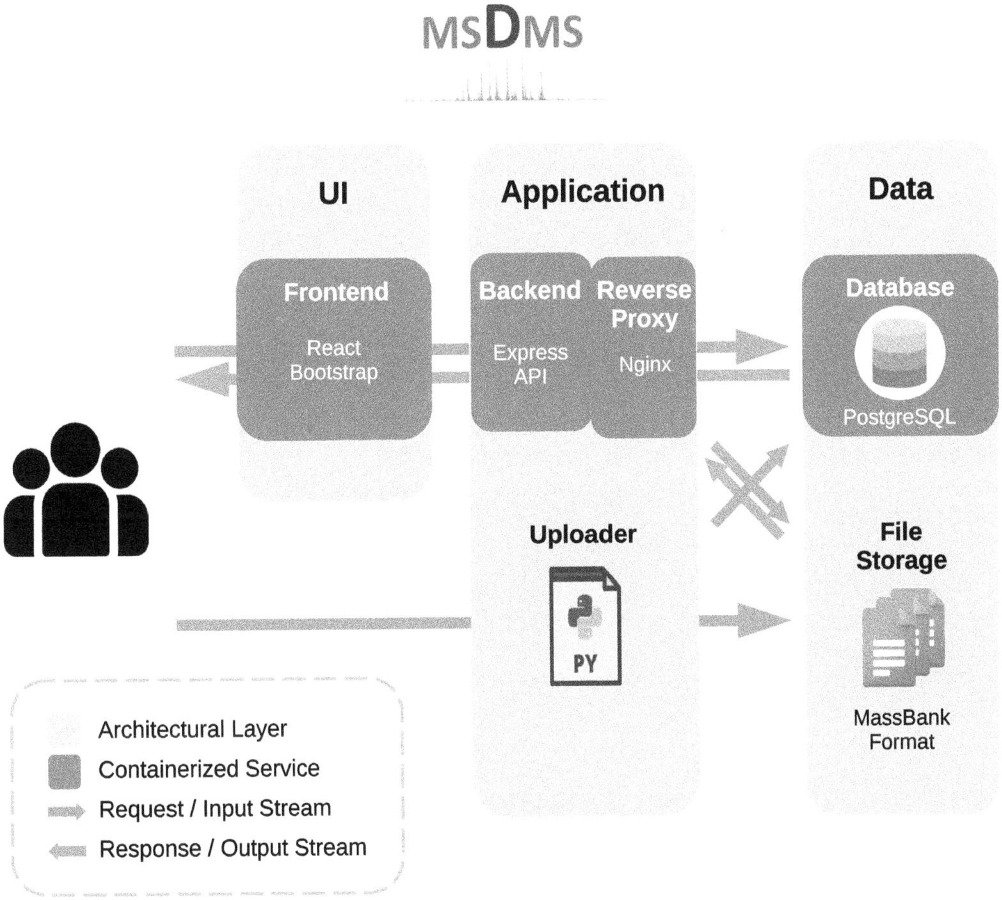

Fig. 1 MSDMS architecture. The system is organized into a user interface; an application layer with an API, a reverse proxy, and a Python uploader, and a data layer that contains a PostgreSQL database and file storage. All components are deployed as Docker containers

configured and parameterized via a docker-compose file. This modular design enables MSDMS to be deployed either as a public web service for sharing data or as a local tool for managing libraries that contain both public and private datasets. The frontend and backend are provided as Docker images to ensure stable and reproducible deployments across systems and prevent code alterations while still enabling local configuration and full access to the stored files and database.

The full details of the tool are presented in this chapter and illustrated with examples of urine sample analysis.

2 Materials

MSDMS can be installed on a local computer server (e.g., Debian 12). An alternative is to set it up on a virtual machine such as VMware. VMware offers the flexibility to easily clone the VM machine to set up both exploratory and production MSDMS VMs.

The technologies used for MSDMS are: (1) Frontend: React.js (JavaScriptJavaScript library) used to build and structure the user interface; (2) Backend: Express.js (JavaScript framework) used to implement the API and application logic that interacts with the database; (3) Database: PostgreSQL (relational database) used to store metabolite and spectrum data. MB files are stored in a local volume; (4) Data Uploader: Python scripts used to programmatically import metabolite and spectrum data; (5) Web server/reverse proxy: Nginx used to serve the frontend and route API requests to the backend; (6) Containerization: Docker used to containerize the application's services and ensure reproducible deployments.

Table 1
Mandatory and optional fields of the MB file format for MSDMS

Mandatory fields	Optional fields
ACCESSION:	DATE:
CH$NAME:	CH$LINK: CAS
CH$FORMULA:	CH$LINK: CHEBI
CH$EXACT_MASS:	CH$LINK: CHEMSPIDER
CH$SMILES: SMILES	CH$LINK: PUBCHEM ID.
CH$IUPAC: InChI=	AC$INSTRUMENT_TYPE:
CH$LINK: INCHIKEY	AC$MASS_SPECTROMETRY: ION_MODE
AC$MASS_SPECTROMETRY: MS_TYPE	AC$MASS_SPECTROMETRY: FRAGMENTATION_MODE
AC$MASS_SPECTROMETRY: ION_MODE	AC$MASS_SPECTROMETRY: COLLISION_ENERGY
PK$PEAK:	AC$CHROMATOGRAPHY: RETENTION_TIME
	MS$FOCUSED_ION: CCS
	MS$FOCUSED_ION: DMS
	PK$SPLASH: Splash ID

The software uses MB format files for import and requires the input data to include all the mandatory fields (*see* **Note 1**) presented in Table 1. For successful import, a record needs to have a unique accession. The database records selected to build a custom library can be exported in MB format with all fields. Various summary files can be generated to monitor the data import and export, as well as the library import and export processes. In addition to the common MB fields, drift time ion mobility and differential ion mobility fields are implemented.

Data can easily be imported using the "Upload New Data" page by drag and drop or by selecting the files with the browse option. As a test for importing public databases, we used the latest MassBank Europe release, which contains 58 different data folders [17]. Data files can also be uploaded using a Python script (*see* **Note 2**).

The following molecular notation fields are used in MSDMS, SMILES (Simplified Molecular Input Line Entry System) [18], InChI (International Chemical Identifier) [19], and InChIKey (Hashed, fixed-length representation of InChI) [20]. In MSDMS the molecules are visualized using SmilesDrawer [21].

3 Methods

3.1 Data Import

For data import and export in MSDMS, we rely exclusively on the MB format. The *Upload New Data* page supports drag-and-drop or file browsing to upload: (1) one or multiple records contained in a single MB file, (2) multiple MB files each containing one or multiple records, or (3) a folder containing one or multiple MB files. Custom source names can be defined based on the folder structure of the uploaded MB files by enabling the subfolder option (*see* **Note 3**).

During the upload process, MSDMS checks if all mandatory fields (Table 1) are present and provides a summary *that* can be downloaded (*uploaded_entries.txt*) before the data are written. The following status flags are defined: NEW, STRUCTURE, DUPLICATE, and FORMAT.

While most of the 58 folders of MassBank/MassBank-data: Release 2025.10 [17] can be uploaded without any problem, some sources show the flag FORMAT, and the records are not imported. Table 2 shows an excerpt of the import summary file (*uploaded_entries.txt*) for the ACES_SU [17] source, which contains 1779 spectra/records. About 1490 spectra are flagged as NEW, 17 as STRUCTURE, and 271 as FORMAT. Here, 271 spectra with format issues are not imported, because the mandatory field CH$LINK: INCHIKEY is empty. STRUCTURE is a warning flag that reports the presence of spectra of the same metabolite documented under different metabolite names. Records with this flag will still be imported. DUPLICATE, which is not present here, means that a spectrum with the same accession is already in the database and, therefore, the

Table 2
Extract of the import summary file for source ACES_SU

Accession	Name	Formula	InChI	InChIKey	Smiles	Status
AS000001	2,4-Dinitrophenol	C6H4N2O5	InChI=1S/ C6H4N2O5/ c9-6-2-1-4(7		Oc1ccc(cc1N(= O)..	FORMAT
AS000002	2-Nitrophenol	C6H5NO3	InChI=1S/ C6H5NO3/ c8-6-4-2-1-3...		Oc1ccccc1[N+]([..	FORMAT
AS000272	Theobromine	C7H8N4O2	InChI=1S/ C7H8N4O2/ c1-10-3-8-5...	YAPQBXQYLJRXSA...	CN1C=NC2= C1C...	NEW
AS000273	Acetopromazine	C19H22N2OS	InChI=1S/ C19H22N2OS/ c1-14(22)...	NOSIYYJFMPDDSA...	CC(=O)C1= CC2=...	NEW
AS000274	Benzethonium	[C27H42NO2]+	InChI=1S/ C27H42NO2/ c1-26(2,3)...	SIYLLGKDQZGJHK..	CC(C)(C)CC©...	NEW
AS000289	(R)-(+)-Atenolol	C14H22N2O3	InChI=1S/ C14H22N2O3/ c1-10(2)16..	METKIMKYRPQLGS...	CC(C) NCC(COC1...	STRUCTURE
AS000558	(R)-Propranolol	C16H21NO2	InChI=1S/ C16H21NO2/ c1-12(2)17..	AQHHHDLHHXJYJD	CC(C) NCC(COC1=...	STRUCTURE

corresponding record will not be imported. Some public or in-house datasets may have missing information. In that case, it is necessary to add N/A or any other comment to the mandatory fields, and the record will be imported. This may also be the case when creating a library with completely unknown spectra.

3.2 Library Search and Library Builder

The homepage of MSDMS is presented in Fig. 2 with a menu to access five different pages. Once the data are imported, they are stored as individual spectra, which can be searched using various filters. The main filters are metabolite name and SMILES representation, both allowing the use of regex code (regular expression pattern matching function) to define search patterns. The search is organized in positive and negative modes. The number of metabolites and spectra obtained from a search is also shown. The metabolite count is based on the InChIKey field. The Query Mass selection is defined as the exact mass, and any adduct can be applied. The chromatographic retention time can also be applied as a search criterion. Upon clicking the "More filters" button, additional filter options are displayed, such as MS/MS fragments (Peaks *m/z*, drift time ion mobility (CCS), or differential mobility spectrometry (SV, CoV, modifier). A summary of the search results can be exported in a tab-separated text file using the ↓ button.

The Library Builder page (Fig. 3) is used to select spectra to generate a custom library. The list can be reviewed, edited, and emptied from there and from the *Export Cart* and Manage Cart

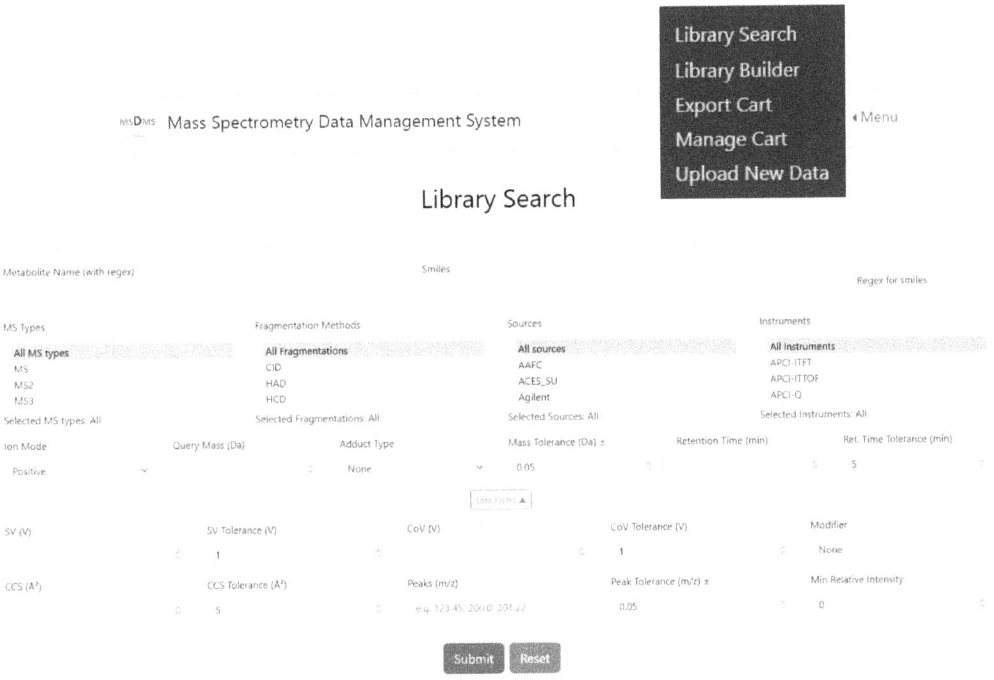

Fig. 2 Home page and menu of MSDMS

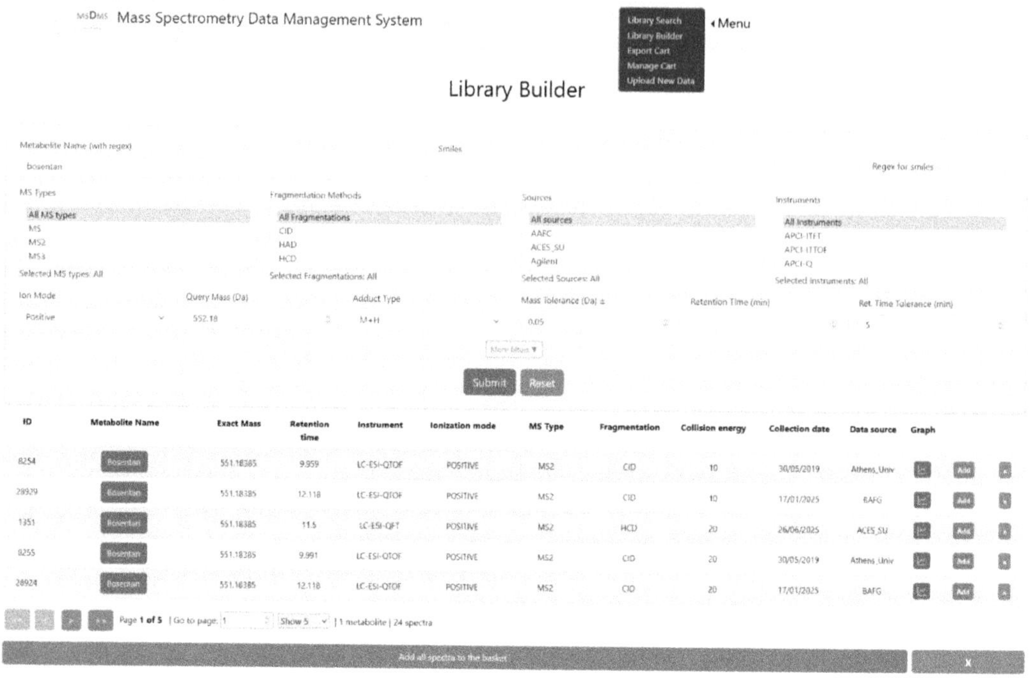

Fig. 3 Library builder page of MSDMS. Searching for positive mode spectra of bosentan in 58 different sources containing 19,998 metabolites and 93,198 spectra. The search returned 24 spectra

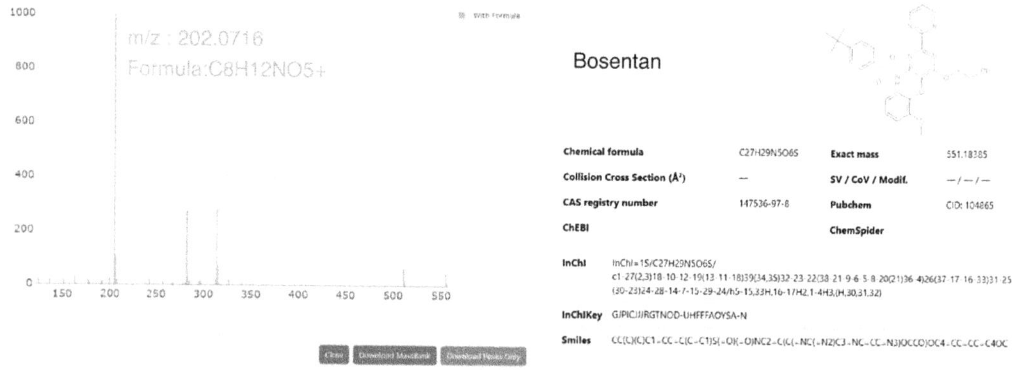

Fig. 4 Spectrum, metabolite structure, and metadata display in MSDMS. If the spectrum is annotated with formulas, they will be shown upon hovering over the peaks

pages. As shown in Fig. 4, the spectrum, the metabolite structure, and metadata can be displayed. For individual records, users may download the complete MB file or export only the spectrum peaks (*m/z* and intensity) as a .txt file. Single entries, a single source, or all data sources can be deleted from the *Library Builder* page.

3.3 Export Cart and Manage Cart

Using the Add button on the *Library Builder* page places the selected spectra into the cart. Different search filters can be applied, and the cart can be filled with records from different sources (see examples

Table 3

Exampyle of spectra selected to create a custom library

ID	Metabolite name	Instrument	Ionization mode	Collision energy (Volt)	Collection date	Data source	
28922	Bosentan	LC-ESI-QTOF	POSITIVE	40	17/01/2025	BAFG	X
16317	Reserpine	LC-ESI-QTOF	POSITIVE	40	17/01/2025	BAFG	X
4668	Diclofenac	LC-ESI-QTOF	POSITIVE	30	23/09/2015	Athens_Univ	X
6412	Acetamiprid	LC-ESI-QTOF	POSITIVE	40	29/05/2019	Athens_Univ	X
6416	Acetamiprid	LC-ESI-QTOF	NEGATIVE	20	28/03/2019	Athens_Univ	X

in Table 3). The cart content can then be reviewed and exported as a compressed file from the Export Cart page. The compressed file will contain all selected spectra as MB text records merged into a single text file, with a maximum of 25,000 records per file.

The *Manage Spectra Cart* page enables users to reload any previously exported cart and to add and remove spectra from it. If one tries to reimport the spectra via the *Upload New Spectra* page, the software will label the records DUPLICATE, and no import will be performed, as the software only accepts unique accession numbers.

3.4 General Screening with MS-Dial 5.0 for Drug of Abuse in Human Urine

Data Dependent Acquisition (DDA) and SWATH data-independent acquisition (DIA) mass spectrometry (MS) has become established techniques in MS-based 'omics' research used for the screening of xenobiotics (e.g., drugs, drug metabolites, pesticides, contaminants, toxicants). Such xenobiotic screening methods are mostly applied for tentative compound identification purposes based on spectral library searching. DIA workflow outperforms DDA in so far as for any precursor ion range, a product ion spectrum is recorded so that the analysis can be reprocessed at any time.

Figure 5 shows the TIC of the SWATH LC–MS/MS analysis of two human urine samples tested positive for THC and cocaine (for details see [22]). The MS/MS spectra were recorded with a collision energy range from 20 to 50 eV (*see* **Note 4**).

Using the MS-DIAL 5.0 [23] software for data analysis, several thousand features (*m/z*, retention time intensity) over the mass range *m/z* 100–1000 can be detected as illustrated in Fig. 6, and MS/MS spectra are available for the mass range *m/z* 50–600. MS-DIAL offers

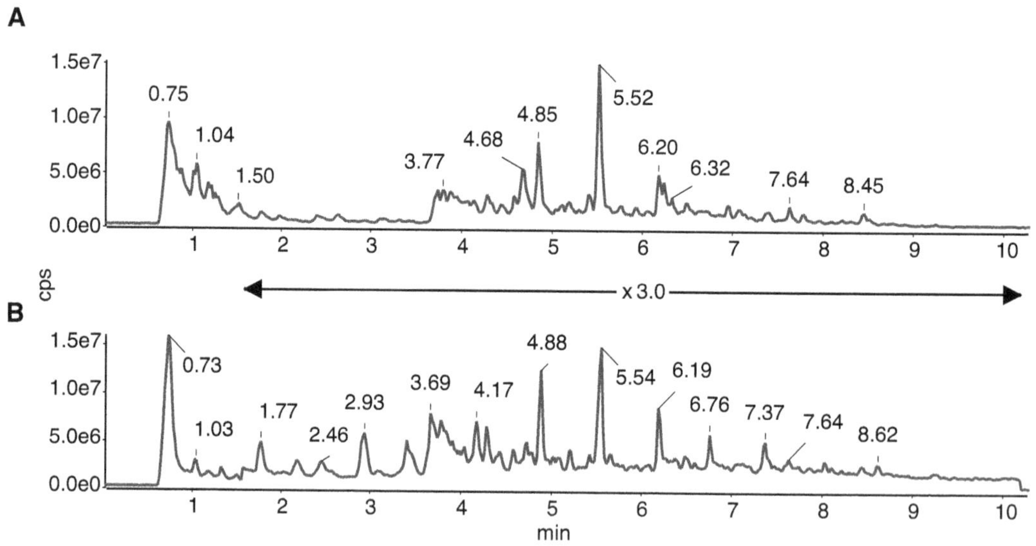

Fig. 5 TIC of the SWATH LC-MS/MS analysis of human urine sample (**A**) TC007 and (**B**) TC013 of subjects tested positive for cocaine and THC

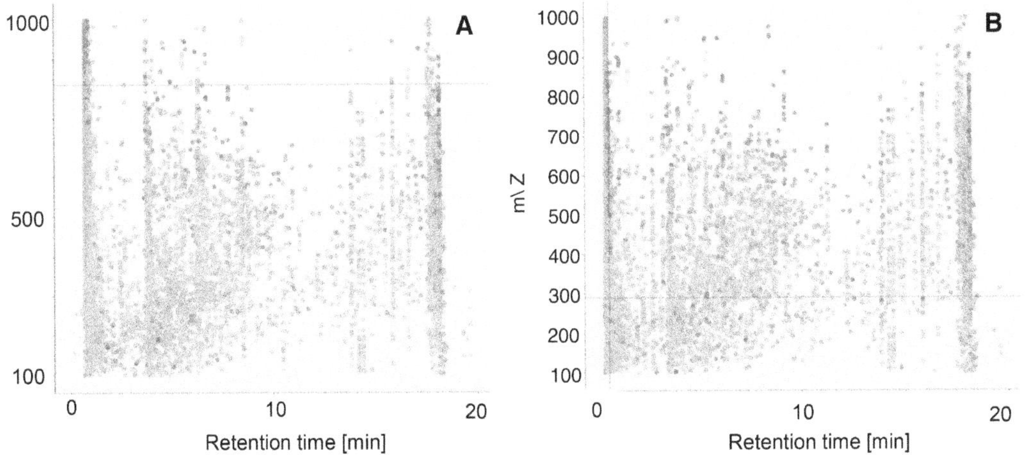

Fig. 6 MS-Dial peak spot viewer of MS1 retention time, *m/z*, intensity. (**A**) TC007 and (**B**) TC013 human urine samples: Red dots represent high-intensity features (strong peak), green dots indicate intermediate-intensity peaks, and blue dots denote low-intensity features

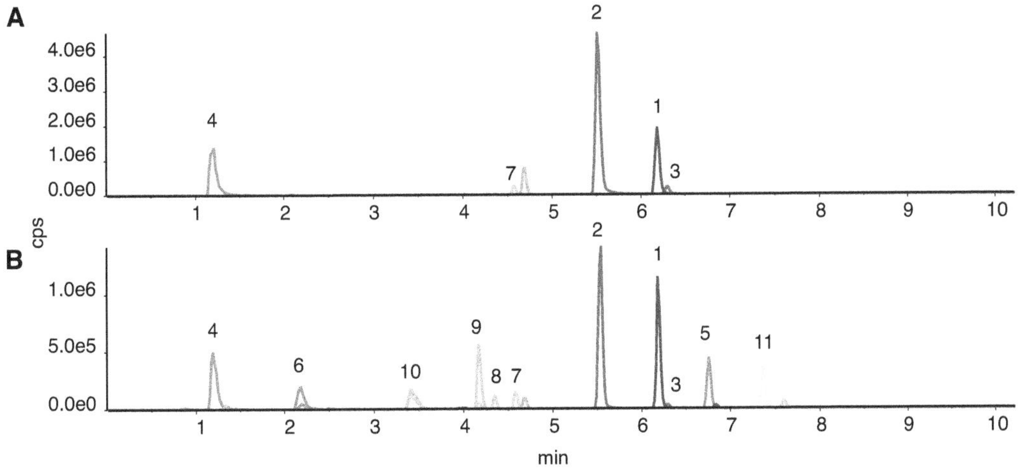

Fig. 7 Extracted ion current of human urine samples (**A**) TC007 and (**B**) TC013: (1) cocaine, (2) benzoylecgonine, (3) norcocaine, (4) ecgonine methyl ester, (5) cocaethylene, (6) ecgonine ethyl ester, (7) levamisole, (8) acetaminophen, (9) acetaminophen sulfate, (10) acetaminophen glucuronide, (11) phenacetin. Phenacetin and levamisole and cutting agents of cocaine preparation and paracetamol, paracetamol glucuronide, and sulfate are metabolites of phenacetin

the option to load libraries in various formats (*see* **Note 5**). With the MassBank release (2025.10), MSDMS shows 16,603 metabolites and 93,198 spectra in positive mode.

MS-DIAL suggested 51 and 39 metabolites with good reference matches and 2636 and 2885 with lower confidence matches. In fact, both samples contain 6 and 11 compounds related to consumption of cocaine and smoking. (Fig. 7). With a limited sample set, manual

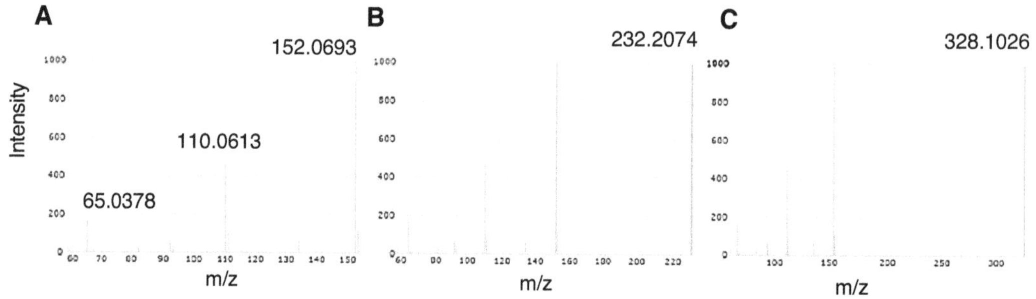

Fig. 8 (**A**) Library spectrum of acetaminophen. (**B**) Reconstructed spectrum of acetaminophen sulfate. (**C**) Reconstructed spectrum of acetaminophen glucuronide

data reduction may be considered, but this becomes impossible with a large set of samples. A substantial portion of the library content is redundant and not relevant to the conditions investigated (instrument, CE, fragmentation type, retention time). Using the MSDMS cart export menu, a custom library was exported from MassBank with records only produced with LC-ESI-QTOF selected in the instrument field. The library still contained 4000 metabolites and 33,500 spectra in positive mode. MS-DIAL proposes 45 and 25 metabolites with good reference matches and 1553 and 1580 with lower confidence.

Of course, numerous endogenous metabolites in urine such as hippuric are detected. Ideally for a forensic screening, one would like to build a library reflecting the specificity of the MS instrument, as well as MS and LC conditions. This task is possible but tedious and can be expensive. The generation of custom libraries from public sources becomes interesting to build a general screening workflow, but requires a library management tool. As MSDMS uses a MassBank text format, any record from any custom library can be further modified or completed using R or Python scripts. As an example using MassBank (release 2025.10), we generated a custom library containing 18 metabolites related to cocaine with 95 spectra (*see* **Note 6**). We also created a theoretical spectrum for acetaminophen sulfate and glucuronide metabolites based on the acetaminophen spectrum (Fig. 8). With this approach, we rapidly identified the cocaine metabolites and some additives. For additional samples, the retention time can also be customized for improved identification. Through an iterative process, libraries specific to lab conditions or projects can be generated from public libraries without having to run reference compounds. Of course, a library can also be built, including a combination of public MS/MS, spectra, in-house recorded spectra, and/or theoretical spectra.

4 Notes

1) The content of the mandatory fields can be defined by the user; only the accession number needs to be unique (e.g., AS000672). In some cases, it is of interest to create libraries of intense signal from unknown compounds. In this case only a few fields can be completed with information, while all the others will be, for example, NA.
2) The software, release version 2.4.3, and installation instructions can be downloaded from https://github.com/lsmsgeneva/MSDMS.
3) The following sequence is required to define DoA Library as source: Library->DoA_Library -> MB file(s)
4) The original.wiff files can be download from 10.26037/yareta:zus3usfesvd5hc67ctawgelcym.
5) MS-Dial cannot use MB library files directly. But conversion tools such MassBank to NIST MSP converter can be used to convert MB.txt to .msp files. (https://systemsomicslab.github.io/compms/others/main.html#Massbank2msp).
6) The custom library can be downloaded from 10.26037/yareta:zus3usfesvd5hc67ctawgelcym.

Conflict of Interest Statement

The authors declare no conflicts of interest that are relevant to the content of this chapter.

Funding SNF Grant 200021_192306

References

1. Maciel EVS, Pereira dos Santos NG, Vargas Medina DA, Lanças FM (2022) Electron ionization mass spectrometry: Quo vadis. Electrophoresis 43(15):1587–1600. https://doi.org/10.1002/elps.202100392
2. Covey TR, Thomson BA, Schneider BB (2009) Atmospheric pressure ion sources. Mass Spectrom Rev 28(6):870–897. https://doi.org/10.1002/mas.20246
3. De Vijlder T, Valkenborg D, Lemière F, Romijn EP, Laukens K, Cuyckens F (2018) A tutorial in small molecule identification via electrospray ionization-mass spectrometry: the practical art of structural elucidation. Mass Spectrom Rev 37(5):607–629. https://doi.org/10.1002/mas.21551
4. Raffaelli A, Saba A (2003) Atmospheric pressure photoionization mass spectrometry. Mass Spectrom Rev 22(5):318–331. https://doi.org/10.1002/mas.10060
5. Sleno L, Volmer DA (2004) Ion activation methods for tandem mass spectrometry. J Mass Spectrom 39(10):1091–1112. https://doi.org/10.1002/jms.703
6. McLuckey SA, Mentinova M (2011) Ion/neutral, ion/electron, ion/photon, and ion/ion interactions in tandem mass spectrometry: do we need them all? Are they enough. J Am Soc Mass Spectrom 22(1):3–12. https://doi.org/10.1007/s13361-010-0004-9
7. Glish GL, Burinsky DJ (2008) Hybrid mass spectrometers for tandem mass spectrometry. J Am

Soc Mass Spectrom 19(2):161–172. https://doi.org/10.1016/j.jasms.2007.11.013
8. Bittremieux W, Wang M, Dorrestein PC (2022) The critical role that spectral libraries play in capturing the metabolomics community knowledge. Metabolomics 18(12):94. https://doi.org/10.1007/s11306-022-01947-y
9. 2023 N (2025) NIST23: updates to the NIST tandem and electron ionization spectral libraries. https://chemdata.nist.gov/dokuwiki/lib/exe/fetch.php?media=chemdata:asms2023:asms2023_nist23_features.pdf. Accessed 12 Jan 2026
10. Ichou F, Schwarzenberg A, Lesage D, Alves S, Junot C, Machuron-Mandard X et al. (2014) Comparison of the activation time effects and the internal energy distributions for the CID, PQD and HCD excitation modes. J Mass Spectrom 49(6):498–508. https://doi.org/10.1002/jms.3365
11. Wang M, Carver JJ, Phelan VV, Sanchez LM, Garg N, Peng Y et al. (2016) Sharing and community curation of mass spectrometry data with Global Natural Products Social Molecular Networking. Nat Biotechnol 34(8):828–837. https://doi.org/10.1038/nbt.3597
12. Horai H, Arita M, Kanaya S, Nihei Y, Ikeda T, Suwa K et al. (2010) MassBank: a public repository for sharing mass spectral data for life sciences. J Mass Spectrom 45(7):703–714. https://doi.org/10.1002/jms.1777
13. Neumann S, Meier R, Wenk M, Elapavalore A, Nishioka T, Schulze T et al. (2025) MassBank: an open and FAIR mass spectral data resource. Nucleic Acids Res https://doi.org/10.1093/nar/gkaf1193
14. Bruderer T, Varesio E, Hidasi AO, Duchoslav E, Burton L, Bonner R et al. (2018) Metabolomic spectral libraries for data-independent SWATH liquid chromatography mass spectrometry acquisition. Anal Bioanal Chem 410(7):1873–1884. https://doi.org/10.1007/s00216-018-0860-x
15. da Silva KM, van de Lavoir M, Robeyns R, Iturrospe E, Verheggen L, Covaci A et al. (2022) Guidelines and considerations for building multidimensional libraries for untargeted MS-based metabolomics. Metabolomics 19(1):4. https://doi.org/10.1007/s11306-022-01965-w
16. Padilla-Gonzalez F, Rizzo S, Dirks C, Bergkamp W, Rasker S, Aloisi I (2025) Creation of an open-access high-resolution tandem mass spectral library of 1000 food toxicants. Anal Chem 97(43):23822–23830. https://doi.org/10.1021/acs.analchem.5c03020
17. Consortium M (2025) MassBank/MassBank-data: Release 2025. 10 Accessed 12 Jan 2026
18. SMILE, S. https://www.daylight.com/dayhtml/doc/theory/theory.smiles.html, https://opensmiles.org/opensmiles.html Accessed 12 Jan 2026
19. InChI. https://www.inchi-trust.org/technology/, https://www.inchi-trust.org/downloads/ Accessed 12 Jan 2026
20. InChIKey. https://www.inchi-trust.org/technology/inchikey/, https://www.inchi-trust.org/downloads Accessed 12 Jan 2026
21. Probst D, Reymond JL (2018) SmilesDrawer: parsing and drawing SMILES-encoded molecular structures using client-side JavaScript. J Chem Inf Model 58(1):1–7. https://doi.org/10.1021/acs.jcim.7b00425
22. Klont F, Jahn S, Grivet C, Konig S, Bonner R, Hopfgartner G (2020) SWATH data independent acquisition mass spectrometry for screening of xenobiotics in biological fluids: opportunities and challenges for data processing. Talanta 211:120747. https://doi.org/10.1016/j.talanta.2020.120747
23. Takeda H, Matsuzawa Y, Takeuchi M, Takahashi M, Nishida K, Harayama T et al. (2024) MS-DIAL 5 multimodal mass spectrometry data mining unveils lipidome complexities. Nat Commun 15(1):9903. https://doi.org/10.1038/s41467-024-54137-w

Check for updates

Chapter 2

Standardized Annotation of Quantitative Proteomics Datasets Using SDRF-Proteomics

Lev Levitsky, Tine Claeys and Veit Schwämmle

Abstract

Proteomics datasets require detailed and standardized metadata annotation to ensure reproducibility, reusability, and alignment with FAIR (Findable, Accessible, Interoperable, Reusable) principles. This chapter presents a practical, hands-on workflow for annotating a label-free proteomics dataset using the Sample and Data Relationship Format for Proteomics (SDRF-Proteomics), a HUPO-PSI and EuBIC-MS standard that has been adopted by major proteomics data repositories. We demonstrate the process on a COVID-19 nasal-swab dataset (PXD020394), showing how to derive sample and assay metadata from the study files, encode them in SDRF with the lesSDRF web application, and validate the result. Although the worked example is label-free, the approach can be generalized to other quantification strategies.

Keywords Proteomics, SDRF, Annotation, COVID-19, Sample metadata, lesSDRF, FAIR, Mass spectrometry

1 Introduction

Large-scale proteomics experiments generate not only vast quantities of mass spectrometry data but also complex metadata describing biological samples, experimental conditions, and optionally data processing steps. When this metadata is poorly documented, it severely hinders reproducibility, downstream reanalysis, and integration into public repositories. Annotation, in this context, refers to the structured documentation of experimental metadata that links biological samples to data acquisition and the resulting analytical data files.

To address these challenges, the Sample and Data Relationship Format for Proteomics (SDRF-Proteomics) [1] was developed as a standardized, tab-delimited format for capturing sample metadata and experimental design information in a machine-readable yet human-interpretable form. Building on the broader MAGE-TAB [2] and ISA [3] standards developed at EMBL-EB, SDRF-Proteomics is tailored for proteomics-specific use cases, such as isobaric labeling

Frédérique Lisacek (ed.), *Bioinformatics for Mass Spectrometry*, Methods in Molecular Biology, vol. 3071,
https://doi.org/10.1007/978-1-0716-5519-1_2,

experiments and LC–MS/MS workflow. Unlike *ad hoc* spreadsheets, SDRF explicitly encodes (1) sample characteristics, (2) experimental factors used in statistical analysis, and (3) the mapping of samples to raw data files, fractions, and technical replicates. This explicit mapping is essential for transparent quality control, differential analysis, and meta-analysis across cohorts.

SDRF-Proteomics has shown its strength by enabling direct reanalysis of public data using scalable workflow [4,5] and statistical analysis [6] without the necessity for further configuration or manual annotation. In addition, there is an increasing number of proteomics software that deliver SDRF templates from the given experimental design [7,8], facilitating the completion of a full data set annotation. As adoption grows across public repositories such as PRIDE [9], SDRF-Proteomics will enable large-scale reanalysis and harmonization at an unprecedented scale. This approach is crucial, as retrospectively annotating complex public datasets presents significant challenges (*see* **Note 1**).

Beyond structured organization of metadata, standardization depends on controlled vocabularies and ontologies to minimize ambiguity and enable automated processing. SDRF-Proteomics leverages key ontologies such as PSI-MS (instrument models, file types, and acquisition settings) [10], UNIMOD (modifications) [11], EFO (biological characteristics and disease terms) [12], among others. Using ontology-backed terms rather than free text improves FAIRness [13], making data Findable, Accessible, Interoperable, and Reusable, and directly supports programmatic validation and semantic searches across resources. Within the proteomics domain, this interoperability is particularly critical for data deposition and reuse via the ProteomeXchange consortium [12,14], where standardized metadata determine whether a dataset can be reliably reanalyzed or integrated into larger studies.

We also introduce lesSDRF [15], a lightweight, browser-based environment for authoring and validating SDRF-Proteomics tables. The tool provides guided column templates, ontology-aware fields, and built-in checks for common inconsistencies (e.g., missing factor values, duplicated assay names, or mismatched technical replicate indices). While SDRF-Proteomics files can be authored in any spreadsheet editor, lesSDRF reduces common errors and accelerates convergence to a validated submission that is ready for repository deposition while reducing the requirement to know about metadata by removing the need for detailed familiarity with the format.

This chapter demonstrates end-to-end SDRF annotation for the COVID-19 dataset PXD020394 [15,16], which compares 5 COVID-19-positive and 5 negative subjects, each measured in technical duplicate (20 RAW files). The goal is a validated SDRF that is both human-interpretable and computationally robust for downstream reuse.

2 Materials

2.1 Dataset

- Dataset identifier: PXD020394.
- Study: COVID-19 nasal swab proteomics comparing five positive and five negative samples.
- Data files: 20 raw MS files (10 biological samples with two technical replicates each).
- File naming pattern: NEG1.raw through NEG5rep.raw (negative samples) and POS1.raw through POS5rep.raw (positive samples).
- Quantification approach: Label-free quantification.
- Publication: "Quantitative proteomic dataset from oro- and naso-pharyngeal swabs used for COVID-19 diagnosis: detection of viral proteins and host's biological processes altered by the infection." (Data Brief)[ ].

2.2 Software and Tools

- lesSDRF web application: https://lessdrf.streamlit.app/.
  - Browser-based tool for creating and validating SDRF files.
  - Provides templates for different organisms.
  - Guides users through required and optional annotation fields.
- Alternative: Spreadsheet editor (e.g., Microsoft Excel, LibreOffice Calc) for manual SDRF creation.

2.3 Reference Resources

- SDRF-Proteomics specification: https://github.com/bigbio/proteomics-sample-metadata/tree/master/sdrf-proteomics.
- Experimental Factor Ontology (EFO): https://bioportal.bioontology.org/ontologies/EFO.
- PSI-MS ontology: https://www.ebi.ac.uk/ols/ontologies/ms.
- PRIDE repository: https://www.ebi.ac.uk/pride/.
- Dataset README file: https://ftp.pride.ebi.ac.uk/pride/data/archive/2020/07/PXD020394/README.txt.

3 Methods

3.1 Understanding the Experimental Design

Before beginning the annotation process, thoroughly review the experimental design and study context. The PXD020394 dataset collected nasal swab samples from 10 individuals, including 5 COVID-19-negative (control) and 5 COVID-19-positive patients. Each sample was analyzed using liquid chromatography–tandem mass spectrometry (LC–MS/MS) with technical replicates, resulting in 20 raw data files total.

Understanding the experimental design is crucial for proper SDRF annotation, as it determines which sample characteristics and experimental factors must be captured in the metadata. Key design elements to identify include:

- Biological samples: Number and source of samples.
- Experimental factors: Primary variables being investigated (e.g., disease status).
- Replication strategy: Technical and biological replicates.
- Sample processing: Extraction, digestion, and modification protocols.
- MS acquisition: Instrument type and acquisition parameters.

3.2 SDRF Structure and Information Categories

An SDRF file is fundamentally a tab-separated values (TSV) table where each row represents a data file (typically one MS run) and columns describe sample characteristics and experimental parameters. The general flow of information follows the experimental workflow: from biological samples through sample processing and mass spectrometry acquisition to data analysis parameters.

SDRF captures several key categories of information (see Fig. 1):

1. **Sample characteristics**: organism, tissue type, disease state, demographic information.

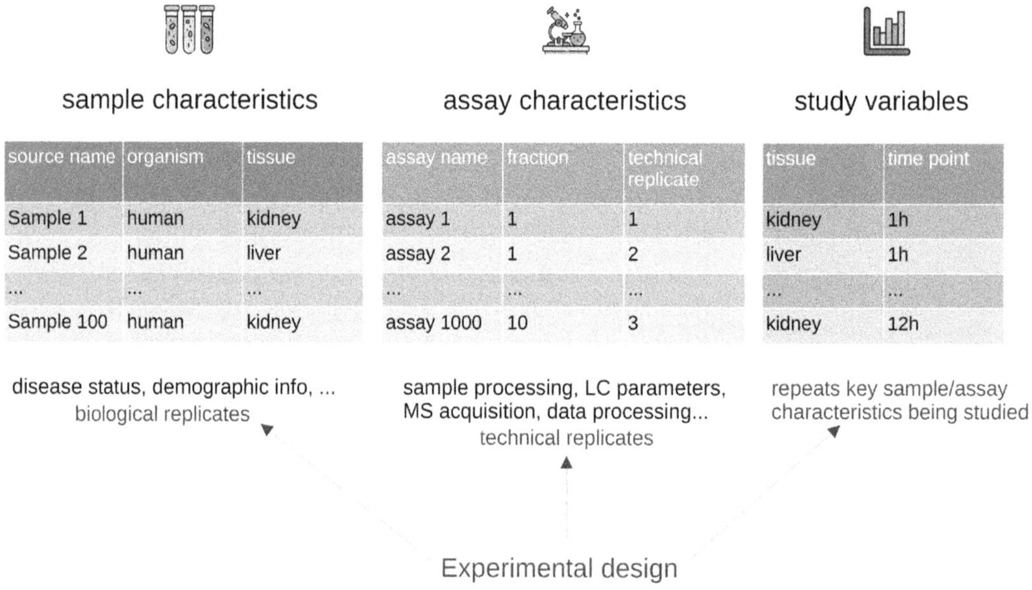

Fig. 1 Schematic of SDRF format

2. **Experimental design**: biological and technical replicates, experimental factors.
3. **Sample processing**: extraction methods, chemical modifications, enzymatic digestion.
4. **MS acquisition**: instrument details, acquisition methods, mass tolerances.
5. **Data processing**: search parameters, modifications, quantification methods.

Column names and values follow controlled vocabularies and ontologies to ensure machine readability. Adherence to these standards is essential for all attribute values (*see* **Note 2**). Most sample characteristics are defined in the Experimental Factor Ontology (EFO), while instrument models are defined in the PSI-MS ontology. However, the full list of supported ontologies is significantly longer and may change with the continued development of the specification. The minimum set of columns is defined by the applicable SDRF template. The specification includes templates for most common sample types, e.g., human samples, other vertebrate animals, plants, etc.

3.3 Step-by-Step Annotation Workflow Using lesSDRF

Step 1: Initialize the SDRF File

1. Access the lesSDRF web application at https://lessdrf.streamlit.app/.
2. Under "Start here with a completely new SDRF file," click the dropdown menu.
3. Select "human" as the organism template.
4. Obtain the list of data file names:
   - Download the README file from the PRIDE repository.
   - Open in a spreadsheet editor and filter by TYPE = RAW.
   - Copy all file names from the NAME column (20 RAW files).
5. Paste the file names into the input field and press Enter.
6. Verify the preview table shows 20 rows.

Step 2: Specify Quantification Strategy

1. Navigate to "2. Labeling" in the left sidebar.
2. Under the SDRF preview, check "label free sample."
3. Click "Submit selection."
4. Choose "ALL" from the dropdown menu to apply to all files.
5. Check the "Ready?" box to proceed.

For label-free data, each raw file represents a separate acquisition with one row per file in the SDRF. For labeled approaches (TMT, iTRAQ), *see* **Note 3**.

Step 3: Annotate Sample Characteristics

Navigate to "3. Required columns" to begin filling required metadata fields.

Source Name and Sample Identification

1. Select "source name" from the column list.
2. Enter sample identifiers that are identical for all files from the same sample.
3. Use patient designations: NEG1 through NEG5 and POS1 through POS5.
4. Each identifier should appear exactly twice (for the two technical replicates).
5. Double-click "Update" to save.

A clear understanding of the replication strategy is critical for successful annotation (*see* **Note 4**).

Organism Information

1. Select characteristics[organism].
2. Specify that data does NOT have multiple organisms.
3. Select that only model organisms are present.
4. Choose "Homo sapiens" from the option.
5. Click "Ready for input?"

For host–pathogen studies like this one, it is possible to annotate multiple organisms (see Fig. 2 and **Note 5**).

Anatomical Location

1. Select characteristics[organism part].
2. Enter "nasal cavity mucosa" as the anatomical source.
3. This specificity is important for biological context and cross-study comparisons.

Demographic Information

For this dataset, demographic details are not available in the publication. Handle missing information as follows:

1. Fill characteristics[age] with "Not available."
2. Fill characteristics[sex] with "Not available."
3. Fill characteristics[ancestry category] with "Not available."
4. Fill characteristics[cell type] with "Not available."

Disease State

1. Select characteristics[disease].
2. Annotate negative samples (NEG1–NEG5) as "normal".
3. Annotate positive samples (POS1–POS5) as "COVID-19".

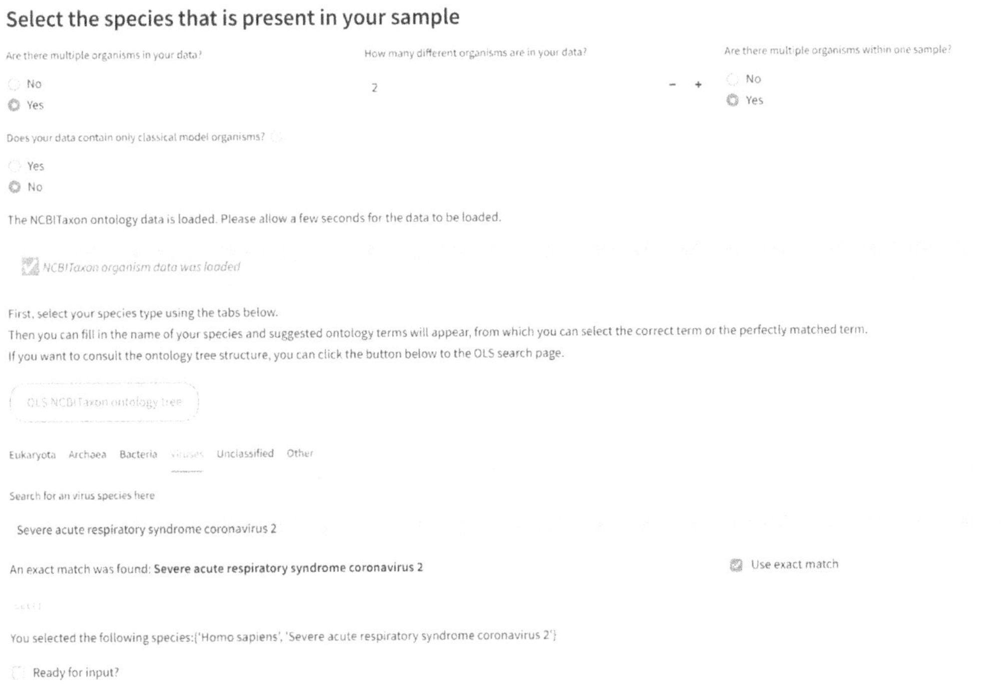

Fig. 2 Annotation of multiple organisms present in a sample using lesSDRF

4. This classification is the primary experimental factor.

Individual and Biological Replicate Identifiers

1. For characteristics[individual], use numbers 1–10 to identify each patient.
2. For characteristics[biological replicate], use numbers 1–5 for both positive and negative sample groups.
3. Context-dependent annotation: consider the study design when defining biological replicates.

Step 4: Annotate Technical and Analytical Parameters.

Assay Name and Technology Type

1. For assay name, create unique identifiers for each experimental run.
2. Technical replicates have different assay names but the same source name.
3. Fill technology type with "proteomic profiling by mass spectrometry."

Experimental Design Metadata

1. For comment[technical replicate], use numbers 1 and 2 for the two replicates of each patient.

2. For comment[fraction identifier], enter “1” for all rows (no fractionation used).

Sample Processing Details

1. Fill comment[cleavage agent details] with “NT=Trypsin”.
2. Select comment[modification parameters].
3. Configure carbamidomethylation:
   - Choose “Carbamidomethyl”
   - Type: “Fixed”
   - Position: “Anywhere”
   - Target: “C”
   - Result: NT=Carbamidomethyl; AC=4; MT=Fixed; PP=Anywhere; TA=C

This five-key format is mandatory for all modification parameters in SDRF-Proteomics (*see* **Note 6**).”

Mass Spectrometry Parameters

1. Fill comment[instrument] with “Q Exactive Plus”.
2. Fill comment[precursor mass tolerance] with “30 ppm”.
3. Fill comment[fragment mass tolerance] with “0.05 Da”.
4. Navigate to “4. Additional columns”.
5. Choose comment[dissociation method].
6. Search for and select “higher energy beam-type collision-induced dissociation (HCD)”.

Step 5: Define Experimental Factors

1. Return to required columns.
2. Select factor value[disease].
3. Link to characteristics[disease] to duplicate values.
4. This captures the primary experimental variable driving the biological investigation.

3.4 Validation and Export

1. Review the completed SDRF table for consistency.
2. Verify all required columns are filled.
3. Check that technical replicates are properly annotated.
4. Export the SDRF file in TSV format.
5. The resulting file contains 20 rows and all annotated columns.

Users interested in practicing on more complex, labeled datasets can find additional resources (*see* **Note 7**).

3.5 Best Practices for SDRF Annotation

Annotate Early

Create SDRF files during or immediately after data generation, not during publication preparation. Early annotation ensures accurate capture of experimental details. When working with external collaborators, such as submitting samples to a core facility, we recommend sharing your information immediately in a structured manner.

Consult Multiple Sources

When (re-)annotating public datasets, examine:

- Publication methods sections.
- Supplementary materials.
- README files from repositories.
- Raw file metadata (instrument-dependent).

Use Specific Terms and Minimize Free Text

Prefer specific anatomical and technical terms over general descriptions. This improves discoverability and enables accurate cross-study comparisons, e.g., not using "cell line" but rather the cell line name.

Document Missing Information

Use "Not available" rather than leaving cells empty when information cannot be determined. This is distinct from "Not applicable" when information is required in the standard format but does not apply to the specific experimental setup. Using standardized placeholders for missing data improves machine readability of the format.

Validate Ontology Terms

Verify that terms exist in the appropriate controlled vocabularies (EFO, PSI-MS, etc.) and use the community-supported SDRF-validators to verify file structure.

Consider the Audience

Remember that SDRF files serve both humans reviewing the data and computational tools are performing automated analysis. Contribute with examples of well-annotated projects (*see* **Note 8**).

4 Notes

1. **Challenges with Public Datasets**. Retroactive annotation of publicly available data remains difficult as key details (file conventions, technical parameters, or clinical information) are often incomplete, ambiguous, or absent from the associated publication. When annotating such datasets, rely on the raw data files for clues (e.g., file names and information embedded in the raw file) and acknowledge any missing information.

2. **Use of Controlled Vocabularies (CV) and Ontologies**. To ensure machine readability and reusability, all attribute values must be selected from standardized lists (ontologies or CVs). Always verify the term with the accession number. Key resources include the Experimental Factor Ontology (EFO) for sample characteristics, the PSI-MS ontology for instruments and methods, and UNIMOD for modifications. The full and updated list of supported ontologies, as this is an evolving resource, is available in the SDRF specification at https://github.com/bigbio/proteomics-sample-metadata/tree/master/sdrf-proteomics.
3. **Beyond Label-Free Quantification**. While the primary protocol covers label-free, SDRF fully supports other quantification strategies, including isobaric labeling (TMT/iTRAQ), metabolic labeling (SILAC), and chemical labeling. These methods require specific, often complex, annotation patterns for mapping the multiplexed samples to the raw data files.
4. **Distinguishing Replicates**. Correctly annotating replicates is crucial for downstream analysis. Technical replicates are repeat measurements of the same sample. In label-free experiments, they share the same "Source Name" but different "Assay Name". Biological replicates are independent samples from different biological sources and thus have a different "Source Name". For multiplexed (TMT/iTRAQ) data, different samples in the same run share the same "Assay Name" but have unique "Source Name" entries.
5. **Handling Multiple Organisms**. For studies where both host and pathogen proteins are analyzed (such as the COVID-19 example), you can list both organisms if the viral data are relevant for processing or discovery by creating a column for each organism. Ensure you use the exact ontology term for the pathogen, e.g., "Severe acute respiratory syndrome coronavirus 2" for the virus.
6. **Modification Key-Value Format**. When annotating modifications, you must use the five-key system within the value field. These five keys are: **NT** (Name of the term), **AC** (Accession number, e.g., UNIMOD), **MT** (Modification type: Fixed/Variable), **PP** (Position), and **TA** (Target amino acid). Failure to include all five keys will result in validation failure when using the SDRF-Parser.
7. **Additional Learning Resources**. For those working with more complex, TMT-labeled datasets, additional practice and advanced challenges in sample-to-file mapping can be found at the Clinical Proteomics course: https://hds-sandbox.github.io/proteomics-sandbox/TeachingModule/teachingmodule.

html. This resource provides practical experience that complements the basic protocol.

8. **Repository Support**. To maximize the visibility and reusability of your data, always submit your SDRF with your final raw data. PRIDE Archive encourages SDRF use, and the wider ProteomeXchange consortium recognizes it as the standard metadata format. Community contributions of high-quality SDRF examples are welcomed on GitHub: https://github.com/bigbio/proteomics-sample-metadata/tree/master/annotated-projects.

Conflict of Interest

The authors declare no conflicts of interest that are relevant to the content of this chapter.

References

1. Dai C, Füllgrabe A, Pfeuffer J, Solovyeva EM, Deng J, Moreno P et al. (2021) A proteomics sample metadata representation for multiomics integration and big data analysis. Nat Commun 12: 5854
2. Rayner TF, Rocca-Serra P, Spellman PT, Causton HC, Farne A, Holloway E et al. (2006) A simple spreadsheet-based, MIAME-supportive format for microarray data: MAGE-TAB. BMC Bioinform 7: 489
3. Johnson D, Batista D, Cochrane K, Davey RP, Etuk A, Gonzalez-Beltran A et al (2021) ISA API: an open platform for interoperable life science experimental metadata. Gigascience 10. https://doi.org/10.1093/gigascience/giab060
4. Dai C, Pfeuffer J, Wang H, Zheng P, Käll L, Sachsenberg T et al. (2024) quantms: a cloud-based pipeline for quantitative proteomics enables the reanalysis of public proteomics data. Nat Methods 21:1603–1607
5. Bouyssié D, Altıner P, Capella-Gutierrez S, Fernández JM, Hagemeijer YP, Horvatovich P et al. (2024) WOMBAT-P: benchmarking label-free proteomics data analysis workflows. J Proteome Res 23:418–429
6. Kohler D, Staniak M, Yu F, Nesvizhskii AI, Vitek O (2024) An MSstats workflow for detecting differentially abundant proteins in large-scale data-independent acquisition mass spectrometry experiments with FragPipe processing. Nat Protoc 19:2915–2938
7. Viegener W, Urazbakhtin S, Ferretti D, Cox J, Xiao J (2025) Facilitating analysis and dissemination of proteomics data through metadata integration in MaxQuant. Nat Commun 16:1–5
8. Chang H-Y, Deng Y, Li R, Avtonomov D, Wen B, Haynes SE, et al. (2026) Analysis of isobaric quantitative proteomic data using TMT-Integrator and FragPipe computational platform. Nat Commun 17:4010. https://doi.org/10.1038/s41467-026-70118-7
9. Perez-Riverol Y, Bandla C, Kundu DJ, Kamatchinathan S, Bai J, Hewapathirana S et al. (2025) The PRIDE database at 20 years: 2025 update. Nucleic Acids Res 53:D543–D553
10. Mayer G, Montecchi-Palazzi L, Ovelleiro D, Jones AR, Binz P-A, Deutsch EW et al. (2013) The HUPO proteomics standards initiative-mass spectrometry controlled vocabulary. Database (Oxford) 2013:bat009
11. Creasy DM, Cottrell JS (2004) Unimod: protein modifications for mass spectrometry. Proteomics 4:1534–1536
12. Malone J, Holloway E, Adamusiak T, Kapushesky M, Zheng J, Kolesnikov N et al. (2010) Modeling sample variables with an experimental factor ontology. Bioinformatics 26:1112–1118
13. Wilkinson MD, Dumontier M, Aalbersberg IJJ, Appleton G, Axton M, Baak A et al. (2016) The FAIR Guiding Principles for scientific data management and stewardship. Sci Data 3: 160018
14. Deutsch EW, Bandeira N, Perez-Riverol Y, Sharma V, Carver JJ, Mendoza L et al. (2023) The ProteomeXchange consortium

at 10 years: 2023 update. Nucleic Acids Res 51:D1539–D1548
15. Claeys T, Van Den Bossche T, Perez-Riverol Y, Gevaert K, Vizcaíno JA, Martens L (2023) lesSDRF is more: maximizing the value of proteomics data through streamlined metadata annotation. Nat Commun 14: 6743
16. Rivera B, Leyva A, Portela MM, Moratorio G, Moreno P, Durán R et al. (2020) Quantitative proteomic dataset from oro- and naso-pharyngeal swabs used for COVID-19 diagnosis: detection of viral proteins and host's biological processes altered by the infection. Data Brief 32:106121

Chapter 3

An Open Infrastructure for Mass Spectrometry Data in R

Johannes Rainer, Marilyn De Graeve, Philippine Louail, Sebastien Gibb and Laurent Gatto

Abstract

We present the *Spectra* software, an R/Bioconductor package, defining an efficient infrastructure for storing, handling, and visualizing mass spectrometry data. We describe how *Spectra* adapts to different data types, sizes, and use cases, addressing scalability while keeping a consistent user interface. We also demonstrate how it can integrate with third-party Python tools and follow modern tidyomics principles, seamlessly integrating with more general tidy data science tools.

Keywords Mass spectrometry, Proteomics, Metabolomics, Bioconductor, R, Reproducible research, Software

1 Introduction

Mass spectrometry (MS) is a powerful analytical technique widely used in life sciences and related fields to detect, identify, and quantify small molecules. These include metabolites, peptides, and other biomolecules present in complex biological samples such as blood, tissue extracts and plant materials. MS plays a crucial role in various applications, ranging from monitoring metabolic pathways and studying biochemical processes to support production workflows and ensuring strict quality control in both research and industrial environments. Its high sensitivity, specificity, and ability to analyze diverse compounds make it an indispensable tool in modern analytical laboratories.

MS data are complex, large, and heterogeneous. The data volume depends on the resolution of the MS instruments and the size of the metabolomics or proteomics experiments, both of which have steadily increased in recent years. With further technological advancements, including the ability to measure additional properties of the quantified ions, such as the ion mobility, the dimensionality and thus the sizes and complexity of the data sets continue to increase.

Frédérique Lisacek (ed.), *Bioinformatics for Mass Spectrometry*, Methods in Molecular Biology, vol. 3071,
https://doi.org/10.1007/978-1-0716-5519-1_3,

The software infrastructure to process such data must thus be able to handle large amounts of data and also adapt to the new properties (e.g., ion mobility), file types, sizes, etc. Moreover, different MS vendors have their own data file formats that contain the acquired MS data. Converting these files to a standard open file format (i.e., mzML [1]) allows MS analysis to be performed and streamlined [2]. However, information loss can occur during such file format conversion.

As one of the first, and still actively maintained, software packages in R, the *xcms* package [3] supported import and processing of liquid chromatography (LC)–MS data for untargeted metabolomics data analysis. This package was also part of Bioconductor version 1.6, released in 2005. Bioconductor is a free, open-source, and open-development software project for the analysis and understanding of biology data from large-scale experiments [4,5]. It is primarily based on the statistical programming language R but also includes contributions in and interfaces to other programming languages.

To handle the, at the time, large MS data volumes, early R packages such as the *xcms* Bioconductor package only loaded data from one sample at a time into memory. This concept was then also adopted and formalized by the *MSnbase* Bioconductor package [6]. Through this *on-disk* data model, mass peaks data were retrieved and imported only upon demand. This provides even better support for large-scale data analysis. Initial concepts for large-scale data support, implemented in the *MSnbase* Bioconductor package, were reused in other R packages for MS data analysis, including the *xcms* package. This infrastructure was then expanded and further generalized in the *Spectra* package (Fig. 1), the current standard infrastructure for MS data analysis in Bioconductor.

The *Spectra* package is developed as part of the *R for Mass Spectrometry* initiative, an effort to harmonize and orchestrate software development for MS data analysis in R, launched in 2019. A goal of this initiative is to define a robust, flexible, and expandable

Fig. 1 The *Spectra* package sticker

infrastructure for MS-based proteomics and metabolomics data analysis. This makes MS data analysis more efficient, robust, and powerful, and provides a reproducible framework for representation and analysis of MS data in R. Efficiency is guaranteed by providing support for large-scale data analysis, but also connectors to other data processing strategies (such as the tidyverse or Python MS packages), make the software infrastructure of the R for Mass Spectrometry initiative extremely powerful.

The *Spectra* package plays a central role in the *R for Mass Spectrometry* ecosystem. It offers an efficient and consistent interface to manipulate and process raw MS data and is adaptable to various experiment sizes. Many other packages depend on `Spectra` as the raw MS data source and provide additional, specialized functionality. Several such *backend* packages extend how MS data is stored and accessed. Packages such as *CompoundDb*, *MetaboAnnotation*, and *MsFeatures* offer dedicated functionalities to extract MS features and annotate metabolites. The *PSMatch* [7] package handles peptide spectrum matching results and allows to associate and visualize these in conjunction with MS spectra. Two additional packages play a special role in the MS software ecosystem, namely *MsCoreUtils* and *MetaboCoreUtils*, that provide low-level functions that are used by other packages to process MS data. Together, *Spectra* and the other packages of the *R for Mass Spectrometry* initiative offer a sound ground for large-scale MS data processing and analysis.

In this chapter, we explore the *Spectra* infrastructure available to represent and analyze MS data within R. We begin by introducing the conceptual representation of MS data, followed by an overview of the different data structures and formats commonly used to store and manage MS data. Next, we discuss approaches and tools that support large-scale MS data analysis, enabling efficient handling of extensive experiments. The chapter then illustrates how `Spectra` objects can seamlessly integrate into the tidyverse and tidyomics [8] tools for streamlined, reproducible workflows. We further extend the discussion beyond R, presenting methods to integrate MS data analysis pipelines with Python using the new *SpectriPy* package [9]. Finally, we cover practical aspects of processing and visualizing MS data, ensuring readers are equipped with both conceptual understanding and actionable tools to analyze their mass spectrometry experiments effectively.

2 Materials

The *Spectra* Bioconductor package provides a flexible and expandable infrastructure for MS data representation and analysis in R. The main data object is the `Spectra` class. Users will operate the MS data through this object, as it provides all the necessary functionality to access, modify, subset, and analyze the data. The code required for import, export, and *representation* of the MS data are implemented separately from the `Spectra` class, in dedicated data *backend* classes (see below for details). This separation guarantees expandability of the infrastructure to support additional file formats or storage modes, as well as a flexibility to change between data representations for, or even during, a data analysis depending on needs and properties of the experiment.

2.1 Representation of MS Data

The typical data generated by an MS instrument consists of several mass spectra acquired for each sample. In its simplest form, a mass spectrum is composed of two paired numeric vectors of *m/z* and intensity values. Each such pair of *m/z* and intensity values represents one mass peak. The number of mass peaks in one spectrum can vary between spectra and very much depends on the type of data acquired and the MS instrument used. For each spectrum, additional metadata information such as the MS level of a spectrum or its retention time is typically available.

The MS data in a `Spectra` object are organized linearly, i.e., one spectrum after the other. The length of a `Spectra` object matches the number of MS `spectra` it represents. The `Spectra` object provides access to the spectrum metadata (the so-called *spectra variables* stored in the `spectraData` table), as well as its mass peaks data (the *m/z* and intensity *peaks variables*, stored in `peaksData` tables).

2.2 Installation

The *Spectra* package, as well as all other packages used in conjunction with *Spectra,* is available from Bioconductor and should thus be installed with the `install()` function from the `BiocManager` package, available from CRAN.

```
if (!require("BiocManager"))
    install.packages("BiocManager")
BiocManager::install("Spectra")
```

Session Information

A complete list and version of R and packages used to run this complete analysis and produce the results is provided below. **Note 1** provides important details about the R and Bioconductor versions compatibility.

- R version 4.5.0 (2025-04-11), `x86_64-pc-linux-gnu`.
- Running under: `Ubuntu 24.04.2 LTS`.
- Matrix products: default.
- BLAS: `/opt/R-4.5/lib/R/lib/libRblas.so`.
- LAPACK: `/opt/R-4.5/lib/R/lib/libRlapack.so`; LAPACK version3.12.1.

- Base packages: base, datasets, graphics, grDevices, methods, stats, stats4, utils.
- Other packages: AnnotationFilter 1.32.0, AnnotationHub 3.16.0, BiocFileCache 2.16.0, BiocGenerics 0.54.0, BiocParallel 1.42.0, BiocStyle 2.36.0, CompoundDb 1.12.1, dbplyr 2.5.0, generics 0.1.4, ggplot2 3.5.2, knitr 1.50, MsBackendSql 1.8.0, MsDataHub 1.8.0, pheatmap 1.0.12, RColorBrewer 1.1-3, reticulate 1.42.0, RSQLite 2.3.11, S4Vectors 0.46.0, Spectra 1.18.2, SpectriPy 0.99.5.
- Loaded via a namespace (and not attached): AnnotationDbi 1.70.0, base64enc 0.1-3, Biobase 2.68.0, BiocManager 1.30.25, BiocVersion 3.21.1, Biostrings 2.76.0, bit 4.6.0, bit64 4.6.0-1, bitops 1.0-9, blob 1.2.4, cachem 1.1.0, ChemmineR 3.60.0, cli 3.6.5, clue 0.3-66, cluster 2.1.8.1, codetools 0.2-20, colorspace 2.1-1, compiler 4.5.0, crayon 1.5.3, curl 6.2.3, data.table 1.17.4, DBI 1.2.3, digest 0.6.37, dplyr 1.1.4, DT 0.33, evaluate 1.0.3, ExperimentHub 2.16.0, farver 2.1.2, fastmap 1.2.0, fastmatch 1.1-6, filelock 1.0.3, fs 1.6.6, GenomeInfoDb 1.44.0, GenomeInfoDbData 1.2.14, GenomicRanges 1.60.0, glue 1.8.0, grid 4.5.0, gridExtra 2.3, gtable 0.3.6, hms 1.1.3, htmltools 0.5.8.1, htmlwidgets 1.6.4, httr 1.4.7, IRanges 2.42.0, jsonlite 2.0.0, KEGGREST 1.48.0, labeling 0.4.3, lattice 0.22-6, lazyeval 0.2.2, lifecycle 1.0.4, magrittr 2.0.3, MASS 7.3-65, Matrix 1.7-3, memoise 2.0.1, MetaboCoreUtils 1.16.1, mime 0.13, MsCoreUtils 1.20.0, mzR 2.42.0, ncdf4 1.24, parallel 4.5.0, pillar 1.10.2, pkgconfig 2.0.3, png 0.1-8, prettyunits 1.2.0, progress 1.2.3, ProtGenerics 1.40.0, purrr 1.0.4, R6 2.6.1, rappdirs 0.3.3, Rcpp 1.1.0, RCurl 1.98-1.17, rjson 0.2.23, rlang 1.1.6, rmarkdown 2.29, rsvg 2.6.2, scales 1.4.0, stringi 1.8.7, tibble 3.2.1, tidyselect 1.2.1, tinytex 0.57, tools 4.5.0, UCSC.utils 1.4.0, vctrs 0.6.5, withr 3.0.2, xfun 0.52, xml2 1.3.8, XVector 0.48.0, yaml 2.3.10.

3 Methods

3.1 Getting Started

Below we create a `Spectra` object from MS data stored in an mzML file. This file, provided through the *MsDataHub* Bioconductor package, contains LC–MS/MS data measured from a pesticide mixture using a data-dependent acquisition (DDA) method.

```
library(MsDataHub)
fl <- PestMix1_DDA.mzML()
```

We can now load the file with the `Spectra()` constructor to create an object of class `Spectra`.

```
library(Spectra)
sp <- Spectra(fl)
sp

MSn data (Spectra) with 7602 spectra in a MsBackendMzR
backend:
       msLevel     rtime scanIndex
     <integer> <numeric> <integer>
1            1     0.231         1
2            1     0.351         2
3            1     0.471         3
4            1     0.591         4
5            1     0.711         5
...        ...       ...       ...
7598         1   899.491      7598
7599         1   899.613      7599
7600         1   899.747      7600
7601         1   899.872      7601
7602         1   899.993      7602
 ...34 more variables/columns.

file(s):
baf70544dba2e_7861
```

The Spectra object assigned to the sp variable represents the MS data from the mzML file. It would also have been possible to directly import data from multiple files at once. The number of spectra in the object is:

```
length(sp)
[1]  7602
```

A Spectra object can be subset using [x:y]. In addition, a number of dedicated *filter* functions are available to subset a Spectra object, based on certain criteria. These will be discussed in Sect. 3.4. Below, we subset the Spectra object to the first ten spectra and determine the length of the resulting object. We use the R pipe operator |> concatenating otherwise nested function calls for better readability.

```
sp[1:10] |>
      length()

[1]  10
```

Which spectrum metadata (i.e., *spectra variables*) are available depends on what was defined in the original MS data file. The names of the available spectra variables of a Spectra object can be extracted with the spectraVariables() function:

```
spectraVariables(sp)

[1] "msLevel"                                    "rtime"
[3] "acquisitionNum"                         "scanIndex"
[5] "dataStorage"                           "dataOrigin"
```

```
[7] "centroided"                              "smoothed"
[9] "polarity"                                "precScanNum
[11] "precursorMz"                    "precursorIntensity"
[13] "precursorCharge"                   "collisionEnergy"
[15] "isolationWindowLowerMz" "isolationWindowTargetMz"
[17] "isolationWindowUpperMz"                 "peaksCount"
[19] "totIonCurrent"                         "basePeakMZ"
[21] "basePeakIntensity"               "ionisationEnergy"
[23] "lowMZ"                                      "highMZ"
[25] "mergedScan"                    "mergedResultScanNum"
[27] "mergedResultStartScanNum" "mergedResultEndScanNum"
[29] "injectionTime"                        "filterString"
[31] "spectrumId"                   "ionMobilityDriftTime"
[33] "scanWindowLowerLimit"         "scanWindowUpperLimit"
[35] "electronBeamEnergy"
```

These are the union of the `coreSpectraVariables()` and the spectrum metadata provided in the input data file. The core spectra variables, which include, for example, the MS level of a spectrum or its retention time, are variables with defined data types that can always be extracted from a `Spectra` object. If no values for these are defined in the `Spectra` object, `NA` is returned. Generally, spectra variables have a single value per spectrum. Thus, the full-spectrum metadata of a `Spectra` can be thought of as being a table (or more specifically, a data frame), with one row per spectrum and each column representing one of the spectra variables. Similar to a R's `data.frame`, it is possible to access metadata columns using the $ operator or use dedicated accessors. Below we extract the retention time values for the first ten spectra of the data set.

```
rtime(sp[1:10])
[1]  0.231 0.351 0.471 0.591 0.711 0.831 0.951 1.071
1.191
[10] 1.311
```

In addition, new spectra variables can be added using the $< - operator. Below, we add a new variable "`experiment`" to the `Spectra` object and assign the value "`Pest Mix 1`" to all spectra.

```
sp$experiment <- "Pest Mix 1"
```

Values for existing spectra variables can be replaced with the same $<- operator. Below, we replace the precursor's intensity (stored in the precursorIntensity spectra variable) by the estimated precursor ion intensity of MS2 spectra based on the MS1 data.

```
sp$precursorIntensity <- estimatePrecursorIntensity(sp)
```

MS data consist, next to the above-discussed spectra metadata, of the mass peaks data. In general, this is composed of paired *m/z* and intensity values of the individual mass peaks and represents a much larger amount of data than the spectra variables do. While for spectra variables a single value is defined for one spectrum, the number of *m/z* and intensity values can differ between spectra and range from 1 to several thousands. The number of mass peaks (hence *m/z* and intensity pairs) per spectrum can be determined with the `lengths()` method. Below we list the number of peaks of the first ten spectra.

```
sp[1:10] |>
   lengths()

[1]  223 211 227 210 220 228 201 215 214 211
```

The *m/z* or intensity values can be extracted, similar to the spectra variables above, using the `mz()` function (or `$mz`). Note that *m/z* values within each spectrum are by definition sorted increasingly.

```
mz(sp)

NumericList of length 7602
[[1]] 50.0157928466797… 610.188232421875
[[2]] 51.9395599365234 53.002872467041…
      610.180847167969
[[3]] 52.0061492919922 ... 685.213562011719
[[4]] 51.9393882751465 ... 610.178771972656
[[5]] 51.9397811889648 ... 834.235107421875
[[6]] 51.9395523071289 ... 612.173400878906
[[7]] 51.9389915466309 ... 743.192993164062
[[8]] 51.9403991699219 ... 611.171997070312
[[9]] 50.0138053894043 ... 758.209777832031
[[10]] 52.0183029174805 ... 758.214477539062
...
<7592 more elements>
```

While it is possible to extract *m/z* and intensity values separately, in most cases it is more efficient to extract the combined mass peaks data using the `peaksData()` function. This function returns a list of numeric matrices, with each column in a peaks matrix being one of the peaks variables (such as *m/z* or intensity) and each row representing one mass peak. The number of rows of these peaks matrices differs depending on the number of mass peaks present for one spectrum.

3.2 Different MS Data Representations

MS data can be very large, depending on the MS instrument properties and settings or the experiment setup. Thus, in most cases it is not advised, or even possible, to have the full MS data loaded into the computer's main memory. The MS infrastructure defined by the *Spectra* package is based on a clear separation between the data analysis code, implemented for the `Spectra` class, and the functionality required to handle and *represent* the MS data. Different MS backend implementations allow to adapt to the available computational resources or experiment properties, thus guaranteeing maximal flexibility of the infrastructure.

The default data backend to represent MS data imported from raw MS data files in mzML, mzXML or netCDF format is `MsBackendMzR`, based on the *mzR* package [10]. This backend, which was also used in the example above, loads all spectra metadata into memory, while the MS peaks data (*m/z* and intensity values) get loaded and hence *realised* into memory only on demand. This reduction in memory usage allows analyzing larger data sets, requiring, however, that the original MS data files remain available throughout the analysis.

The information on the backend used by a `Spectra` object is printed on the first line with the default `show()` method:

```
sp
```

```
MSn data (Spectra) with 7602 spectra in a MsBackendMzR
backend:
msLevel rtime scanIndex
<integer> <numeric> <integer>
1             1      0.231          1
2             1      0.351          2
3             1      0.471          3
4             1      0.591          4
5             1      0.711          5
... ... ... ...
7598          1    899.491       7598
7599          1    899.613       7599
7600          1    899.747       7600
7601          1    899.872       7601
7602          1    899.993       7602
... 35 more variables/columns.

file(s):
baf70544dba2e_7861
```

The size of this data in memory is:

```
print(object.size(sp), units = "MB")
```

```
2.4 Mb
```

A call to `mz()` (`$mz`), `intensity()` (`$intensity`), or `peaksData()` will load the respective data from the original MS data file(s) and return it to the user.

An alternative data representation would be the `MsBackendMemory`. This backend keeps all data, peaks and spectra metadata, in memory. It has thus a larger memory footprint but a faster access to its peaks data, and it is independent of the original MS data files. The *Spectra* package allows changing between data representations at any time, which adds to the flexibility of the infrastructure: original, raw data can be loaded using the `MsBackendMzR` and after initial cleaning and filtering, the backend could be changed to a `MsBackendMemory` for faster processing. Below, we change the backend of our `Spectra` object from the `MsBackendMzR` to the `MsBackendMemory` using the `setBackend()` function.

```
sp <- setBackend(sp, MsBackendMemory())
sp
MSn data (Spectra) with 7602 spectra in a
MsBackendMemory backend:
msLevel rtime scanIndex
<integer> <numeric> <integer>
1            1     0.231     1
2            1     0.351     2
3            1     0.471     3
4            1     0.591     4
5            1     0.711     5
... ... ... ...
7598         1   899.491  7598
7599         1   899.613  7599
7600         1   899.747  7600
7601         1   899.872  7601
7602         1   899.993  7602
... 35 more variables/columns.
Processing:
Switch backend from MsBackendMzR to MsBackendMemory
[Tue Jul 22 18:54:38 2025]
```

The size in memory of the sp data variable is now larger, because also the MS peaks data was loaded into memory.

```
print(object.size(sp), units = "MB")

25.3 Mb
```

Additional `MsBackend` implementations are defined in other R packages. The *MsBackendSql* package provides, for example, an MS data backend that uses an SQL database to store all data. Any SQL database supported by the *DBI* R package [11] can be used. Below we change the data representation to a backend from this package, storing all MS data into an SQLite database file, *sp.sqlite*.

```
library(MsBackendSql)
library(RSQLite)
sp <- setBackend(sp, MsBackendOfflineSql(),
     dbname = "sp.sqlite", drv = SQLite())

Importing data...

[========================================] 1/1
(100%) in 1s
Creating indices.... Done

sp

MSn data (Spectra) with 7602 spectra in a
MsBackendOfflineSql backend:
     msLevel precursorMz polarity
<integer> <numeric> <integer>
1            1     NA     1
```

```
2            1     NA     1
3            1     NA     1
4            1     NA     1
5            1     NA     1
... ... ... ...
7598            1     NA     1
7599            1     NA     1
7600            1     NA     1
7601            1     NA     1
7602            1     NA     1
... 36 more variables/columns.
Use 'spectraVariables' to list all of them.
Database: /home/lgatto/wrk/2025-Spectra-MiMB/sp.sqlite
Processing:
Switch backend from MsBackendMzR to MsBackendMemory
[Tue Jul 22 18:54:38 2025]
Switch backend from MsBackendMemory to
MsBackendOfflineSql [Tue Jul 22 18:54:39 2025]
```

Now, a `MsBackendOfflineSql` backend is used to represent the data. This backend keeps only the primary keys of the spectra in memory, and all MS data, metadata, or peaks data are retrieved from the database on-the-fly when requested. Due to its minimal memory footprint, this backend is best used to represent very large MS data sets.

```
print(object.size(sp), units = "MB")

0.1 Mb
```

The file size of the related SQLite database file is about 27 M:

```
file.size("sp.sqlite")
[1] 27336704
```

Specific backend implementations allow to add support for alternative MS data file formats as well, such as, for example, provided through the *MsBackendMgf* (support for files in the MGF Mascot Generic File format) or *MsBackendMsp* (support for files in MSP format).

3.3 Large-Scale Data Analysis Support

As discussed in the previous sections, data representations using a small memory footprint are ideally used with large data sets. Still, even if the backend does not occupy much memory, the MS data still needs to be loaded into memory for the data analysis.

The *Spectra* package uses two strategies to keep memory demand to a minimum for data analysis: any data analysis operation is not applied immediately to the data but is cached internally and only applied to the data upon request. This data access is in most cases done on data subsets, such as on spectra of one sample at a time, keeping the memory demand manageable. In addition, `Spectra` supports *chunk-wise* data processing. Using the `processingChunkSize()` function, it is possible to configure and set the size of the data chunks of a `Spectra` object. For any data analysis operation, the Spectra will then only load the data from the number of spectra defined with `processingChunkSize()` at a time. In addition to processing data in chunks, this also allows to perform the data operation in parallel (*see* **Note 2**).

Below we set the chunk size of our `Spectra` object to 1000 and use the `estimatePrecursorMz()` function. This will load and process the MS data for 1000 spectra at a time and return an *estimated* precursor *m/z* for MS2 spectra based on the signal of the previously measured MS1 spectrum.

```
processingChunkSize(sp) <- 1000
pmz <- estimatePrecursorMz(sp)
```

3.4 Tidy Spectra Processing

Tidy data processing [12], popularized by the *tidyverse*, has become a new standard in data analysis, and has been applied to omics data, under the tidyomics term [8]. In a nutshell, tidy data analysis revolves around two main principles, namely endomorphism, i.e., the transformation of an object to another instance of the same class, and the definition of standardized verbs (actions) to act on these objects. These two principles allows to chain operations using the pipe operator.

These principles have been adopted for the `Spectra` package. Spectra objects can be piped into various processing functions that will themselves return updated `Spectra` instances. The package defines for instances over 20 filtering functions, such as `filterMsLevel()`, `filterRt()`, `filterEmptySpectra()`, ... that can be used to filter the MS data.

In the example below, we keep MS level 1 scans acquired between 5 and 10 min, filtering on the retention time and MS level spectra variables.

```
sp |>
filterRt(rt = c(300, 600)) |>
filterMsLevel(1L)

MSn data (Spectra) with 1003 spectra in a
MsBackendOfflineSql backend:
msLevel precursorMz polarity
<integer> <numeric> <integer>
1           1      NA      1
2           1      NA      1
3           1      NA      1
4           1      NA      1
5           1      NA      1
... ... ... ...
999         1      NA      1
1000        1      NA      1
1001        1      NA      1
1002        1      NA      1
1003        1      NA      1
... 36 more variables/columns.
```

```
Use 'spectraVariables' to list all of them.
Database: /home/lgatto/wrk/2025-Spectra-MiMB/sp.sqlite
Processing:
Switch backend from MsBackendMzR to MsBackendMemory
[Tue Jul22 18:54:38 2025]
Switch backend from MsBackendMemory to
MsBackendOfflineSql [Tue Jul 22 18:54:39 2025]
Filter: select retention time [300..600] on MS
level(s) [Tue Jul 22 18:56:01 2025]
...1 more processings. Use'processingLog' to list all.
```

It is also possible to filter on the peaks data to keep mass peaks with intensities greater than 100 and m/z between 200 and 800:

```
sp |>
filterIntensity(intensity = c(100, Inf)) |>
filterMzRange(mz = c(200, 800))

MSn data (Spectra) with 7602 spectra in a
MsBackendOfflineSql backend:
msLevel precursorMz polarity
<integer> <numeric> <integer>
1            1      NA      1
2            1      NA      1
3            1      NA      1
4            1      NA      1
5            1      NA      1
... ... ... ...
7598         1      NA      1
7599         1      NA      1
7600         1      NA      1
7601         1      NA      1
7602         1      NA      1
... 36 more variables/columns.
Use 'spectraVariables' to list all of them.
Database: /home/lgatto/wrk/2025-Spectra-MiMB/sp.sqlite
Lazy evaluation queue: 2 processing step(s)
Processing:
Switch backend from MsBackendMzR to MsBackendMemory
[Tue Jul 22 18:54:38 2025]
Switch backend from MsBackendMemory to
MsBackendOfflineSql [Tue Jul 22 18:54:39 2025]
Remove peaks with intensities outside [100, Inf]
in spectra of MS level(s) 1, 2. [Tue Jul...1 more
processings. Use'processingLog' to list all.
```

It is of course possible to extend the chaining of operations outside of `Spectra` and pipe the spectra variables into the `ggplot()` tidyverse function to visualize the total ion chromatogram (Fig. 2).

```
library(ggplot2)
sp |>
filterMsLevel(1L) |>
spectraData() |>
```

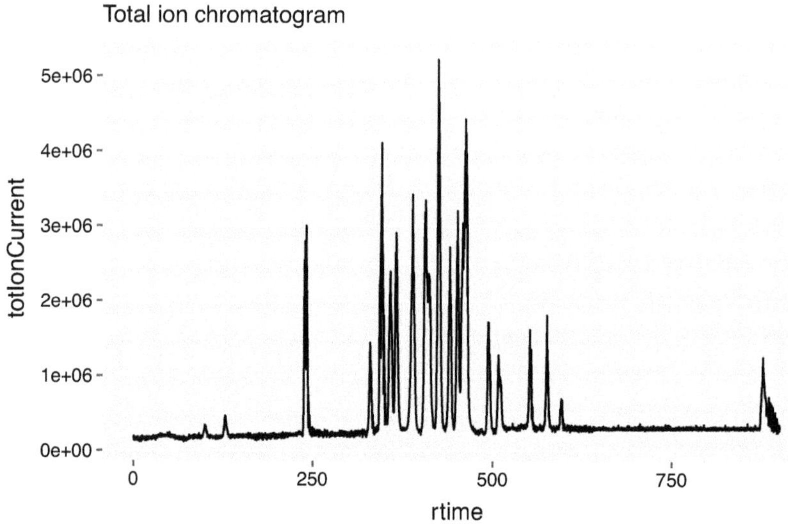

Fig. 2 Total ion chromatogram of MS level 1 scans

```
ggplot(aes(x = rtime, y = totIonCurrent)) +
geom_line() +
ggtitle("Total ion chromatogram")
```

For the subsequent sections, we focus on the MS2 scans in the object and update the sp variable to contain only MS level 2 scans. In addition, we remove empty spectra, i.e., spectra without mass peaks.

```
sp <- sp |>
filterMsLevel(2L) |>
filterEmptySpectra()
```

3.5 Data Analysis Beyond R: Python Integrating with SpectriPy

The *Spectra* package provides a powerful infrastructure for MS data handling and analysis in R. In Python, specific libraries such as *matchms* [13] or *spectrum_utils* [14] offer useful features that would ideally complement *Spectra*.

The *SpectriPy* package graeve_spectripy_2025 [15] bridges these two worlds, allowing access to functionality from both programming languages and thus avoiding the need to re-implement algorithms missing in the other language. *SpectriPy* builds upon the R *reticulate* package [16] that translates base R data types between R and Python, opening the way to translate MS data structures between both programming languages. In addition, *SpectriPy* defines a dedicated MS data backend to represent MS data stored in an associated Python environment, which enables a more elegant and seamless integration of functionalities from MS libraries, and combine these in mixed R-Python analysis workflows.

As an example, we change the backend of our sp `Spectra` data to an `MsBackendPy` backend, as defined in *SpectriPy*. This will convert the full MS data to respective MS data structures in Python (by default to `matchms.Spectrum` objects from the *matchms* Python library) and store it in a new variable (defined by the `pythonVariableName parameter`) in the associated Python environment. We first set the `processingChunkSize` to Inf to disable chunk-wise processing when switching to the new backend.

```
library(SpectriPy)

Loading required package: reticulate

Using virtual environment'r-spectripy'

processingChunkSize(sp) <- Inf
sp <- setBackend(sp, MsBackendPy(), pythonVariableName
= "pysp")
sp

MSn data (Spectra) with 2756 spectra in a MsBackendPy
backend:
Data stored in the "pysp" variable in Python
Processing:
Switch backend from MsBackendMzR to MsBackendMemory
[Tue Jul 22 18:54:38 2025]
Switch backend from MsBackendMemory to
MsBackendOfflineSql [Tue Jul 22 18:54:39 2025]
Filter: select MS level(s) 2 [Tue Jul 22 18:56:02
2025]
...2 more processings. Use 'processingLog' to list all.
```

The `MsBackendPy` keeps only a reference to the data that is now stored in the associated Python environment; the size on the R side is thus negligiible.

```
print(object.size(sp), units = "MB")
0 Mb
```

The data are now stored as a list of `matchms.Spectrum` objects in Python, but we can still access and extract data in R. The peak intensities for the first ten spectra are:

```
sp[1:10] |>
intensity()

NumericList of length 10
[[1]] 0.0482765696942806 0.0972539931535721
      0.0576995946466923
[[2]] 0.032534446567297 0.00529412413015962
      0.199114724993706
[[3]] 0.0443157367408276 0.0587722808122635
      0.00599231710657477
[[4]] 0.0401345416903496 0.0546061582863331
      0.048479650169611
[[5]] intensity=0.0218546539545059
[[6]] 0.0801825001835823 ... 0.0308814756572247
[[7]] 0.0179884266108274 ... 0.106236264109612
```

```
[[8]] 0.077180951833725 ... 0.0169821456074715
[[9]] 0.00520043354481459 ... 0.00520535605028272
[[10]] 0.00510971155017614 0.0114039815962315
       0.0265298280864954
```

Python code blocks are labeled with #' PythonProgramming languagePython session to discriminate them from the R code blocks in this chapter.

In the following Python code block, we view the number of spectra and access the first spectrum of the `pysp` `matchms.Spectrum` object using basic Python commands.

```
#' Python session

#' Number of spectra
len(pysp)

2756

#' Access the first spectrum
pysp[0]

Spectrum(precursor m/z=137.96, 3 fragments between
55.9 and 138.0)
```

Next, we apply some filtering on the `matchms.Spectrum` object using available libraries from *matchms*. Default filtering was performed to standardize ion mode, correct charge, and more, as documented in the matchms filtering documentation. In addition, we scale the peak intensities to a maximum of one in each spectrum.

```
 #' Python session

from matchms.filtering import default_filters,
normalize_intensities

# Apply filters to clean and enhance each spectrum
for i in range(len(pysp)):
# Apply default filter to standardize ion mode, correct
charge and
# more. Default filter is fully explained at

# https://matchms.readthedocs.io/en/latest/api/
matchms.filtering.html

spectrum = default_filters(pysp[i])
# Scale peak intensities to maximum of 1
pysp[i] = normalize_intensities(spectrum)
```

The `MsBackendPy` backend used by the `sp` `Spectra` object refers to the variable pysp in the associated Python session and does not contain a copy of the MS data. The above applied data manipulations are thus immediately accessible in R:

```
sp[1:10] |>
intensity()
```

```
NumericList of length 10
[[1]] 0.496396786690783 1 0.593287666405428
[[2]] 0.163395482520569 0.0265883104844555 1
[[3]] 0.754024450444343 1 0.101958219483026
[[4]] 0.734981968149085 1 0.887805545949651
[[5]] intensity=1
[[6]] 0.312258644632166 ...0.12026324585649
[[7]] 0.111564910914543 1 ...0.658881379550376
[[8]] 0.890933866704096 ...0.196032418511653
[[9]] 0.0768728714428873 1 0.230786693452863
      0.0769456359781453
[[10]] 0.192602512670527 0.429855088357564 1
```

```
The maximum intensity is thus 1 for each spectrum.
```

```
sp[1:10] |>
intensity() |>
max()
```

```
[1] 1 1 1 1 1 1 1 1 1 1
```

The *SpectriPy* package enables to combine MS data analysis routines and functions from Python libraries and R packages, unlocking the way to more powerful and complete cross-language data analysis.

3.6 Data Processing and Visualization

We covered previously the infrastructure of a Spectra object and how to inspect it, as well as performing some direct data manipulations using Python libraries. We now turn our attention to visualization, a crucial aspect of mass spectrometry analysis. Despite working in a command-line environment, rich and informative visual exploration of the data remains entirely possible. The *Spectra* package offers a wide array of functions for both processing and visualizing spectral data.

We will begin by loading a publicly available spectral library, namely *MassBank* [17], using the *AnnotationHub* package [18]. MassBank is a large collection of mass spectra focused primarily on small molecules. It is widely used in metabolomics and environmental sciences for spectral matching and compound identification, offering a curated reference set against which experimental spectra can be compared.

```
library(AnnotationHub)

Loading required package: BiocFileCache

Loading required package: dbplyr

library(CompoundDb)

Loading required package: AnnotationFilter

ah <- AnnotationHub()
```

```
#' List available MassBank data sets
query(ah, "MassBank")

AnnotationHub with 8 records
# snapshotDate(): 2025-04-08
# $dataprovider: MassBank
# $species: NA
# $rdataclass: CompDb
# additional mcols(): taxonomyid, genome,
# description, coordinate_1_based, maintainer,
# rdatadateadded, preparerclass, tags, rdatapath,
# sourceurl, sourcetype
# retrieve records with, e.g., 'object[["AH107048"]]'

  AH107048 | MassBank CompDb for release 2021.03
  AH107049 | MassBank CompDb for release 2022.06
  AH111334 | MassBank CompDb for release 2022.12.1
  AH116164 | MassBank CompDb for release 2023.06
  AH116165 | MassBank CompDb for release 2023.09
  AH116166 | MassBank CompDb for release 2023.11
  AH119518 | MassBank CompDb for release 2024.06
  AH119519 | MassBank CompDb for release 2024.11
```

Let us load the latest MassBank release.

```
mb <- ah[["AH119519"]]
loading from cache

mb

class: CompDb
data source: MassBank
version: 2024.11
organism: NA
compound count: 122512
MS/MS spectra count: 122512
```

The MassBank data are provided as a self-contained SQLite database and includes metadata on small molecules (e.g., names, chemical formula, exact monoisotopic mass, InChI, etc.), as well as their associated fragment spectra.

Since we are only interested in the spectra, we extract them and filter to retain only MS2-level spectra (some entries may be NA or MS level > 2).

```
mb_sp <- Spectra(mb) |>
filterMsLevel(2L)
```

Next, we scale the intensities of these reference spectra with the same approach applied previously, so that the intensity of each spectrum's highest peak is set to 1. This scaling makes spectra easier to compare and visualise.

```
mb_sp <- scalePeaks(mb_sp, by = max)
```

Here, `scalePeaks(…, by = max)` sets the peak with the highest intensity in each spectrum to 1 and scales all other peaks proportionally. We then remove peaks with an intensity below 5% of the maximum intensity, which we consider to be noise. Applying the same preprocessing steps to both experimental and reference spectra ensures fair comparison.

```
#' remove peaks below 0.05 of the max (1L) peak.
low_int <- function(x,...) x > 0.05
mb_sp <- filterIntensity(mb_sp, intensity = low_int)
```

We can check the scaled intensities for the first 10 MassBank spectra:

```
mb_sp[1:10] |>
intensity()

NumericList of length 10
[[1]] 0.226978413480687 … 0.751891512076341
[[2]] 0.194635909762203 … 0.760492840055249
[[3]] 0.240220272080312 ... 0.578481255079962
[[4]] 0.0651552850616005 ... 0.33119738639121
[[5]] 0.423181771259126 1 0.709196486100968
0.0572011624793549
[[6]] intensity=1
[[7]] 0.246862101303013 0.419460453597735 ... 1
[[8]] 0.0502634970728531 ... 0.0516241159091104
31
[[9]] 0.0815629737381796 ... 0.48514919810953
[[10]] 0.372648818851002 ... 0.245518575569845
```

We can now see that the peak intensities are now scaled to a maximum of 1.

Before performing spectral matching, we can further filter the reference spectra: restrict to spectra with positive polarity (our experimental data were acquired in ESI+ mode) and keep only spectra with more than three peaks, as very sparse spectra may be uninformative.

```
mb_sp <- mb_sp[lengths(mb_sp) > 3] |>
filterPolarity(polarity = 1)

sp <- sp[lengths(sp) > 3]
```

Multiple plotting functions are available in `Spectra` for data visualization. Below, we display the first and tenth spectrum from the MassBank database using the `plotSpectra()` function (Fig. 3).

```
plotSpectra(mb_sp[c(1, 10)],
   main = c("1st MassBank spectrum", "10th MassBank
   spectrum"))
```

It is also possible to overlay multiple spectra in one plot using `plotSpectraOverlay()` (Fig. 4).

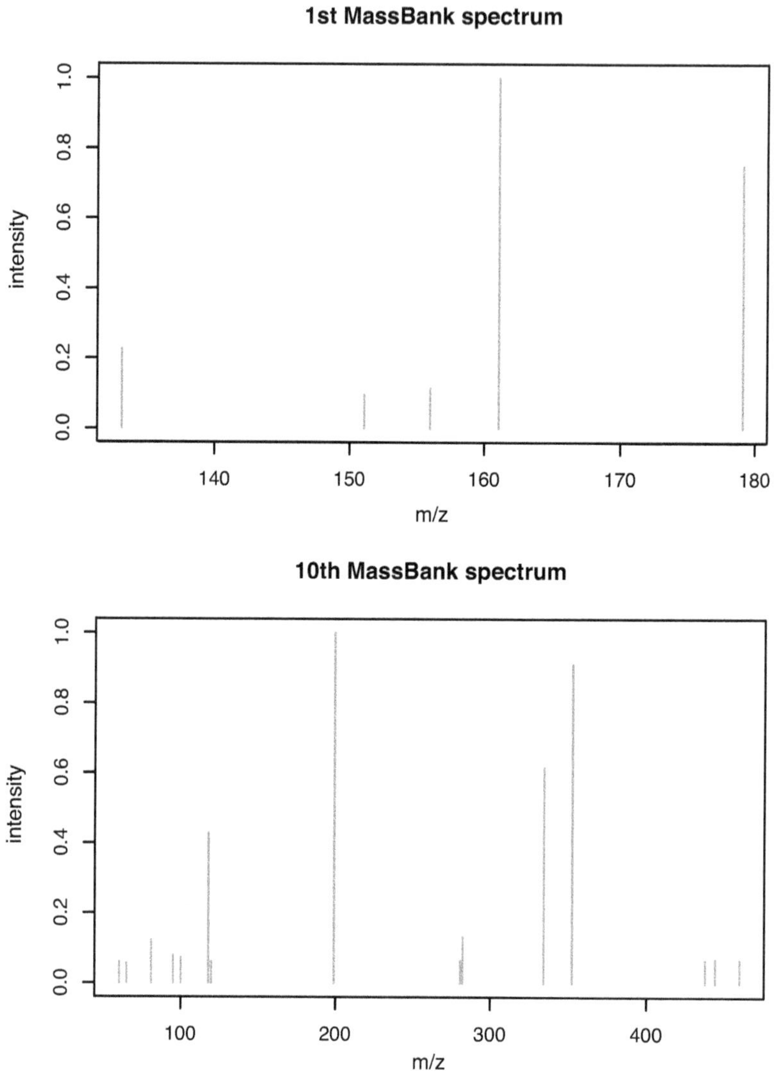

Fig. 3 First and tenth MassBank spectra

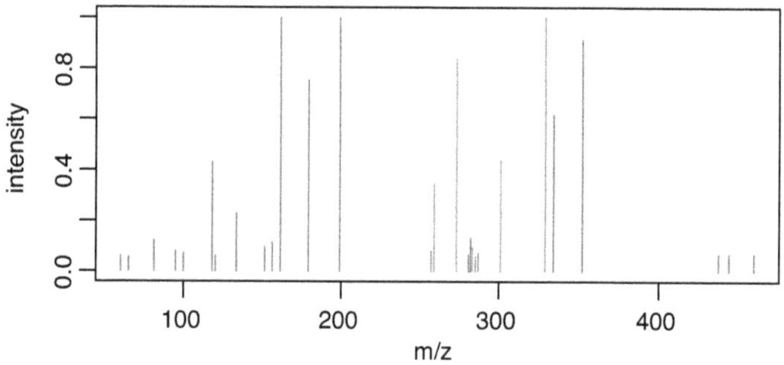

Fig. 4 Overlay of 1st, 10th, and 20th MassBank spectra

```
library(RColorBrewer)
tencol <- brewer.pal(3, "Set1")
plotSpectraOverlay(mb_sp[c(1,10, 20)], col = tencol,
main = "")
```

Spectral similarity calculations can be computationally intensive. To reduce the number of comparisons, we filter the MassBank database using `filterPrecursorMz()` to retain only spectra whose precursor *m/z* values are close to those in our experimental dataset:

```
mb sp <- filterPrecursorMzValues(mb_sp, mz =
sp$precursorMz, ppm = 20)
mb_sp

MSn data (Spectra) with 3484 spectra in a
MsBackendCompDb backend:
msLevel precursorMz polarity
<integer> <numeric> <integer>
1          2      529.269      1
2          2      529.269      1
3          2      529.269      1
4          2      529.269      1
5          2      529.269      1
... ... ... ...
3480       2      283.096      1
3481       2      429.100      1
3482       2      429.100      1
3483       2      253.000      1
3484       2      253.000      1
...    46 more variables/columns.
Use 'spectraVariables' to list all of them.
data source: MassBank
version: 2024.11
organism: NA
Lazy evaluation queue: 2 processing step(s)
Processing:
Filter: select MS level(s) 2 [Tue Jul 22 18:56:26
2025]
Scale peak intensities in spectra of MS level(s) 2.
[Tue Jul 22 18:56:27 2025]
Remove peaks based on their intensities and a user-
provided function in spectra of MS level(...2 more
processings. Use'processingLog' to list all.
```

Now we have 3484 scans, a more manageable number of reference spectra. The lazy evaluation queue in the `Spectra` print output indicates the processing steps (e.g., scaling, filtering) that have yet to been applied to the underlying data when effectively is accessed. This has the advantage that data manipulation operations that affect mass peak data can also be used with read-only mass spectrometry data—such as the SQLite database with the reference spectra used in our example. In addition, since data manipulations are in most cases applied only data subsets, it is computationally more efficient.

Since we now intend to use the full data for comparison, it makes sense to apply the processing steps explicitly. Additionally, we will move the spectra to memory (via `MsBackendMemory`) to avoid latency caused by accessing data from disk or database.

```
mb_sp <- setBackend(mb_sp, MsBackendMemory())
```

```
'MsBackendCompDb' does not support parallel
processing.
Switching to serial processing.
```

```
mb_sp <- applyProcessing(mb_sp)
```

We now have both the reference and experimental spectra ready for comparison using `compareSpectra()`. This function computes similarity scores using a user-defined function. Here, we use the normalized dot product defined in `MsCoreUtils::ndotproduct()`. More information of how spectra are compared can be found in the documentation of `?compareSpectra`.

```
mtch <- compareSpectra(sp, mb_sp)
dim(mtch)
```

```
[1] 2038 3484
```

The spectra comparison returns a similarity matrix. To visualize the top matches, we subset the matrix to keep only spectra with at least one similarity score above 0.6 and visualize the similarity matrix subset as a heatmap (Fig. 5).

```
submtch <- mtch[, apply(mtch, 2, function(x) any(x >
0.6))]
```

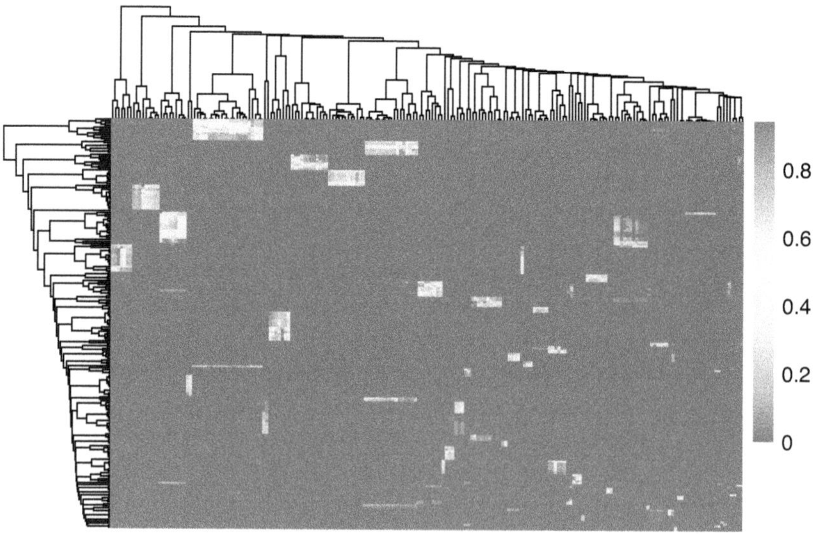

Fig. 5 Heatmap of experimental vs MassBank spectra similarity scores

```
submtch <- submtch[apply(submtch, 1, function(x) any(x
> 0.6)),]

library(pheatmap)
pheatmap(submtch,
      show rownames = FALSE,
      show colnames = FALSE)
```

To explore specific matches in more detail, we use mirror plots, which are a standard visualization tool in mass spectrometry.

First, we plot the best match (highest similarity score, Fig. 6):

```
#' Get coordinate of best match
idx_mx <- which(mtch == max(mtch), arr.ind = TRUE)
plotSpectraMirror(

sp[idx_mx[1]], mb_sp[idx_mx[2]],
main = c(mb_sp[idx_mx[2] ]$name, "score: ",
mtch[idx_mx]))
grid()
```

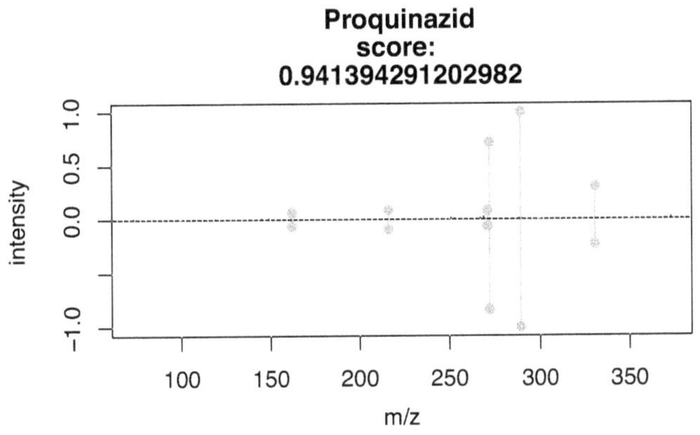

Fig. 6 Mirror plot of the best match

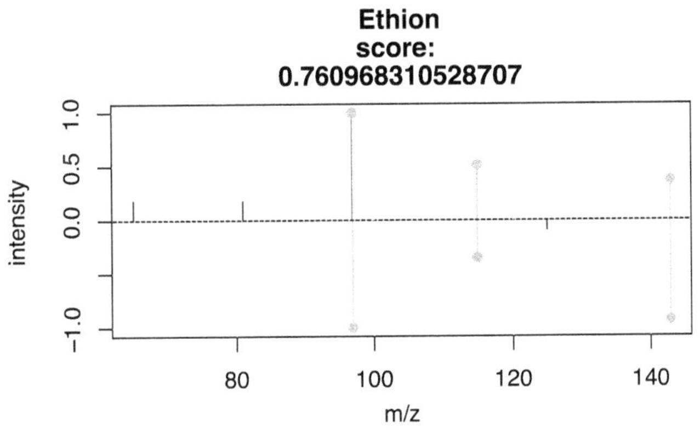

Fig. 7 Mirror plot of the 200th best match

We observe that both *m/z* and intensity values match extremely well. Next, let's look at a less perfect match (Fig. 7):

```
#' Get coordinate of the 200th best match
idx_mx <- which(mtch == sort(mtch, decreasing = TRUE)
[200], arr.ind = TRUE)
plotSpectraMirror(


sp[idx mx[1]], mb sp[idx mx[2]],
main = c(mb sp[idx mx[2] ]$name, "score: ", mtch[idx
mx]))
grid()
```

The two compared spectra share some of their fragments, but the intensities also differ, which resulted in a lower similarity score. See Note 3 for additional plotting functions.

The *Spectra* package therefore does not limit itself to comprehensive and efficient infrastructure for mass spectrometry data but also allows for processing, comparison, and visualization.

4 Notes

1. The versions of R and the specific Bioconductor release are tight. It is thus necessary to use a recent version of R itself for `BiocManager::install()` to install the latest versions of Bioconductor packages.
2. The user should, however, be aware that in such cases, the data for `processingChunkSize * no cpu cores` spectra is loaded in memory, and that collecting and combining the results from the parallel processes will require an additional, of not negligible, amount of memory.
3. The `PSMatch` package [7] provides additional plotting capabilities, including support for the visualization of posttranslational modification with `plotSpectraPTM()`.

Funding

PL is funded by an European Union HORIZON-MSCA-2021 project, Grant No. 101073062: HUMAN—Harmonising and Unifying Blood Metabolic Analysis Networks.

Data and Software Availability All data needed to reproduce the code are directly available through data-dedicated packages. All software packages are open-source and freely available as part of the

Bioconductor project as well as in their respective GitHub repositories (https://github.com/RforMassSpectrometry/Spectra).

Competing Interests The authors report no competing interests.

References

1. Martens L, Chambers M, Sturm M, Kessner D, Levander F, Shofstahl J, Tang WH, R¨ompp A, Neumann S, Pizarro AD, Montecchi-Palazzi L, Tasman N, Coleman M, Reisinger F, Souda P, Hermjakob H, Binz P-A, Deutsch EW (2011) mzML–a community standard for mass spectrometry data. Mol Cell Proteomics 10(1):110
2. Holman JD, Tabb DL, Mallick P (2014) Employing ProteoWizard to convert raw mass spectrometry data. Curr Protoc Bioinform 46:13.24.1–13.24.9. https://doi.org/10.1002/robo0471250953.bi1324s46
3. Louail P, Brunius C, Garcia-Aloy M, Kumler W, Storz N, Stanstrup J, Treutler H, Vangeenderhuysen P, Witting M, Neumann S, Rainer J (2025) Xcms in peak form: now anchoring a complete metabolomics data preprocessing and analysis software ecosystem. Anal Chem 97(50):27639–27645. https://doi.org/10.1021/acs.analchem.5c04338. Epub 2025 Dec 8. PMID: 41359826; PMCID: PMC12750407
4. Gentleman RC, Carey VJ, Bates DM, Bolstad B, Dettling M, Dudoit S, Ellis B, Gautier L, Ge Y, Gentry J, Hornik K, Hothorn T, Huber W, Iacus S, Irizarry R, Leisch F, Li C, Maechler M, Rossini AJ, Sawitzki G, Smith C, Smyth G, Tierney L, Yang JYH, Zhang J (2004) Bioconductor: open software development for computational biology and bioinformatics. Genome Biol 5(10):80. https://doi.org/10.1186/gb-2004-5-10-r80
5. Huber W, Carey VJ, Gentleman R, Anders S, Carlson M, Carvalho BS, Bravo HC, Davis S, Gatto L, Girke T, Gottardo R, Hahne F, Hansen KD, Irizarry RA, Lawrence M, Love MI, MacDonald J, Obenchain V, Ole´s AK, Pag'es H, Reyes A, Shannon P, Smyth GK, Tenenbaum D, Waldron L, Morgan M (2015) Orchestrating high-throughput genomic analysis with bioconductor. Nat Methods 12(2):115–121. https://doi.org/10.1038/nmeth.3252
6. Gatto L, Gibb S, Rainer J (2021) MSnbase, efficient and elegant R-based processing and visualization of raw mass spectrometry data. J Proteome Res 20(1):1063–1069. https://doi.org/10.1021/acs.jproteome.0c00313 Epub 2020 Sep 28. PMID: 32902283
7. Deflandre G, Gibb S, Gatto L (2025) PSMatch: an R/Bioconductor package to explore proteomics identification data. OSF Preprints. https://doi.org/10.31219/osf.io/62v9pv1. osf.io/62v9pv1
8. Hutchison WJ, Keyes TJ, tidyomics Consortium, Crowell HL, Serizay J, Soneson C, Davis ES, Sato N, Moses L, Tarlinton B, Nahid AA, Kosmac M, Clayssen Q, Yuan V, Mu W, Park J-E, Mamede I, Ryu MH, Axisa P-P, Paiz P, Poon C-L, Tang M, Gottardo R, Morgan M, Lee S, Lawrence M, Hicks SC, Nolan GP, Davis KL, Papenfuss AT, Love MI, Mangiola S (2024) The tidyomics ecosystem: enhancing omic data analyses. Nat Methods 21(7):1166–1170
9. Graeve MD, Bittremieux W, Naake T, Huber C, Anagho-Mattanovich M, Hoffmann N, Marchal P, Chrone V, Louail P, Hecht H, Witting M, Rainer J (2025) SpectriPy: enhancing cross-language mass spectrometry data analysis with R and Python. J Open Source Softw 10(109):8070. https://doi.org/10.21105/joss.08070
10. Chambers MC, Maclean B, Burke R, Amodei D, Ruderman DL, Neumann S, Gatto L, Fischer B, Pratt B, Egertson J, Hoff K, Kessner D, Tasman N, Shulman N, Frewen B, Baker TA, Brusniak M-Y, Paulse C, Creasy D, Flashner L, Kani K, Moulding C, Seymour SL, Nuwaysir LM, Lefebvre B, Kuhlmann F, Roark J, Rainer P, Detlev S, Hemenway T, Huhmer A, Langridge J, Connolly B, Chadick T, Holly K, Eckels J, Deutsch EW, Moritz RL, Katz JE, Agus DB, MacCoss M, Tabb DL, Mallick P (2012) A cross-platform toolkit for mass spectrometry and proteomics. Nat Biotechnol 30(10):918–920
11. Databases (R-SIG-DB), RSIG, Wickham H, M¨uller K (2024) DBI: R database interface. https://doi.org/10.32614/CRAN.package.DBI . R package version 1.2.3. https://CRAN.R-project.org/package=DBI
12. Wickham H (2014) Tidy data. J Stat Softw 59(10):1–23
13. Huber F, Verhoeven S, Meijer C, Spreeuw H, Castilla E, Geng C, Van Der Hooft J, Rogers S, Belloum A, Diblen F, Spaaks J (2020) matchms— processing and similarity evaluation

of mass spectrometry data. J Open Source Softw 5(52):2411. https://doi.org/10.21105/joss.02411
14. Bittremieux W (2020) spectrum utils: a Python package for mass spectrometry data processing and visualization. Anal Chem 92(1):659–661. https://doi.org/10.1021/acs.analchem.9b04884
15. De Graeve et al (2025) SpectriPy: Enhancing cross-language mass spectrometry data analysis with R and Python. J. Open Source Softw 10(109):8070. https://doi.org/10.21105/joss.08070
16. Ushey K, Allaire J, Tang Y (2025) Reticulate: interface to 'Python'. https://doi.org/10.32614/CRAN.package.reticulate. R package version 1.41.0. https://github.com/rstudio/reticulate
17. Elapavalore A, Kondi´c T, Singh RR, Shoemaker BA, Thiessen PA, Zhang J, Bolton EE, Schymanski EL (2023) Adding open spectral data to massbank and pubchem using open source tools to support non-targeted exposomics of mixtures. Environ Sci Process Impacts 25(11):1788–1801. https://doi.org/10.1039/d3em00181d
18. Morgan M, Shepherd L (2025) AnnotationHub: client to access AnnotationHub resources. https://doi.org/10.18129/B9.bioc.AnnotationHub. R package version 3.16.0. https://bioconductor.org/packages/AnnotationHub

Part II

Mass Spectrometry Data Analysis

Check for updates

Chapter 4

Preprocessing Large-Scale Proteomics Data

Xindong Sun, Markus Ralser and Christoph B. Messner

Abstract

Recent advances in mass spectrometry instrumentation and computational analysis have established label-free data-independent acquisition (DIA) as a powerful tool in quantitative proteomics, offering depth, throughput, and reproducibility. These developments now enable large-scale projects with hundreds or even thousands of samples, opening new applications in systems biology and biomedicine. However, large-scale datasets require special consideration during data preprocessing. Here, we present a practical guide for preprocessing large-scale DIA datasets, illustrated using a public dataset. We cover key steps, including database searching, normalization, batch correction, missing value imputation, and evaluation of preprocessing strategies, and provide R code together with practical considerations.

Keywords Large-scale proteomics, Data-independent acquisition (DIA), Label-free quantification, Data preprocessing, Normalization, Batch effect correction, Missing value imputation

1 Introduction

Mass spectrometry (MS)-based proteomics enables the simultaneous identification and quantification of large numbers of proteins in complex biological samples. Over the past decade, major advances in instrumentation and data analysis have increased throughput, depth, and sensitivity. In particular, label-free data-independent acquisition (DIA) methods have been widely adopted for quantitative proteomics and are particularly powerful for large-scale studies [1–3].

With increased throughput, it is now possible to run hundreds or even thousands of samples within a reasonable timeframe. This enables many new applications and insights into systems biology, large clinical studies, and even population-wide studies [3–11].

However, performing sample preparation, data acquisition, and database searching at scale introduces challenges for data analysis, including increased batch effects and missing values (Fig. 1).

Frédérique Lisacek (ed.), *Bioinformatics for Mass Spectrometry*, Methods in Molecular Biology, vol. 3071,
https://doi.org/10.1007/978-1-0716-5519-1_4,

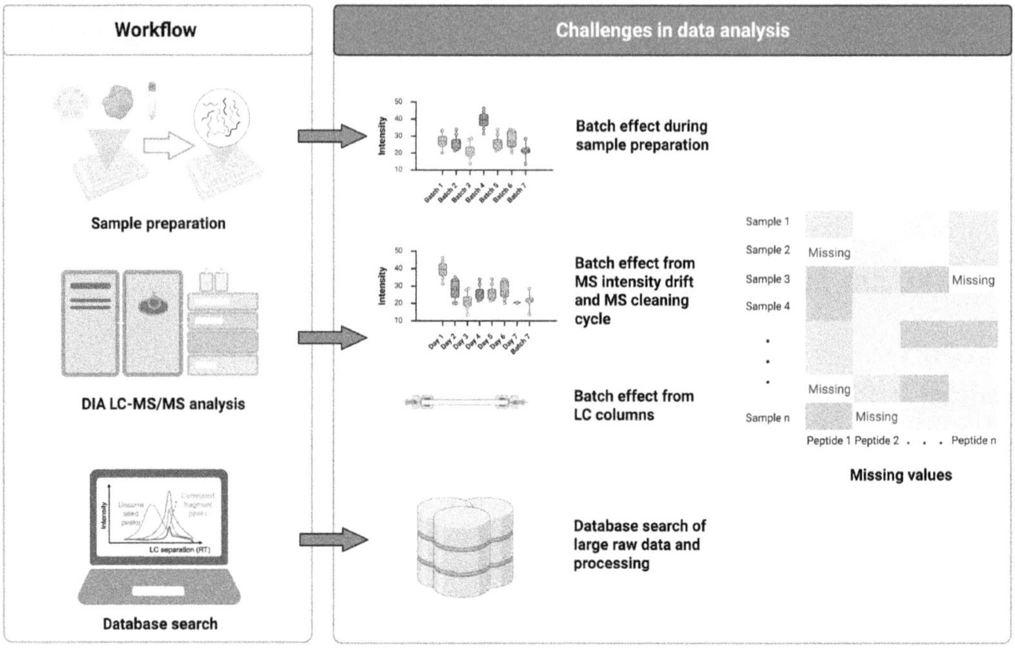

Fig. 1 Challenges in raw data processing and preprocessing in large-scale DIA experiments. The left panel (workflow) highlights key steps in the workflow, while the right panel illustrates common challenges. The schematic diagram was created using BioRender.com

1.1 Raw Data Processing for Large-Scale DIA Datasets

DIA generates highly multiplexed spectra containing fragment ions from multiple coeluting peptides [12]. This complexity requires a sophisticated algorithm to reliably identify peptides and quantify them based on extracted ion chromatogram. Several software platforms are now available for DIA analysis, which have been compared and benchmark [13–16]. Widely used DIA analysis tools include DIA-NN [17,18], Spectronaut, OpenSWATH [19], fragpipe (MSfragger-DIA) [20], Skyline [21], maxDIA [22], EncyclopeDIA [23], and CHIMERYS [24].

Computation time is a critical consideration in large-scale DIA data processing and depends strongly on the applied algorithms. In particular, processing speed differs between library-based and library-free workflow. Library-based methods rely on preconstructed spectral libraries containing information such as retention times, fragment ion spectra, and peptide intensities, enabling targeted signal extraction. In contrast, library-free approaches analyze DIA data without prior libraries, either by using in silico-generated spectral libraries [18] or by applying algorithms that directly perform database matching [25]. While library-free methods avoid dependence on preexisting libraries, they are often more computationally demanding due to the size of in silico libraries or the complexity of the algorithm used.

To increase scalability, DIA processing can be deployed in cloud or high-performance computing (HPC) environments, allowing parallelized data analysis for thousands of samples [26]. Scalable "match-between-runs" (MBR) algorithm have been further implemented for such distributed workflow [26].

1.2 Normalization and Protein Summarization

Normalization is a critical step to correct for systematic technical variations and ensure comparability across samples [27,28]. It is typically integrated into DIA processing software, which provides a built-in algorithm that can be adapted to different experimental setups. Common normalization strategies include global normalization methods, such as median or quantile normalization, which assume that the overall protein abundance distribution remains constant across samples [27,28]. More sophisticated approaches apply retention time-dependent normalization to account for signal drift over the course of chromatographic separation, which is particularly relevant for longer LC gradients [18]. The choice of normalization strategy should consider the specific characteristics of the dataset, including gradient length, instrument stability, and the expected biological variation [29]. In addition to normalization, DIA processing software performs protein summarization methods, such as topN summarization [12] or the MaxLFQ algorithm [22,30], which aggregate peptide-level quantification to protein-level abundances.

1.3 Sources of Batch Effects in Large-Scale DIA Proteomics

Batch effects are one of the biggest challenges in large-scale DIA proteomics, as they can mask, or in cases of poor experimental design, mimic biological signals. While batch effects should be minimized wherever possible, they cannot be entirely eliminated. Therefore, diagnosing and correcting batch effects are essential steps to improve the biological signal in the data [31].

Sample preparation introduces batch effects through differences in sample handling, protein extraction efficiency, digestion completeness, or peptide recovery. Additional sources of batches can result from reagent lot variation or laboratory personnel. Standardization is therefore critical. Automated workflow, standardized protocols, robust quality management systems, and system suitability testing all help reduce variability [3,32–35].

Although improvements in instrument design have reduced contamination, leading to fewer cleaning cycles, and advances in chromatography have increased robustness [36–39], even well-maintained instruments can show gradual intensity drifts. Additionally, instrument maintenance or column changes can introduce batch effects. Further, environmental factors such as laboratory temperature or barometric pressure can further affect data consistency over time [40]. Regular system suitability tests and QC sample injections are essential for diagnosing and mitigating instrument-related batch effects [41,42].

For clinical samples, consistency in preanalytical steps (e.g., sample collection, storage, transport) must be considered. While these factors are often not modifiable, they should be accounted for during experimental design.

1.4 Batch Effect Correction and Drift Correction

Though good standardization and quality management can minimize systematic variation, batch effects will remain and must be addressed computationally. While batches can be adjusted using classical statistical approaches (e.g., including batch as a covariate in linear models), most proteomics studies first apply batch correction and then use a "batch-free" dataset for downstream analysis [27]. Algorithm such as median centering, ComBat, and RUV-III-C are commonly employed [3,27,43–50]. Depending on dataset size and sample heterogeneity, correction can be guided by quality control (QC) samples or applied across all samples.

In addition to discrete batch effects, continuous instrumental drift can occur over time, often correlating with acquisition order. Regular acquisition of QC samples and inspection of total ion chromatogram (TICs) provide essential diagnostic information [41]. Drift can subsequently be corrected computationally using linear or nonlinear models, such as LOESS regression [3,27,51].

While batch correction is an important step, overcorrection must also be considered, as it may remove biological variation, particularly when biological signals are confounded with batch structure. The appropriateness of batch correction should therefore be carefully assessed, for example, by visualization and by evaluating whether biological heterogeneity is preserved [41,52–54].

Computational correction can only partially mitigate the effects of poor experimental design. Careful planning remains essential for the success of large-scale proteomics studies. Best practices include proper randomization and blocking of samples to minimize systematic biases [3,33,55–57].

1.5 Missing Values and Imputation

Missing values represent another challenge that becomes increasingly problematic as experiments grow in scale. In proteomics data, missingness generally arises from two sources: (1) missing not at random (MNAR), typically reflecting true low-abundance proteins that fall below the detection limit, and (2) missing at random (MAR), which is generally less prominent in DIA than in data-dependent acquisition methods [58–60].

MNAR values are commonly imputed using left-censored imputation (replacing missing entries with low or minimum values), whereas MAR values are imputed using algorithm such as k-nearest neighbors (KNN) or random forest-based approaches [61,62]. The relative probability of MAR versus MNAR can be estimated from the data, for example, by using entropy-based metrics [60].

In large-scale datasets, left-censored imputation can exacerbate batch effects, especially when the missingness rate differs systematically between batches. To address this, mixed imputation strategies have been proposed, using low-value imputation when the missingness rate is low within a batch and MAR-type imputation otherwise [43,63]. More recent work has used self-supervised deep learning to more accurately handle missing values in label-free quantitative proteomics [62].

1.6 Evaluating Data Preprocessing: Strategies and Considerations

Data preprocessing can inadvertently lead to overcorrection or the introduction of unwanted artifacts. Therefore, the data workflow must be carefully evaluated. However, evaluating data preprocessing remains challenging, as no widely accepted standards exist for assessing steps such as batch correction [53,54].

Common evaluation strategies include principal component analysis (PCA) to assess variance before and after preprocessing, calculation of coefficients of variation (%CV) across replicates or QC samples, and correlation analysis between replicates [27,33,47,50,64]. In addition, known biological relationships can be examined.

In this chapter, we set out to provide a practical guide for (pre) processing large-scale proteomics datasets. Using a large-scale yeast knockout dataset [63] that presents both technical challenges (multiple batches, missing values) and built-in validation features (knockout proteins as ground truth), we demonstrate a comprehensive preprocessing workflow (Fig. 2). Below, we describe a step-by-step protocol of the previously published workflow [63], which consists of two main parts: (1) raw data processing (database search and quantification) and (2) downstream data preprocessing, including batch correction, filtering, and imputation strategies detailed in the following subsections.

2 Materials

2.1 Example Dataset

As an illustrative example of a large-scale study, we use a published DIA proteomics dataset [63]. With approximately 5000 measured samples, including quality controls, this study is among the largest single DIA datasets to date. The dataset quantified protein abundances across a genome-scale knockout library in *Saccharomyces cerevisiae*. By systematically mapping gene deletions to proteomic signatures, it connects genetic perturbations with protein-level changes [63].

The dataset has been used to uncover proteome changes related to global biological properties (e.g., overall protein abundance changes, protein half-life, growth rates, effects on protein complexes, copy number variation, and paralogs) as well as functional relationships of proteins and genes (e.g., connectivity in protein–protein interaction (PPI) networks). Furthermore, it demonstrated how functional proteomics can complement gene annotation by analyses such as proteome profile similarity, protein covariation, and reverse proteome profiling [63].

The scale and complexity of this dataset make it particularly suitable for demonstration here: its size introduces batch correction challenges, while the systematic knockouts enable distinction between true biological absence and technical missingness, enabling the testing of imputation strategies.

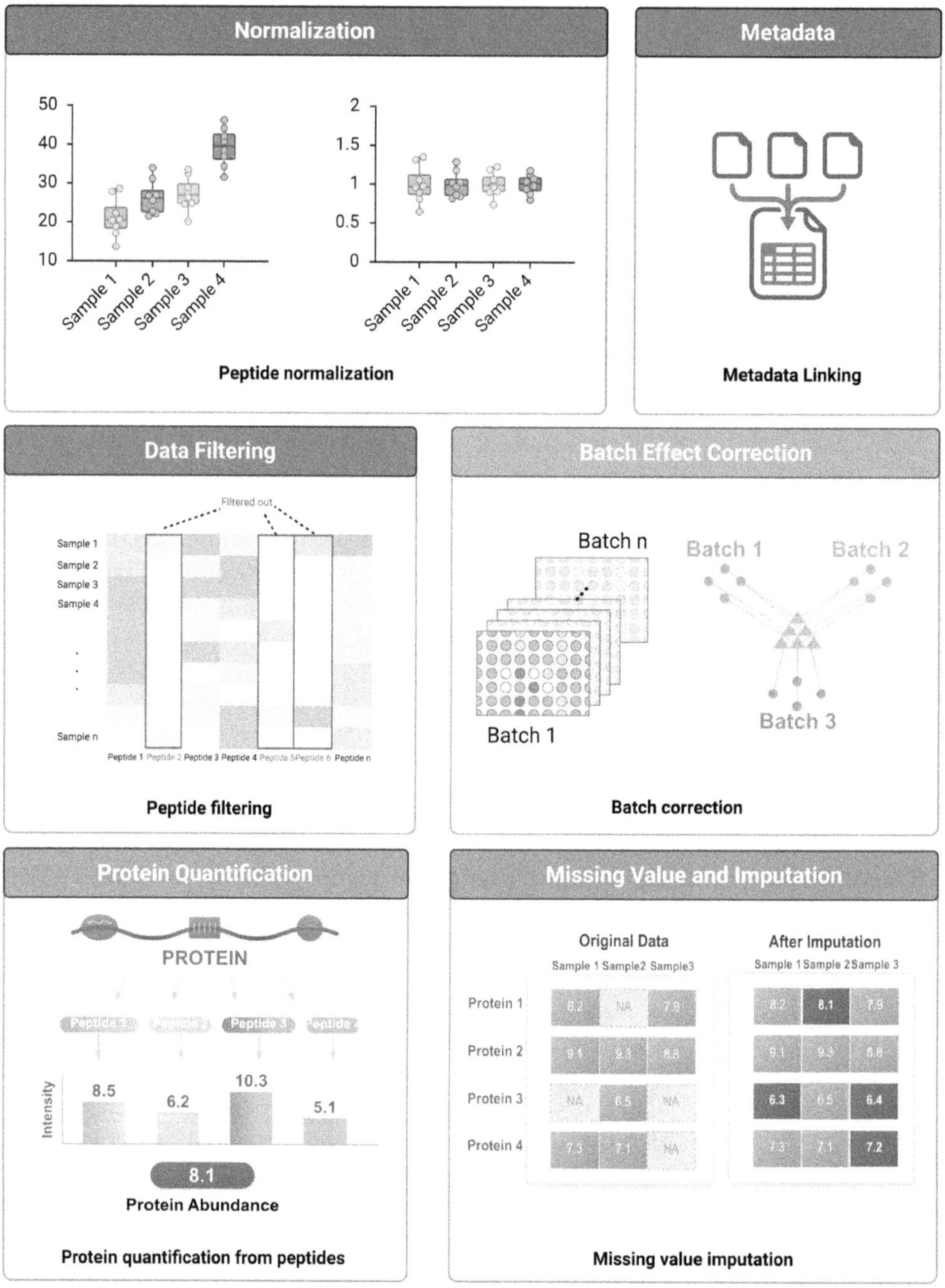

Fig. 2 Overview of an example preprocessing workflow for large-scale proteomics data. The workflow includes six steps: (1) Normalization of peptide intensities to correct for sample loading; (2) Linking data with metadata; (3) Data Filtering to remove low-quality features; (4) Batch Effect and Drift Correction to address systematic technical variation; (5) Protein Quantification to aggregate peptide-level data to protein abundances; and (6) Missing Value Imputation to infer missing measurements. The schematic diagram was created using BioRender.com

2.2 Input Data and Metadata

- DIA-NN output quantification data (20210208_diannoutput.tsv) were obtained from ProteomeXchange (PXD036062) [65].
- Protein Database (annotation) from UniProt [66]

2.3 Software and Algorithms

R [67] (version ≥4.3) within RStudio, utilizing the following packages:

- tidyverse, version 2.0.0 [68].
- impute, version 1.76.0 [69].
- DIA-NN R package, version 1.0.1 (https://github.com/vdemichev/diann-rpackage) [18].
- data.table, version 1.17.8 (https://CRAN.R-project.org/package=data.table).

2.4 Computational Resources

- An operating system compatible with R (version ≥4.3). We recommend a computer with Windows 10 or later.
- A minimum of 128 GB system memory.
- A minimum of 12 CPU cores.
- A 1 TB SSD is recommended.
- For the example dataset, a network connection with access to (1) MassIVE database and (2) UniProt [66].

3 Methods

The workflow consists of two main stages. First, raw mass spectrometry data undergo database searching, peptide quantification, and normalization (Raw Data Processing). Second, the quantified data are subjected to filtering, batch correction, protein summarization, and missing value imputation (Data Preprocessing).

3.1 Raw Data Processing

Raw data can be downloaded from ProteomeXchange (PXD036062). For downloading large amounts of raw data, using an FTP client is recommended for efficient data transfer. Data were processed with DIA-NN (version 1.7.12) [63]. First, data files were converted to the .dia format using the "convert to .dia function." Database search was performed using a gas-phase fractionation (GPF) spectral library generated from pooled samples [63]. Processing was performed with default settings, with MS2 and MS1 mass accuracies set to 20 ppm and a scan window size of 6 [63].

To address the computational demands of large-scale data analysis, two-stage processing can be used to split the runtime. First, data can be processed in batches corresponding to individual plates for peptide scoring and identification. Then, all batches can be combined in a final quantification step using DIA-NN's "use existing .quant files when available" option to generate the final quantitative data.

Normalization

DIA-NN provides several normalization options (*see* **Note 1**), and default retention time-dependent normalization ("RT-dependent") was used [63]. Figure 3 demonstrates the effectiveness of the normalization approach for a representative experimental plate (hpr35), showing alignment of median precursor intensities across samples following normalization compared to the considerable variation observed before processing. Normalization effectively removed sample-to-sample technical variation while preserving within-sample intensity distributions (*see* **Note 1**).

3.2 Data Preprocessing

Preparation

Data preprocessing was performed using R version 4.3.3 (2024-02-29 ucrt). The DIA-NN output file containing quantification results (*20210208_diannoutput.tsv*) was imported, and standard data manipulation packages (as listed in Sect. 2) were used.

```
#### 3.2.0 Preparation: load the required R packages

require(tidyverse)
require(impute)
require(diann)
require(data.table)
#### Set the working directory
setwd("Your/Working/Directory/")
```

Metadata

To assign relevant annotation (e.g., batch, condition) to each sample, metadata were extracted from the sample names, and the resulting columns were named according to the corresponding information. The sample names followed the structure: injectionnr_wellnumber_platenumber_sampletype_ORF_genename_abs280.

For example, sample 30_28_hpr6_ko_YOR010C_TIR2_0.75 encodes the following information:

- 30: injection number (within plate hpr6)
- 28: well number (counted row-wise)
- hpr6: plate number 6 (hpr = haploid prototroph rearranged)
- ko: sample type (knockout)
- YOR010C: targeted ORF
- TIR2: gene name
- 0.75: measured absorbance at 280 nm (used for adjusting injection volume)

```
#### 3.2.1 Metadata

### Extract and summarise the metadata from file name
## Extract the sample list
sample_list <- data.table::fread("dataset/20210208_diannoutput.tsv", sep = "\t", select = c(2))
```

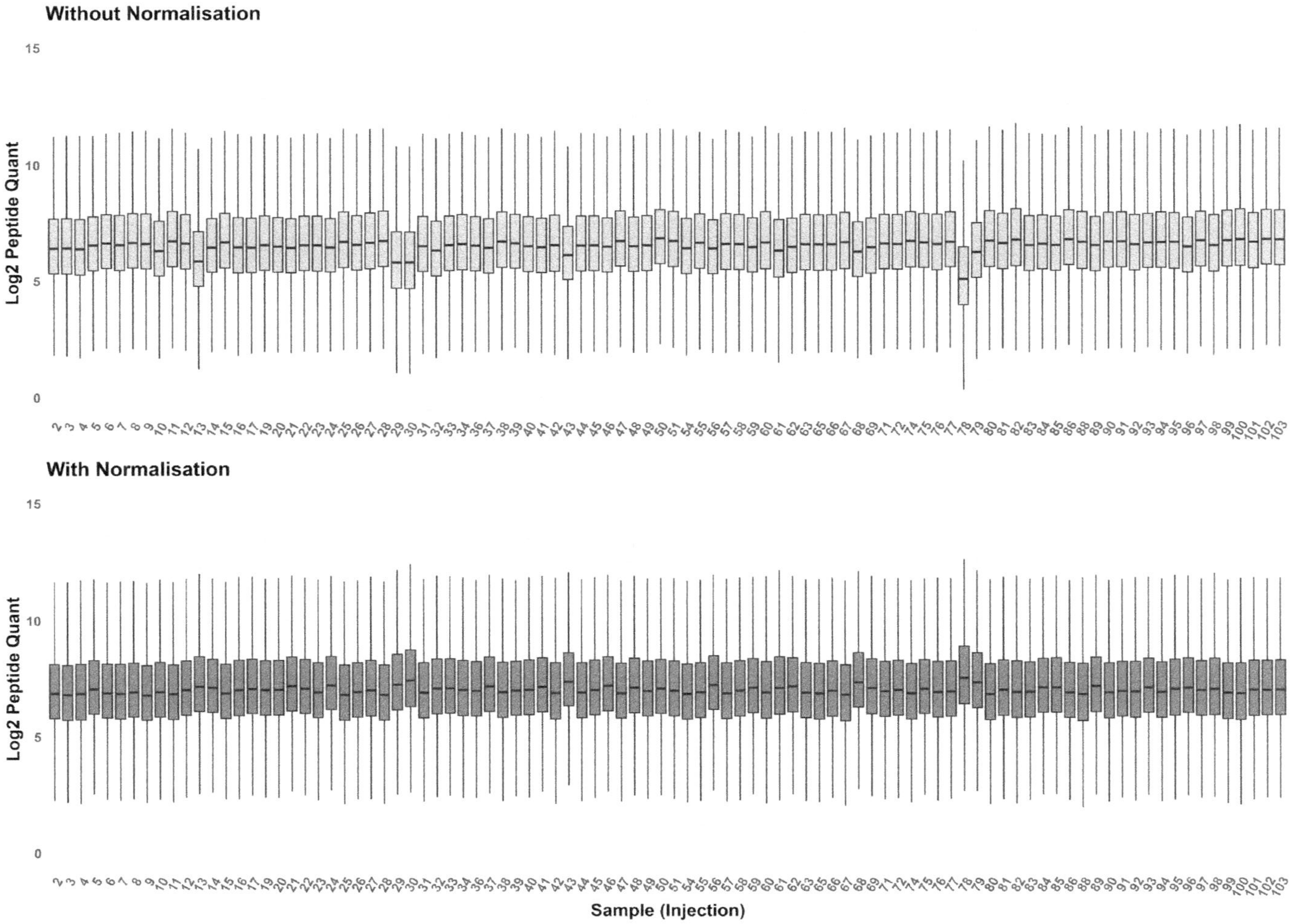

Fig. 3 Normalization of label-free DIA proteomics data based on total precursor quantity. Box plots show the distribution of precursor intensities before (top panel, Precursor.Quantity) and after (bottom panel, Precursor.Normalised) normalization for a representative plate (hpr35) containing both knockout (KO) and wild-type (WT) samples. Normalization reduces sample-to-sample variation in total signal intensity

```
metadata <- data.frame("File.Name" =
unique(sample_list$Run))
## Summarise the sample information into different
columns and arrange by plate
metadata <- metadata %>%
 mutate(File.Name = gsub("\\.wiff", "", File.Name))
%>% ## Remove the.wiff extension in file name
 tidyr::separate(File.Name, into = c("injnr",
"samplenr", "plate", "condition", "orf",
"genename", "a280"), sep = "_", remove = F) ##
Allocate the information into corresponding columns
```

Data Filtering

The filtering criteria are dependent on both the scale of the dataset and the biological questions. We filtered the data to remove unreliable features and samples that are outliers or introduce noise. In this example dataset, we performed data filtering at the precursor level. See in **Note 2** the filtering strategy and filtering level (precursor or protein).

Dataset Loading

```
## Load the example dataset with required columns
including:
## File.Name, name of the sample; Run, name of
the run; Protein.Group, name of the protein group;
Genes, name of the gene; Stripped.Sequence, stipped
peptide sequence; Precursor.Id, peptide sequence;
Q.Value, peptide q value; Protein.Q.Value,
protein q value; Proteotypic; Precursor.Quantity,
unnormalised peptide quantity; Precursor.
Normalised, normalised peptide quantity
col_select <- c(2, 3, 6, 14, 15, 17, 18, 21, 22,
23)
dt <- data.table::fread("dataset/20210208_
diannoutput.tsv", sep = "\t", select = col_select)
```

Q-Value Filtering and Proteotypic Filtering

DIA-NN performs precursor-level q-value filtering by default (< 0.01). Protein-level filtering is not automatically applied; therefore, we use a protein q-value threshold of <0.01 to ensure high-confidence identifications. Additionally, we restricted our analysis to proteotypic peptides only.

```
### 3.2.2.1 Q-value filtering and Proteotypic
filtering
## Protein Q-value filtering and Proteotypic
filtering; Extract the file name without extension
and join with metadata
dt <- dt %>%
 dplyr::filter(Protein.Q.Value < 0.01) %>%
 dplyr::filter(Proteotypic == 1) %>%
 mutate(File.Name = gsub("\\.wiff", "", Run)) %>%
 left_join(metadata, by = c("File.Name"))
```

Completeness Filters

To remove unreliable precursors detected in only a small number of samples, we applied completeness filters. This is effective in homogeneous sample sets; however, in heterogeneous datasets (e.g., studies with expected protein "on/off" behavior), overly stringent completeness filters may remove biologically relevant signals.

For this example dataset, we applied the completeness filter based on the wild-type control samples (a strain in which a his3Δ::kanMX deletion is complemented by heterologous expression of the HIS3 enzyme, hereafter referred to as "HIS3") with a threshold of 80% (requires precursors to be identified in 80% of the samples). As some plates contained more wild-type control samples than others, we included only wild-type samples from the following well positions that were common across all plates: 11, 20, 29, 38, 80, and 89 [63].

```
### 3.2.2.2 Completeness filters
## Generate the peptide list of completeness based
on wild type (wt) HIS3 samples on certain plate
position with precursor q value less than 0.01
wt_completeness_filter <- dt %>%
 dplyr::filter(grepl("HIS3", File.Name)) %>%
 dplyr::filter(samplenr %in% c("11", "20", "29",
"38", "80", "89")) %>%
 group_by(Precursor.Id) %>%
 summarise(count = n())
wt_completeness_filter$completeness <
- wt_completeness_filter$count /
max(wt_completeness_filter$count)
## Filter peptides with completeness threshold at
80%
wt_completeness_filter_completeness80 <- wt_
completeness_filter %>%
 dplyr::filter(completeness >= 0.8)
## Apply the 80% completeness filter to the whole
dataset
dt_completeness80 <- dt %>%
 dplyr::filter(Precursor.Id %in% wt_completeness_
filter_completeness80$Precursor.Id)
```

Remove Outlier Samples

Samples with low numbers of identified precursors are often of poor quality, resulting from factors such as insufficient injection amounts or technical failures. Here, we excluded samples with <80% of the maximum number of identified precursors observed across all samples (*see* **Note 3**).

```
### 3.2.2.3 Remove the outlier samples
## we excluded samples with < 80% of the maximum
number of identified precursors across all samples
sample_outlier <- dt_completeness80 %>%
 group_by(File.Name) %>%
 summarise(count = n()) %>%
```

```
 mutate(sample_ratio = count / max(count)) %>%
 mutate(sample_filter = case_when(
 sample_ratio < 0.8 ~ "Outlier",
 TRUE ~ "Normal"))
## Filter out the outlier samples
dt_outlier_filtered <- dt_completeness80 %>%
 dplyr::filter(File.Name %in% sample_
outlier[sample_outlier$sample_filter ==
"Normal",]$File.Name)
```

CV Filters

To improve quantitative accuracy, precursors showing high variability were excluded using a coefficient of variation (CV) cutoff. We applied a CV filter of <50%, which removed highly variable precursors while minimally impacting overall dataset size (*see* **Note 4**). Since CV values are calculated from normalized precursor intensities, this filtering step was performed after batch correction.

```
### 3.2.2.4 CV filters
## Please note that this step is performed after
data normalization and batch correction in section
3.2.3
## Calculate the CV value of each peptide
cv <- filter(dt_batch_corrected, condition %in%
c("ko", "HIS3")) %>%
 group_by(Precursor.Id) %>%
 summarise(cv = sd(batchcorr.precursornorm) /
mean(batchcorr.precursornorm) * 100)
## Filter the precursor with CV less than 50%
dt_filtered_cv50 <- dt_batch_corrected %>%
 filter(Precursor.Id %in% filter(cv, cv <
50)$Precursor.Id)
```

Batch effect correction

Batch effects inevitably arise in large-scale proteomics experiments and thus require correction. In this dataset, samples were prepared and measured in separate batches corresponding to individual 96-well plates. To correct the resulting batch effects, we applied median centering at the precursor level across all samples within each batch.

```
#### 3.2.3 Batch Effect Correction
### Batch correction with all ko and his3 (wt)
samples
## Here we consider each plate as an individual
batch, and perform plate-wise normalisation for
batch correction
## Calculate the correction factor (center value)
of each peptide across all ko and his3 (wt) samples
on each plate (batch-wise)
dt_batch_corrected <- as.data.
table(dt_outlier_filtered)
```

```
dt_batch_corrected <- dt_batch_corrected[,
batchcorr.center.precursornorm:= Precursor.
Normalised / median(Precursor.Normalised), by =
.(condition %in% c("HIS3", "ko"), Precursor.Id,
plate)]
## Calculate the corrected peptide quantity by
multiply the normalised peptide quantity with the
correction factor
dt_batch_corrected <- dt_batch_corrected[,
batchcorr.precursornorm:= batchcorr.center.
precursornorm * median(Precursor.Normalised), by =
.(condition %in% c("HIS3", "ko"), Precursor.Id)]
```

We also evaluated an alternative approach that uses the medians of wild-type (WT) samples, benchmarked against hold-out QC samples ($n=4$ per plate) excluded from batch correction (*see* **Note 5**). All-sample batch correction demonstrated superior performance (median CV=16.9%) compared to WT-based correction (median CV=19.5%; Fig. 4) and was therefore selected for downstream analysis (*see* **Note 5**).

Protein Summarization

Although protein summarization is implemented in DIA-NN [18], we performed filtering and batch correction at the precursor level in the presented workflow. Protein-level quantification was subsequently performed using the MaxLFQ algorithm [30] via the *maxLFQ* function in the DIANN R package.

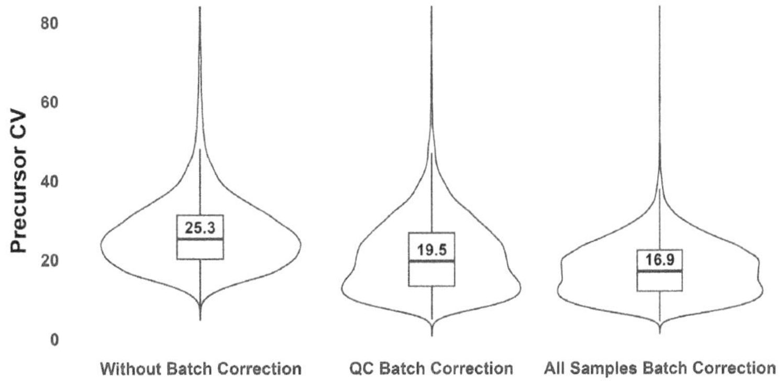

Fig. 4 Comparison of batch correction strategies based on precursor CVs. CVs were calculated for uncorrected data, data corrected using QC (WT) samples, and data corrected using all samples. To avoid bias toward the QC-based correction, CVs were computed on hold-out QC samples. For each precursor, the mean CV across ten iterations (random selection of QC samples for correction and evaluation) was calculated and plotted. Median CV values are shown as numbers. Violin plots show the distribution of precursor CVs; in the overlaid boxplots, the boxes represent the first and third quartiles with the median indicated by a line, and whiskers extend to the most extreme data points within 1.5× the interquartile range

```
#### 3.2.4 Protein summarisation
## We summarized (quantify) the protein quantity
based on the filtered and batch corrected peptide
intensities using maxLFQ function in DIA-NN R
package
## Select the File.Name, Protein.Names, Precursor.
Id and Precursor.Normalised
df <- data.frame(File.Name = dt_batch_
corrected$File.Name, Protein.Names = dt_batch_
corrected$Protein.Group, Precursor.Id = dt_batch_
corrected$Precursor.Id, Precursor.Normalised =
dt_batch_corrected$batchcorr.precursornorm)
## Perform protein quantification with maxLFQ
df_maxlfq <- diann_maxlfq(df)
## Transform the wide maxLFQ output to long data
df_maxlfq_long <- as.data.frame(df_maxlfq) %>%
 mutate(Protein.Group = rownames(df_maxlfq)) %>%
 dplyr::select(Protein.Group, everything()) %>%
 gather(File.Name, "maxLFQ", 2:(ncol(df_maxlfq) +
1))
```

Imputation of Missing Data

Missing values are common in DIA-MS datasets, and in large-scale experiments they can be unevenly distributed across plates (*see* **Note 6**). After protein summarization, missing values remained in the dataset. Simple imputation approaches—such as replacing all missing values with the global minimum—can exacerbate batch effects, especially when entire plates lack specific proteins because of batch effects resulting from different instrument sensitivity.

To address this, we implemented a mixed imputation strategy that combines left-censored imputation for values likely missing not at random (MNAR) with similarity-based imputation for values likely missing due to batch-related sensitivity differences (for assessment, *see* **Note 6**). Specifically:

- When <5% of values for a given protein were missing within a batch, we imputed random values between 0 and half the lowest abundance observed for that protein on that plate.
- When ≥5% of values were missing within a batch, we applied k-nearest neighbor (KNN) imputation.

```
#### 3.2.5 Imputation of missing data
### Here we applied a mixed imputation strategy
based on the percentage of missing
## Label the missing value
set.seed(20210208)
impute_missing <- df_maxlfq_long %>%
 dplyr::select(Protein.Group, File.Name, maxLFQ) %>
%
 as.data.table() %>%
  dcast(Protein.Group ~ File.Name, value.var =
"maxLFQ")
```

```
impute_missing <- impute_missing %>%
   melt(measure.vars = 2:length(impute_missing))
%>%
   dplyr::rename("quant" = value) %>%
   dplyr::rename("File.Name" = variable) %>%
   left_join(metadata, by = c("File.Name")) %>%
   mutate(impute = ifelse(is.na(quant), "impute",
"noimpute"))
## Calculate the percentage of missing
impute_missing_ratio <- impute_missing %>%
 group_by(impute, plate, Protein.Group) %>%
 mutate(impute_count = n()) %>%
 left_join(
  dplyr::select(impute_missing, plate, File.Name)
%>%
   distinct() %>%
   ungroup() %>%
   group_by(plate) %>%
   mutate(sample_nr = n())) %>%
 mutate(ratio = (impute_count) / sample_nr)
```

Impute missing values with random values between 0 and half the minimum quantity observed for the same protein on the same plate, if missingness is below 5% per plate.

```
## Minimum imputation with ratio less than 0.05
impute_missing.min <- impute_missing_ratio %>%
 group_by(Protein.Group) %>%
 mutate(impute_method = impute) %>%
 mutate(impute_method = ifelse(is.na(quant) & ratio
< 0.05, "min_impute", impute_method)) %>%
 ungroup() %>%
 group_by(Protein.Group, plate) %>%
 mutate(quant = ifelse(is.na(quant) & ratio < 0.05,
runif(1, 0, min(quant, na.rm = T) / 2), quant)) %>%
 mutate(impute_method = ifelse(is.na(quant), "knn_
impute", impute_method)) %>%
ungroup()
```

For proteins missing in >5% of samples per plate, values were imputed using k-nearest neighbor averaging (*impute.knn* function from the R package *impute*).

```
## KNN imputation with ratio more than 0.05
## Make data wide and select columns with data
impute_missing.min <- as.data.table(impute_missing.
min)
impute_missing.min.knn <- dcast(impute_missing.min,
Protein.Group ~ File.Name, value.var = "quant")
## Store protein row
impute_missing.min.knn_proteingroups <- impute_
missing.min.knn$Protein.Group
```

```
impute_missing.min.knn <- data.matrix(impute_
missing.min.knn[, 2:ncol(impute_missing.min.knn)])
## Perform KNN imputation
impute_missing.min.knn <- impute.knn(impute_
missing.min.knn, k = 10, rowmax = 0.5, colmax =
0.8, maxp = 1500, rng.seed = 20210208)
## Add protein column and rename columns
impute_missing.min.knn <- as.data.frame(impute_
missing.min.knn[[1]]) %>%
 mutate(Protein.Group = impute_missing.min.knn_
proteingroups) %>%
 dplyr::select(Protein.Group, everything())
## Make data long
impute_missing.min.knn_long <- melt(as.data.
table(impute_missing.min.knn), measure.vars =
2:length(impute_missing.min.knn)) %>%
 dplyr::rename("File.Name" = variable) %>%
 dplyr::rename("quant" = value)
```

4 Notes

1. Normalization with DIA-NN

   DIA-NN offers multiple normalization strategies, including global normalization and retention time-dependent normalization [18]. Selection between these approaches depends on experimental characteristics: retention time-dependent normalization is recommended for datasets acquired with long chromatographic gradients with retention time-dependent drifts, whereas global normalization is more suitable for shorter gradients with stable chromatography. In the example dataset, DIA-NN's default retention time-dependent normalization algorithm was used.

   The effectiveness of this normalization approach is illustrated in Fig. 3, which displays the distribution of precursor intensities before and after normalization for a representative experimental plate (hpr35, randomly selected).

2. Precursor-level versus protein-level filtering

   Filtering criteria depend on both dataset scale and biological questions. In this example, we performed data filtering at the precursor level rather than the protein level. Precursor-level filtering allows identification and removal of individual problematic precursors (e.g., those with excessive missing values, high variability, or potential interferences) while retaining other high-quality precursors from the same protein. This approach ensures only reliable precursors contribute to final protein quantification. For large-scale datasets with substantial missing data or batch effects, precursor-level filtering can better preserve proteome coverage.

3. Identifying and excluding low-quality samples

Samples with low numbers of identified precursors are often of poor quality, resulting from factors such as insufficient injection amounts, technical failures during sample preparation or suboptimal LC–MS performance. To identify such samples, we calculated the number of identified precursors per sample and established a threshold based on the dataset distribution. Here, we excluded samples with <80% of the maximum number of identified precursors observed across all samples. The majority of samples passed the filter, with only a small fraction flagged as outliers (4.65% of total samples, Fig. 5D). This filtering step ensures that poor-quality samples do not compromise the detection of true biological differences. The specific threshold, however, should be adjusted based on dataset characteristics.

4. Combining completeness and CV filtering

Completeness filters were applied to remove unreliable precursors detected in only a small number of samples. A completeness threshold specifies the minimum percentage of samples in which a precursor must be identified to be retained (e.g., 80% completeness requires detection in at least 80% of samples).

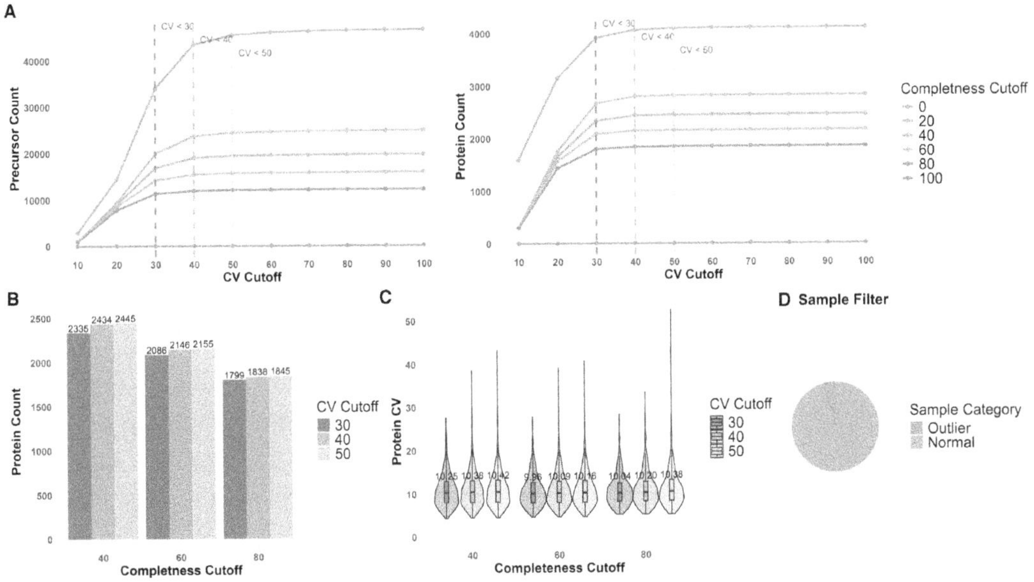

Fig. 5 Evaluation of precursor completeness and coefficient of variation (CV) filters. **(A)** Left: number of precursor identifications across different completeness thresholds (calculated on WT samples) and CV cutoffs. Right: as in the left plot, but for protein identifications. **(B)** Protein identifications across different combinations of completeness and CV filters shown as barplots. **(C)** Distribution of protein-level CV values under the respective filtering conditions. **(D)** Sample outliers with <80% of the maximum number of identified precursors across all samples. In **(C)**, violin plots show the distribution of protein CVs; in the overlaid boxplots, the boxes represent the first and third quartiles with the median indicated by a line, and whiskers extend to the most extreme data points within 1.5× the interquartile range

Further, a CV cutoff was applied to exclude highly variable precursors and improve quantitative accuracy. The CV for filtering can be calculated either across all samples or restricted to WT samples to avoid removing biologically variable precursors. Here, we evaluated different combinations of completeness and CV filters and assessed their impact on the number of retained precursors (Fig. 5). Due to the high quantitative precision of the dataset, only a small number of precursors were removed even with a stringent CV threshold of 30%, retaining 11,298 of 12,217 precursors and 1799 of 1851 proteins with a completeness filter of 80% (Fig. 5A, B). Using a CV filter of < 50%, as applied here, only highly variable precursors were excluded, with minimal impact on the final dataset (retaining 12,114 precursors and 1845 proteins; Fig. 5B). Correspondingly, protein-level CV distributions were comparable across all filter combinations (median CVs 9.96–10.42%; Fig. 5C).

5. Strategies in batch correction

Precursor-level median centering was applied to mitigate batch effects. Centering can be performed either aligning the medians of wild-type (WT) samples or the medians of all samples. The all-sample approach is appropriate only when experimental conditions are balanced across batches, whereas WT-based centering requires a sufficient number of WT replicates per batch. We evaluated both strategies by comparing CV values in hold-out QC samples (=4 per plate) that were excluded from the batch correction. All-sample batch correction achieved better performance (median CV = 16.9%) compared to WT-based batch correction (median CV = 19.5%; Fig. 4) and was therefore applied.

To assess the effectiveness of normalization, filtering, and batch correction, we performed principal component analysis (PCA) on WT and KO samples (Fig. 6). Following these preprocessing steps, the variance of the first two principal components was reduced and the WT samples formed a coherent cluster. This reduction in variance captured by PC1 and PC2 indicates removal of technical noise.

6. Missing value and hybrid imputation strategy

Large-scale DIA-MS datasets commonly exhibit missing values that are unevenly distributed across acquisition batches due to variable instrument performance (Fig. 7A). In the postfiltered dataset of 10,504,435 protein-level quantifications, 268,894 values (2.5%) were missing (Fig. 7B).

Naive imputation strategies, such as uniform replacement with global minimum values, risk amplifying batch effects when entire plates have reduced instrument sensitivity. We therefore employed a conditional imputation framework that distinguishes between values missing not at random (MNAR) and those missing due to technical factors (see "Protein summarization" section).

We validated this strategy by examining knockout-gene proteins, which represent true biological absences requiring left-censored

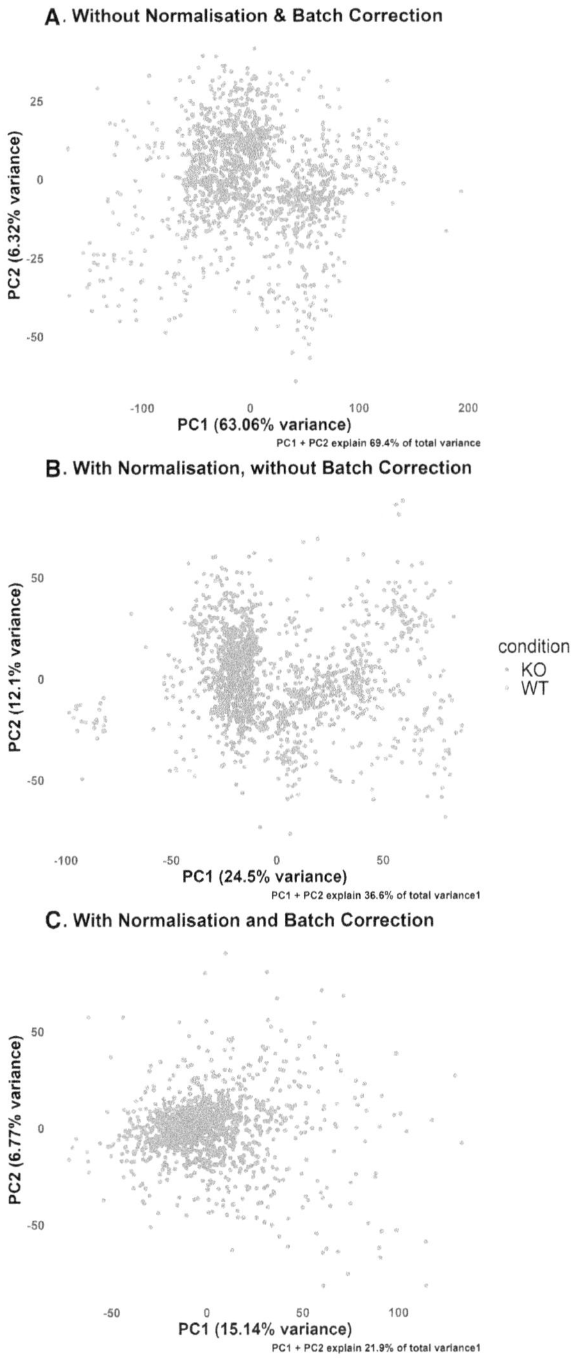

Fig. 6 Normalization and batch correction reduce variability among QC (WT) samples in the first two principal components. (**A**) Without normalization and batch correction, (**B**) normalization only, without batch correction and (**C**) after normalization and batch correction. Different colors indicate KO and WT samples. Principal components were calculated from precursor intensities

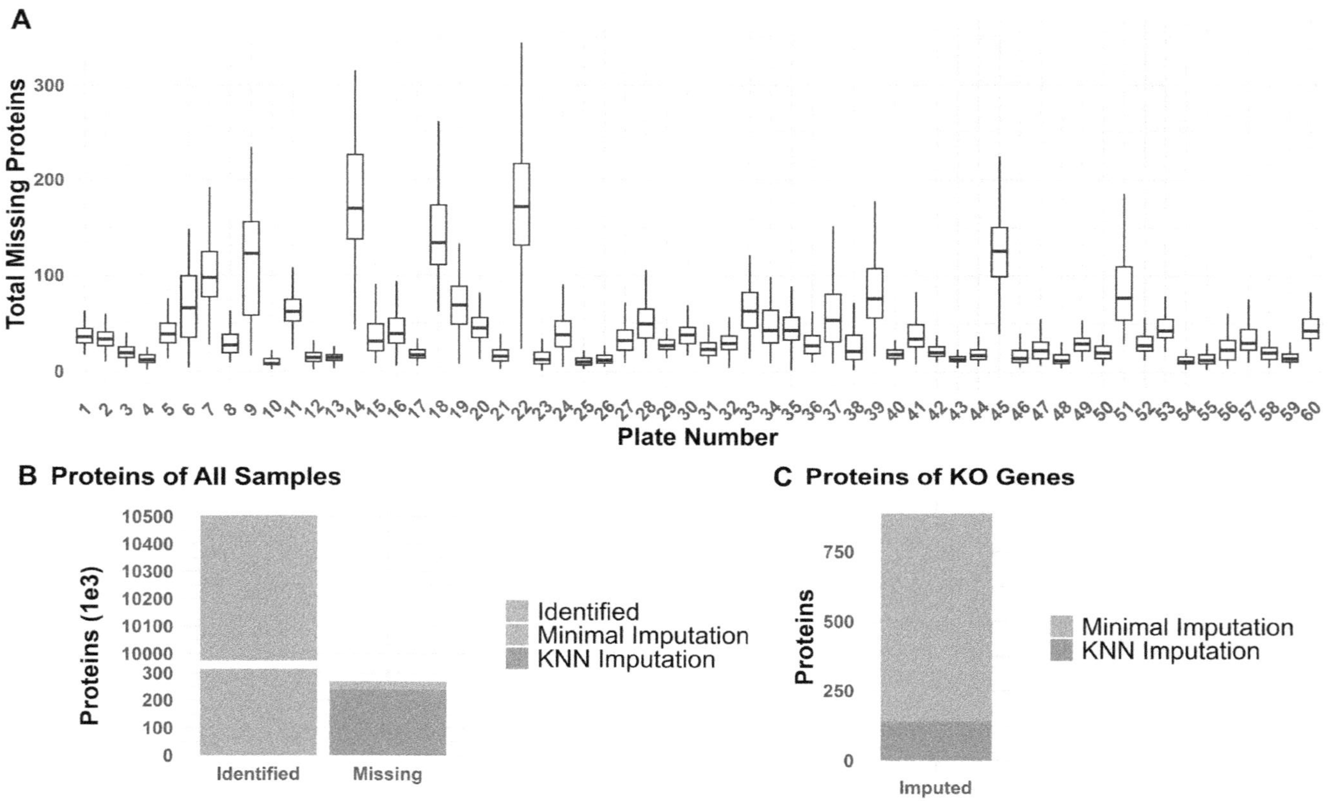

Fig. 7 Missing values and imputation. (**A**) Missing values vary across batches. The number of missing protein values per sample is shown as boxplots. (**B**) Comparison of the number of identified proteins and number of missing proteins imputed with different strategies (minimal imputation and KNN imputation) across all measured samples. (**C**) Proteins corresponding to knocked-out genes predominantly received minimum-value imputation rather than KNN imputation

imputation. The results confirmed appropriate method selection: while KNN imputation was applied to most missing values dataset-wide (89%; Fig. 7B), knockout-gene proteins predominantly received minimum-value imputation (84.0%; Fig. 7C), demonstrating that the algorithm correctly identified biologically absent proteins.

5 Declaration of Generative AI and AI-Assisted also Machine learning Technologies in the Writing Process

During the preparation of this work, the authors used ChatGPT (Versions 5.0 and 5.4, OpenAI) and Claude (Opus 4.6, Anthropic, 2026) to improve the readability and quality of the writing after the initial draft. After using this tool, the authors reviewed and edited the content as needed and took full responsibility for the content of the publication.

Acknowledgments

This work was supported by the Swiss Canton of Grisons, The LOOP Zurich, and the Uniscientia Stiftung. We thank A.S. Egger-Hörschinger, P. Gessner, L. Chatterjee, P. Westermann, and T. Zinsli for helpful discussions about data preprocessing strategies and for critical reading of the manuscript.

Conflict of Interest The authors declare no conflicts of interest that are relevant to the content of this chapter.

References

1. Aebersold R, Mann M (2016) Mass-spectrometric exploration of proteome structure and function. Nature 537:347–355
2. Guo T, Steen JA, Mann M (2025) Mass-spectrometry-based proteomics: from single cells to clinical applications. Nature 638:901–911
3. Messner CB, Demichev V, Wang Z et al (2023) Mass spectrometry-based high-throughput proteomics and its role in biomedical studies and systems biology. Proteomics 23:e2200013
4. Niu L, Stinson SE, Holm LA et al (2025) Plasma proteome variation and its genetic determinants in children and adolescents. Nat Genet 57:635–646
5. Mi Y, Burnham KL, Charles PD et al (2024) High-throughput mass spectrometry maps the sepsis plasma proteome and differences in patient response. Sci Transl Med 16:eadh0185
6. Cai X, Xue Z, Zeng F-F et al (2023) Population serum proteomics uncovers a prognostic protein classifier for metabolic syndrome. Cell Rep Med 4:101172
7. Tang J, Yue L, Xu Y et al (2025) Longitudinal serum proteome mapping reveals biomarkers for healthy ageing and related cardiometabolic diseases. Nat Metab 7:166–181
8. Tüshaus J, Eckert S, Schliemann M et al (2025) Towards routine proteome profiling of FFPE tissue: insights from a 1,220-case pan-cancer study. EMBO J 44:304–329
9. Eckert S, Berner N, Kramer K et al (2024) Decrypting the molecular basis of cellular drug phenotypes by dose-resolved expression

proteomics. Nat Biotechnol https://doi.org/10.1038/s41587-024-02218-y
10. Guzman UH, Martinez-Val A, Ye Z et al (2024) Ultra-fast label-free quantification and comprehensive proteome coverage with narrow-window data-independent acquisition. Nat Biotechnol 42:1855–1866
11. Gonçalves E, Poulos RC, Cai Z et al (2022) Pan-cancer proteomic map of 949 human cell lines. Cancer Cell 40:835–849. e8
12. Ludwig C, Gillet L, Rosenberger G, Amon S, Collins BC, Aebersold R (2018) Data-independent acquisition-based SWATH-MS for quantitative proteomics: a tutorial. Mol Syst Biol 14(8), e8126.
13. Lou R, Cao Y, Li S et al (2023) Benchmarking commonly used software suites and analysis workflows for DIA proteomics and phosphoproteomics. Nat Commun 14: 94
14. Fröhlich K, Brombacher E, Fahrner M et al (2022) Benchmarking of analysis strategies for data-independent acquisition proteomics using a large-scale dataset comprising inter-patient heterogeneity. Nat Commun 13: 2622
15. Staes A, Mendes Maia T, Dufour S et al (2024) Benefit of in silico predicted spectral libraries in data-independent acquisition data analysis workflows. J Proteome Res 23:2078–2089
16. Gotti C, Roux-Dalvai F, Joly-Beauparlant C et al (2021) Extensive and accurate benchmarking of DIA acquisition methods and software tools using a complex proteomic standard. J Proteome Res 20:4801–4814
17. Demichev V, Szyrwiel L, Yu F et al (2022) dia-PASEF data analysis using FragPipe and DIA-NN for deep proteomics of low sample amounts. Nat Commun 13: 3944
18. Demichev V, Messner CB, Vernardis SI et al (2020) DIA-NN: neural networks and interference correction enable deep proteome coverage in high throughput. Nat Methods 17:41–44
19. Röst HL, Rosenberger G, Navarro P et al (2014) OpenSWATH enables automated, targeted analysis of data-independent acquisition MS data. Nat Biotechnol 32:219–223
20. Yu F, Teo GC, Kong AT et al (2023) Analysis of DIA proteomics data using MSFragger-DIA and FragPipe computational platform. Nat Commun 14: 4154
21. Egertson JD, MacLean B, Johnson R et al (2015) Multiplexed peptide analysis using data-independent acquisition and Skyline. Nat Protoc 10:887–903
22. Sinitcyn P, Hamzeiy H, Salinas Soto F et al (2021) MaxDIA enables library-based and library-free data-independent acquisition proteomics. Nat Biotechnol 39:1563–1573
23. Searle BC, Pino LK, Egertson JD et al (2018) Chromatogram libraries improve peptide detection and quantification by data independent acquisition mass spectrometry. Nat Commun 9: 5128
24. Frejno M, Berger MT, Tüshaus J et al (2025) Unifying the analysis of bottom-up proteomics data with CHIMERYS. Nat Methods 22:1017–1027
25. Tsou C-C, Avtonomov D, Larsen B et al (2015) DIA-Umpire: comprehensive computational framework for data-independent acquisition proteomics. Nat Methods 12:258–264. 7 p following 264
26. Guturu H, Nichols A, Cantrell LS et al (2025) Cloud-enabled scalable analysis of large proteomics cohorts. J Proteome Res 24:1462–1469
27. Čuklina J, Lee CH, Williams EG et al (2021) Diagnostics and correction of batch effects in large-scale proteomic studies: a tutorial. Mol Syst Biol 17:e10240
28. Karpievitch YV, Dabney AR, Smith RD (2012) Normalization and missing value imputation for label-free LC-MS analysis. BMC Bioinformatics 13(Suppl 16):S5
29. Graw S, Tang J, Zafar MK et al (2020) ProteiNorm—a user-friendly tool for normalization and analysis of TMT and label-free protein quantification. ACS Omega 5:25625–25633
30. Cox J, Hein MY, Luber CA et al (2014) Accurate proteome-wide label-free quantification by delayed normalization and maximal peptide ratio extraction, termed MaxLFQ. Mol Cell Proteomics 13:2513–2526
31. Čuklina J, Pedrioli PGA, Aebersold R (2020) Review of batch effects prevention, diagnostics, and correction approaches. Methods Mol Biol 2051:373–387
32. Messner CB, Demichev V, Wendisch D et al (2020) Ultra-high-throughput clinical proteomics reveals classifiers of COVID-19 infection. Cell Syst https://doi.org/10.1016/j.cels.2020.05.012
33. Liu X, Sun H, Hou X et al (2025) Standard operating procedure combined with comprehensive quality control system for multiple LC-MS platforms urinary proteomics. Nat Commun 16: 1051
34. Maretty L, Gill D, Simonsen L et al (2025) Proteomic changes upon treatment with semaglutide in individuals with obesity. Nat Med 31:267–277
35. Oliver NC, Choi MJ, Arul AB et al (2024) Establishing quality control metrics for large-scale plasma proteomic sample preparation. ACS Meas Sci 4:442–451
36. Bache N, Geyer PE, Bekker-Jensen DB et al (2018) A novel LC system embeds analytes in pre-formed gradients for rapid, ultra-robust proteomics. Mol Cell Proteomics 17:2284–2296

37. Bekker-Jensen DB, Martínez-Val A, Steigerwald S et al (2020) A compact quadrupole-orbitrap mass spectrometer with FAIMS interface improves proteome coverage in short LC gradients. Mol Cell Proteomics 19:716–729
38. Meier F, Brunner A-D, Koch S, et al (2018) Online parallel accumulation–serial fragmentation (PASEF) with a novel trapped ion mobility mass spectrometer. Mol Cell Proteomics 17:2534–2545.
39. Hsu C, Shulman N, Stewart HI, et al (2025) Evaluation of a prototype Orbitrap Astral Zoom mass spectrometer for quantitative proteomics—Beyond identification lists. J Proteome Res 24:5742–5754.
40. Steidel M, Knecht S, Sweetman G et al (2025) Impact of local air pressure on ion mobilities and data consistency in diaPASEF-based high throughput proteomics. J Proteome Res 24:966–973
41. Tsantilas KA, Merrihew GE, Robbins JE et al (2024) A framework for quality control in quantitative proteomics. J Proteome Res 23:4392–4408
42. Chiva C, Olivella R, Staes A et al (2025) A multiyear longitudinal harmonization study of quality controls in mass spectrometry proteomics core facilities. J Proteome Res 24:397–409
43. Gonidaki C, Latosinska A, Vlahou A, et al (2026) Practical impact of imputation and batch-effect correction for proteomics/peptidomics differential-abundance analysis. Proteomics 26:48–58.
44. Yu Y, Mai Y, Zheng Y, Shi L (2024) Assessing and mitigating batch effects in large-scale omics studies. Genome Biol 25: 254
45. Poulos RC, Hains PG, Shah R et al (2020) Strategies to enable large-scale proteomics for reproducible research. Nat Commun 11: 3793
46. Johnson WE, Li C, Rabinovic A (2007) Adjusting batch effects in microarray expression data using empirical Bayes methods. Biostatistics 8:118–127
47. Yu Y, Zhang N, Mai Y et al (2023) Correcting batch effects in large-scale multiomics studies using a reference-material-based ratio method. Genome Biol 24: 201
48. Leek JT, Johnson WE, Parker HS et al (2012) The SVA package for removing batch effects and other unwanted variation in high-throughput experiments. Bioinformatics 28:882–883
49. Zhu T, Sun R, Zhang F et al (2021) BatchServer: a web server for batch effect evaluation, visualization, and correction. J Proteome Res 20:1079–1086
50. Sundararaman N, Bhat A, Venkatraman V et al (2023) BIRCH: an automated workflow for evaluation, correction, and visualization of batch effect in bottom-up mass spectrometry-based proteomics data. J Proteome Res 22:471–481
51. Tognetti M, Chatterjee L, Beaton N, et al (2025) Serum proteomics reveals survival-associated biomarkers in pancreatic cancer patients treated with chemoimmunotherapy. iScience 28, 112230.
52. Mishra M, Barck L, Moreno P et al (2023) SelectBCM tool: a batch evaluation framework to select the most appropriate batch-correction methods for bulk transcriptome analysis. NAR Genom Bioinform 5:lqad014
53. Salanon E, Comte B, Centeno D et al (2025) A benchmarking workflow for assessing the reliability of batch correction methods. BioRxiv. https://doi.org/10.1101/2025.08.01.668073
54. Hui HWH, Kong W, Goh WWB (2024) Thinking points for effective batch correction on biomedical data. Brief Bioinform 25, bbae515.
55. Burger B, Vaudel M, Barsnes H (2021) Importance of block randomization when designing proteomics experiments. J Proteome Res 20:122–128
56. Krzywinski M, Altman N (2014) Points of significance: analysis of variance and blocking. Nat Methods 11:699–700
57. Oberg AL, Vitek O (2009) Statistical design of quantitative mass spectrometry-based proteomic experiments. J Proteome Res 8:2144–2156
58. Jin L, Bi Y, Hu C et al (2021) A comparative study of evaluating missing value imputation methods in label-free proteomics. Sci Rep 11: 1760
59. Liu M, Dongre A (2021) Proper imputation of missing values in proteomics datasets for differential expression analysis. Brief Bioinform 22, bbaa112.
60. Hediyeh-Zadeh S, Webb AI, Davis MJ (2023) MsImpute: estimation of missing peptide intensity data in label-free quantitative mass spectrometry. Mol Cell Proteomics 22:100558
61. Lazar C, Gatto L, Ferro M et al (2016) Accounting for the multiple natures of missing values in label-free quantitative proteomics data sets to compare imputation strategies. J Proteome Res 15:1116–1125
62. Webel H, Niu L, Nielsen AB et al (2024) Imputation of label-free quantitative mass spectrometry-based proteomics data using self-supervised deep learning. Nat Commun 15: 5405
63. Messner CB, Demichev V, Muenzner J, et al (2023) The proteomic landscape of genome-wide genetic perturbations. Cell 186:2018–2034.

64. Brenes AJ (2024) Calculating and reporting coefficients of variation for DIA-based proteomics. J Proteome Res 23:5274–5278
65. Deutsch EW, Bandeira N, Perez-Riverol Y et al (2023) The ProteomeXchange consortium at 10 years: 2023 update. Nucleic Acids Res 51:D1539–D1548
66. The UniProt Consortium (2017) UniProt: the universal protein knowledgebase. Nucleic Acids Res 45:D158–D169
67. RDc, TRA, 2004. Language and environment for statistical computing. R Foundation for Statistical Computing, Vienna, Austria
68. Wickham H, Averick M, Bryan J et al (2019) Welcome to the tidyverse. J Open Source Softw 4:1686
69. Troyanskaya O, Cantor M, Sherlock G, et al (2001) Missing value estimation methods for DNA microarrays. Bioinformatics 17: 520–525.

Check for updates

Chapter 5

PSMatch: An R/Bioconductor Package to Explore Proteomics Identification Data

Guillaume Deflandre, Sebastien Gibb and Laurent Gatto

Abstract

We present the R/Bioconductor *PSMatch* package, which helps proteomics practitioners to load, manage, and handle peptide spectrum matches. It provides functions to model peptide–protein relations, visualize these as graphs, and make informed decisions about shared peptide filtering. The package also provides functions to calculate fragments and, in conjunction with the *Spectra* package, annotate and visualize MS2 scans.

Keywords Mass spectrometry, Proteomics, Identification, PTM, Visualization, Bioconductor, R, Reproducible research, Software

1 Introduction

Mass spectrometry (MS) is the current state-of-the-art technology to comprehensively study complex proteomes. It is compatible with diverse input sample types and provides a comprehensive proteome coverage without the need for prior knowledge on the proteins present in a sample. The processing of MS-based proteomics data entails two distinct, albeit-related steps, namely, the identification of peptides and their quantitation. Both of these are then combined to produce the quantitative data matrix that is further processed, analyzed, and interpreted. Both are also critical to guarantee biologically accurate interpretation of the proteomics data. However, unlike for quantitative data that benefits from a lot of attention and exploration and analysis tools [1–4] there are relatively few tools and methods to explore identification data [5–8] that tend to be accepted as is without undergoing further exploratory data analysis.

Here, we present *PSMatch* (Fig. 1), an R/Bioconductor [9, 10] package designed to handle identification data in the form of peptide-spectrum matches (PSMs), and offering functionalities to streamline exploration and visualization of PSM data. It provides functions

Frédérique Lisacek (ed.), *Bioinformatics for Mass Spectrometry*, Methods in Molecular Biology, vol. 3071,
https://doi.org/10.1007/978-1-0716-5519-1_5,

Fig. 1 The *PSMatch* package sticker

to load PSM data from *mzIdentML* (*mzID*) [11] or tabular files, generate theoretical fragment ions, model peptide–protein relations and facilitate various visualizations. Recent advancements provide functionalities to study post-translational modifications (PTMs). As the proportion of unidentified spectra in MS-based experiments typically ranges from 50% to 90% [12, 13] , dedicated and flexible tools built for handling the actual identified spectra are essential.

2 Materials

To demonstrate the usefulness of the *PSMatch* package, a publicly available phosphoproteomic dataset [14] was downloaded, and a database search was performed using the search engine Sage [15]. The *mzML* files were downloaded from the PRIDE [16] database (experiment PXD039419) and the `sage-results.tsv` file was generated with Sage *v.0.14.6*. The necessary files to reproduce the code in the paper are available on Zenodo [17]. The versions of the packages used to produce this document are listed in the session information below. Note that *PSMatch* is not dependent on Sage or any other search engine. It can handle any identification data stored in *mzIdentML* or tabular files from any search engine.

2.1 Installation

PSMatch is an open source R/Bioconductor package, openly developed on GitHub. It can be installed from the Bioconductor project [10] using the BiocManager package.

```
if (!require("BiocManager"))
   install.packages("BiocManager")
BiocManager::install("PSMatch")
```

Once installed, the package can be loaded with

```
library(PSMatch)
```

2.2 Documentation

This tutorial showcases how *PSMatch* allows to handle identification data. For a more complete overview of *PSMatch*'s functionalities, please refer to the appropriate vignettes for the respective subjects. The vignettes are available online from the package's webpage or locally in R. To list all available *PSMatch* vignettes, we can use:

```
vignette(package = "PSMatch")

Vignettes in package 'PSMatch':

Fragments MS2 fragment ions (source, html)

AdjacencyMatrix Understanding protein groups with
        adjacency matrices (source, html)
    PSM Working with PSM data (source, html)
```

To open the vignette focusing on "MS2 fragment ions," we add its name as shown in the next code chunk:

```
vignette("Fragments", package = "PSMatch")
```

Session Information

A complete list and version of R and packages used to run this complete analysis and produce the results is provided below. **Note 1** provides important details about the R and Bioconductor versions compatibility.

- R version 4.5.0 (2025-04-11), `x86_64-pc-linux-gnu`.
- Running under: `Ubuntu 24.04.2 LTS`.
- Matrix products: default.
- `BLAS: /opt/R-4.5/lib/R/lib/libRblas.so`.
- `LAPACK: /opt/R-4.5/lib/R/lib/libRlapack.so`; LAPACK version3.12.1.
- Base packages: base, datasets, graphics, grDevices, methods, stats, stats4, utils.
- Other packages: AnnotationFilter 1.32.0, AnnotationHub 3.16.0, BiocFileCache 2.16.0, BiocGenerics 0.54.0, BiocParallel 1.42.0, BiocStyle 2.36.0, CompoundDb 1.12.1, dbplyr 2.5.0, generics 0.1.4, ggplot2 3.5.2, knitr 1.50, MsBackendSql 1.8.0, MsDataHub 1.8.0, pheatmap 1.0.12, RColorBrewer 1.1-3, reticulate 1.42.0, RSQLite 2.3.11, S4Vectors 0.46.0, Spectra 1.18.2, SpectriPy 0.99.5.
- Loaded via a namespace (and not attached): AnnotationDbi 1.70.0, base64enc 0.1-3, Biobase 2.68.0, BiocManager 1.30.25, BiocVersion 3.21.1, Biostrings 2.76.0, bit 4.6.0, bit64 4.6.0-1, bitops 1.0-9, blob 1.2.4, cachem 1.1.0, ChemmineR 3.60.0, cli 3.6.5, clue 0.3-66, cluster 2.1.8.1, codetools 0.2-20, colorspace 2.1-1, compiler 4.5.0, crayon 1.5.3, curl 6.2.3, data.table 1.17.4, DBI 1.2.3, digest 0.6.37, dplyr 1.1.4, DT 0.33,

evaluate 1.0.3, ExperimentHub 2.16.0, farver 2.1.2, fastmap 1.2.0, fastmatch 1.1-6, filelock 1.0.3, fs 1.6.6, GenomeInfoDb 1.44.0, GenomeInfoDbData 1.2.14, GenomicRanges 1.60.0, glue 1.8.0, grid 4.5.0, gridExtra 2.3, gtable 0.3.6, hms 1.1.3, htmltools 0.5.8.1, htmlwidgets 1.6.4, httr 1.4.7, IRanges 2.42.0, jsonlite 2.0.0, KEGGREST 1.48.0, labeling 0.4.3, lattice 0.22-6, lazyeval 0.2.2, lifecycle 1.0.4, magrittr 2.0.3, MASS 7.3-65, Matrix 1.7-3, memoise 2.0.1, MetaboCoreUtils 1.16.1, mime 0.13, MsCoreUtils 1.20.0, mzR 2.42.0, ncdf4 1.24, parallel 4.5.0, pillar 1.10.2, pkgconfig 2.0.3, png 0.1-8, prettyunits 1.2.0, progress 1.2.3, ProtGenerics 1.40.0, purrr 1.0.4, R6 2.6.1, rappdirs 0.3.3, Rcpp 1.1.0, RCurl 1.98-1.17, rjson 0.2.23, rlang 1.1.6, rmarkdown 2.29, rsvg 2.6.2, scales 1.4.0, stringi 1.8.7, tibble 3.2.1, tidyselect 1.2.1, tinytex 0.57, tools 4.5.0, UCSC.utils 1.4.0, vctrs 0.6.5, withr 3.0.2, xfun 0.52, xml2 1.3.8, XVector 0.48.0, yaml 2.3.10.

3 Methods

3.1 Creating PSM Objects

There is a plethora of search engines available, with each their own resulting identification file format with different variable names. Most of them provide variables such as a matched spectrum identifier, a PSM rank, a search engine score, a false discovery rate, a label indicating whether the PSM comes from the decoy or target peptide, a peptide sequence, as well as the protein or proteins the matched peptide sequence stems from. These variables can be defined when creating the `PSM` object that holds the identification results. These variables are then used to filter and analyze the identification data. The other variables are still stored within the object and can be accessed in the same manner as one would be with a `data.frame`.

Let us first load the identification tabular data into a `data.frame` and look at the variable names to identify how these are encoded in the Sage output files. The code below is general and applicable to any search engine, but see **Note 2** on how to simplify importing data from Sage specifically.

```
sage results <- read.delim("./data/PXD039419-sage-
results.tsv")
colnames(sage results)
 [1]    "psm_id"                          "peptide"
 [3]    "proteins"                   "num_proteins"
 [5]    "filename"                         "scannr"
 [7]    "rank"                              "label"
 [9]    "expmass"                        "calcmass"
[11]    "charge"                      "peptide_len"
[13]    "missed_cleavages"          "semi_enzymatic"
```

```
[15]   "isotope_error"                    "precursor_ppm"
[17]   "fragment_ppm"                       "hyperscore"
[19]   "delta_next"                         "delta_best"
[21]   "rt"                                 "aligned_rt"
[23]   "predicted_rt"                   "delta_rt_model"
[25]   "ion_mobility"               "predicted_mobility"
[27]   "delta_mobility"                  "matched_peaks"
[29]   "longest_b"                           "longest_y"
[31]   "longest_y_pct"         "matched_intensity_pct"
[33]   "scored_candidates"                     "poisson"
[35]   "sage_discriminant_score"     "posterior_error"
[37]   "spectrum_q"                          "peptide_q"
[39]   "protein_q"                       "ms2_intensity"
```

We first transform decoy labels into logical (boolean) values and then construct the PSM object from the data.frame setting the functions' arguments with the appropriate column names.

```
sage results[, "label"] <- sage results[, "label"] < 0
psms <- PSM(

     sage results,
     spectrum = "scannr", ## Scan index
     peptide = "peptide", ## Peptide sequence
     protein = "proteins", ## Protein(s) which the
               peptide stems from
decoy = "label", ## Is the PSM a target or a decoy?
rank = "rank", ## Rank of the PSM
score = "sage discriminant score", ## Score given by
        search engine
fdr = "spectrum q") ## Calculated q-value or FDR for
       that PSM
psms

PSM with 562183 rows and 40 columns.
Spectra: 58591 unique
     db: 444298 target, 117885 decoy
     ranks: 1:215239 2:185573 3:161371
names(40): psm_id peptide ... protein_q ms2_intensity
```

The PSM() constructor could also have taken an *mzIdentML* file, rather than a data.frame to create the PSM object.

The psmVariables of the object store the original variable names that will be used later on during the filtering step.

```
psmVariables(psms)
spectrum peptide
"scannr"                                    "peptide"
Protein decoy
"proteins"                                    "label"
rank score
"rank"                    "sage_discriminant_score"
fdr
"spectrum_q"
```

Before proceeding with the PSM data exploration, let us visualize and confirm the separation between target and decoy peptide scores for the nine original raw data acquisitions (defined by their *mzML* file names) (Fig. 2).

```
library(ggplot2)
    ggplot(psms, aes(x = sage discriminant score,
    colour = label)) +
    geom density() +
    facet wrap(~ filename) +
theme(strip.text.x = element text(size = 8))
```

3.2 Filtering Data

We can now filter the PSMs to retain only the targets of rank 1 that have been confidently identified by using the dedicated functions.

```
psms <- psms |>
    filterPsmDecoy() |>
    filterPsmFdr(FDR = 0.01) |>
filterPsmRank()

Removed 117885 decoy hits.

Removed 148826 PSMs with FDR > 0.01.

Removed 155088 PSMs with rank > 1.
```

The filterPSMs() function is a wrapper around the functions above, including also the filterPsmShared() function that keeps only unique peptides.

3.3 Exploring Peptide–Protein Relations

One of the most important aspects of identification data lies in how proteins and peptides are related. Some identified peptides stem from multiple proteins (called *shared peptides*), while others are unique to a single protein or isoform (called *unique* or *proteotypic* peptides). The `describePeptides()` and `describeProteins()` functions provide useful summaries of shared and unique peptides as well as protein and protein groups.

```
describePeptides(psms)

25344 peptides composed of

    unique peptides: 9177

    shared peptides (among protein):

    7649(2) 3983(3) 1912(4) 1099(5) 515(6) 362(7)
    214(8) 148(9) 57(10) 41(11) 39(12) 13(13) 6(14)
    6(15) 35(16) 15(17) 8(18) 5(19) 5(20) 14(21) 10(22)
    9(23) 6(24) 12(32) 1(47) 1(68) 1(113) 1(243)

describeProteins(psms)
```

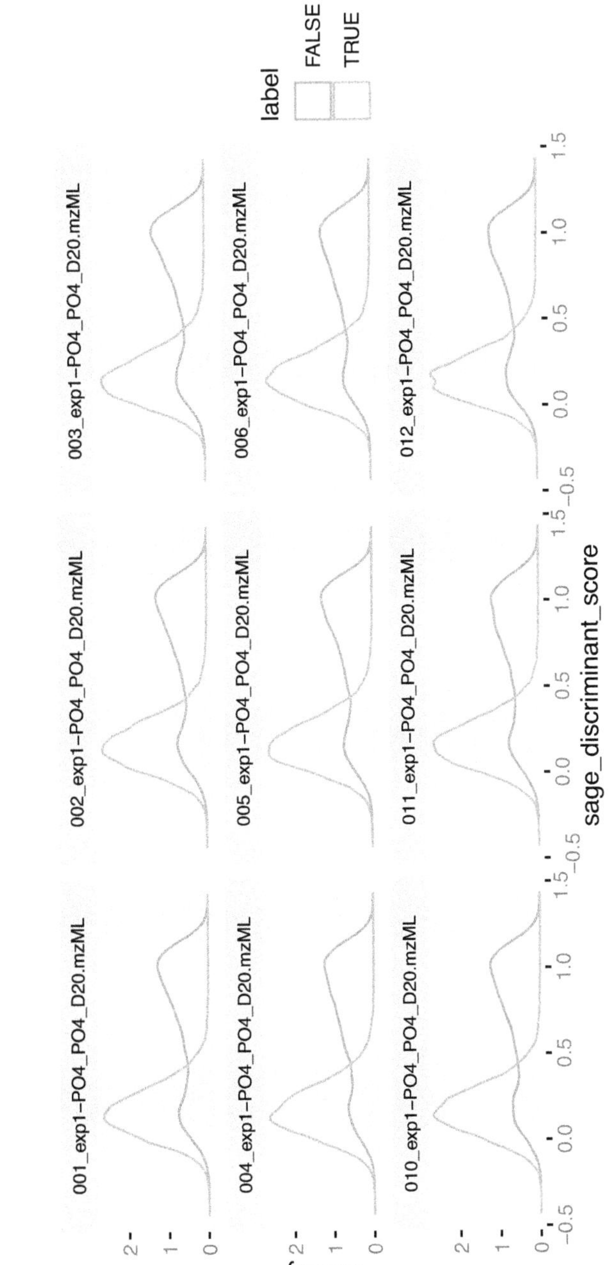

Fig. 2 Target and decoy PSM score distributions

```
10581 proteins composed of

only unique peptides: 1596

only shared peptides: 8437

unique and shared peptides: 548
```

One way to store the peptide–protein relations is through connected components. They represent the relation between proteins and peptides and how shared and unique peptides define groups of proteins. Below, we see that the constituents of the protein groups, as defined in the proteins variable, are separated by a semi-column, which we later set to split the groups and create the connected components.

```
head(psms$proteins)

[1] "sp|Q9H307|PININ_HUMAN"
[2] "sp|P02545-3|LMNA_HUMAN;sp|P02545-4|LMNA_
    HUMAN;sp|P02545-5|LMNA_HUMAN;sp|P02545|LMNA_HUMAN"
[3] "sp|Q13586-2|STIM1_HUMAN;sp|Q13586|STIM1_HUMAN"
[4] "sp|Q7Z417|NUFP2_HUMAN"
[5] "sp|Q7Z2T5|TRM1L_HUMAN"
[6] "sp|Q9NXV6|CARF_HUMAN"

cc <- ConnectedComponents(psms, split = ";")
```

The connected component summary shows the total number of individual proteins and components in the data, and the respective number of single proteins defined by a single unique peptide ($[1 \times 1]$), protein groups defined by a single shared peptide ($[1 \times n]$), single proteins defined by multiple unique peptides ($[n \times 1]$) and protein groups defined by multiple peptides ($[n \times n]$).

```
cc
An instance of class ConnectedComponents
Number of proteins: 10581
Number of components: 4884
     Number of components [peptide x proteins]:
     1476[1 x 1] 411[1 x n] 1127[n x 1] 1870[n x n]
```

Individual peptide-by-protein connected components can be extracted with `connectedComponents()` function, specifying the index of the component.

```
m <- connectedComponents(cc, 22)
m

3 x 3 sparse Matrix of class "dgCMatrix"
sp|A2AJT9-2|BCLA3_HUMAN
SQKLPDVKPS[+79.966]PINLR          4.6364287
RGS[+79.966]EDFETR                0.7867669
GYS[+79.966]PGRGDSNRR             1.0660958
sp|A2AJT9-3|BCLA3_HUMAN
```

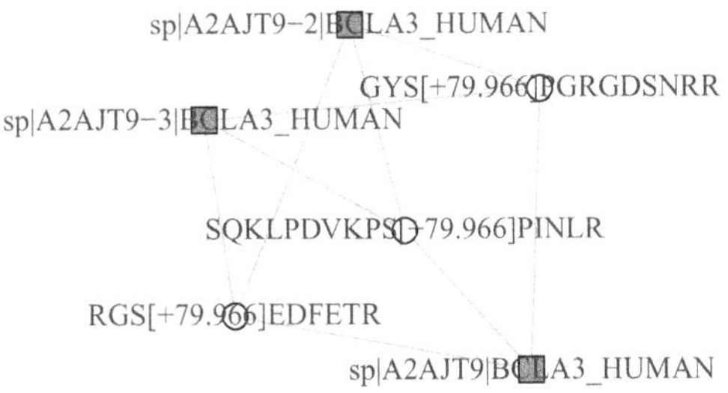

Fig. 3 Network representing the relationships between peptides and proteins within a protein group

```
SQKLPDVKPS[+79.966]PINLR          4.6364287
RGS[+79.966]EDFETR                0.7867669
GYS[+79.966]PGRGDSNRR             1.0660958
sp|A2AJT9|BCLA3_HUMAN
SQKLPDVKPS[+79.966]PINLR          4.6364287
RGS[+79.966]EDFETR                0.7867669
GYS[+79.966]PGRGDSNRR             1.0660958
```

The values in the connected component matrix represent the sum of scores given by the search engine. Some PSMs may have identical peptide sequences but have different charges or modifications sites when these are not explicitly encoded in the sequence. Such instances increase the confidence of the protein that peptide stems from, which is why the scores are summed.

Individual components can be visualized using the `plotAdjacencyMatrix()` function. Figure 3 shows the relation between three peptides and three proteins. Proteins are
represented as colored squares and peptides by circles.

```
plotAdjacencyMatrix(m)
```

In this example, all three peptides are shared among the three isoforms of the BCLAF3 gene.

One way to select relevant connected components is to select those composed by more than one peptide (more than one row) and more than one protein (more than one column), i.e., those annotated as $[n \times n]$ in the textual summary of the object.

```
head(which(nrows(cc) > 1 & ncols(cc) > 1))

[1] 5 6 7 13 18 19

connectedComponents(cc, 7)

4 x 2 sparse Matrix of class "dgCMatrix"
sp|A0A8I5KQE6|RPSA2_HUMAN
EHPWEVM[+15.995]PDLYFYRDPEEIEKEEQAAAEK       2.9922049
FTPGTFTNQIQAAFREPR.
EHPWEVMPDLYFYRDPEEIEKEEQAAAEK                0.8481142
```

```
LLVVTDPR 2.7430056
sp|P08865|RSSA_HUMAN
EHPWEVM[+15.995]PDLYFYRDPEEIEKEEQAAAEK          2.9922049
FTPGTFTNQIQAAFREPR                              1.4500293
EHPWEVMPDLYFYRDPEEIEKEEQAAAEK                   0.8481142
LLVVTDPR                                        2.7430056
```

The vignette "Understanding protein groups with adjacency matrices" from the *PSMatch* package provides further details on prioritizing interesting connected components for further exploration.

3.4 Calculating Fragments Masses

PSMatch also offers functions to compute theoretical fragments, including posttranslational modifications. These can be set as fixed or variable modifications. It is also possible to set the maximum amount of modifications per peptide. Below, we define a single variable phosphorylation on threonines and compute the masses of *b* and *y* fragment ions for a short illustrative peptide.

```
calculateFragments("ATK",
    fixed modifications = NULL,
    variable modifications = c(T = 79.966),
    type = c("b", "y"))

Fixed modifications used: None
Variable modifications used: T=79.966

mz ion type pos z seq peptide
 1      72.04439  b1    b       1   1A          ATK
 2     173.09207  b2    b    2  1   AT          ATK
 3     147.11280  y1    y       1   1K          ATK
 4     248.16048  y2    y    2  1   TK          ATK
 5     129.10224  y1_   y_      1   1K          ATK
 6     230.14992  y2_   y_   2  1   TK          ATK
 7      72.04439  b1    b       1   1A  AT[79.966]K
 8     253.05807  b2    b    2  1   AT  AT[79.966]K
 9     147.11280  y1    y       1   1K  AT[79.966]K
10     328.12648  y2    y    2  1   TK  AT[79.966]K
11     129.10224  y1_   y_      1   1K  AT[79.966]K
12     310.11592  y2_   y_   2  1   TK  AT[79.966]K
```

3.5 Visualizing Annotated Spectra

The theoretical fragments truly shine when combined with the observed peaks from the experimental spectra. The spectra can be loaded using the R/Bioconductor *Spectra* package [18], and the PSMs can be joined to the spectra using a common identifier.

The *Spectra* package can also be installed using `BiocManager::install()`.

```
BiocManager::install("Spectra")
library(Spectra)
```

Let us first load the spectra from the *.mzML* file containing the raw MS data and create an object of class `Spectra`.

```
sp <- Spectra("./data/001 exp1-PO4 PO4 D20.mzML")
sp

MSn data (Spectra) with 62152 spectra in a
MsBackendMzR backend:
msLevel            rtime               canIndex
<integer>          <numeric>           <integer>
1       1          0.0623463               1
2       1          0.4938655               2
3       2          0.5671415               3
4       1          0.9047552               4
5       1          1.3250339               5
… … … …
62148   2          6598.38               62148
62149   1          6598.72               62149
62150   1          6599.11               62150
62151   1          6599.50               62151
62152   1          6599.89               62152
… 34 more variables/columns.

file(s):
001_exp1-PO4_PO4_D20.mzML
```

To match the PSMs from the nine acquisitions to the scans from the first acquisition loaded in sp, we need to define a common identifier. Below, we generate such unique primary keys by combining the *mzML* filenames and the spectrum identifiers. If identification results from a single file were to be matched to a single *mzML* spectrum, the spectrum identifiers would suffice, but we prefer to describe the most general case here.

```
## Create the common identifier for the spectra
head(sp$pkey <- paste0(basename(sp$dataOrigin),
     sub("^.+scan=", "::", sp$spectrumId)))

[1] "001_exp1-PO4_PO4_D20.mzML::1"
[2] "001_exp1-PO4_PO4_D20.mzML::2"
[3] "001_exp1-PO4_PO4_D20.mzML::3"
[4] "001_exp1-PO4_PO4_D20.mzML::4"
[5] "001_exp1-PO4_PO4_D20.mzML::5"
[6] "001_exp1-PO4_PO4_D20.mzML::6"

## Create the common identifier for the PSMs
head(psms$pkey <- paste0(basename(psms$filename),
     sub("^.+scan=", "::", psms$scannr)))

[1] "003_exp1-PO4_PO4_D20.mzML::32424"
[2] "012_exp1-PO4_PO4_D20.mzML::34473"
[3] "011_exp1-PO4_PO4_D20.mzML::26791"
[4] "010_exp1-PO4_PO4_D20.mzML::16044"
[5] "004_exp1-PO4_PO4_D20.mzML::28200"
[6] "003_exp1-PO4_PO4_D20.mzML::20011"
```

We can now join the MS2 scans (using the filterMsLevel() function) from the Spectra object and the identification results from the PSM object into a new annotated Spectra object.

```
sp <- joinSpectraData(
filterMsLevel(sp, 2L),
psms,
by.x = "pkey")
```

As calculateFragments() needs a canonical sequence, we first add a nonmodified sequence variable to the object.

```
sp$sequence <- gsub("[^A-Za-z]", "", sp$peptide)
```

We also remove the unmatched spectra, i.e., spectra that have a missing sequence. Notice that only 31.2% of the experimental MS2 scans have been identified. This low identification rate highlights the relevance of tools that allow proper exploration of identification results.

```
table(is.na(sp$sequence))

FALSE TRUE
17376 38292

sp <- sp[!is.na(sp$sequence),]
```

Now that both PSMs and spectra have been matched, we can visualize the spectra with their matched peaks based on the identified sequences. As we are handling phosphoproteomic data, we are also adding variable modifications for phosphorylations.

The examples in Figs. 4 and 5 show how adding a modification has an impact on the identification and matched peaks.

```
plotSpectraPTM
    (sp[22],
     variable modifications = c(T = 79.966, S = 79.966,
     Y = 79.966),
     USI = FALSE)

plotSpectraPTM(
     sp[1414],
     variable modifications = c(T = 79.966, S = 79.966,
     Y = 79.966),
     USI = FALSE)
```

In this chapter, we introduced *PSMatch*, an open-source R/Bioconductor package that provides a flexible and comprehensive infrastructure for handling mass spectrometry identification data. *PSMatch* facilitates the import, processing, and visualization of PSMs, supporting diverse workflows regardless of the search engine used. Through intuitive functions, users can create structured PSM objects and filter identifications based on confidence measures, explore peptide–protein relationships, generate theoretical fragments including posttranslational modifications, and visualize experimental spectra alongside theoretical predictions in conjunction with the *Spectra* package. Given the significant proportion of spectra that remain unidentified in MS-based experiments, *PSMatch* offers the necessary tools to handle, explore, and understand identification data, from the actual identified spectra to the final inferred protein groups.

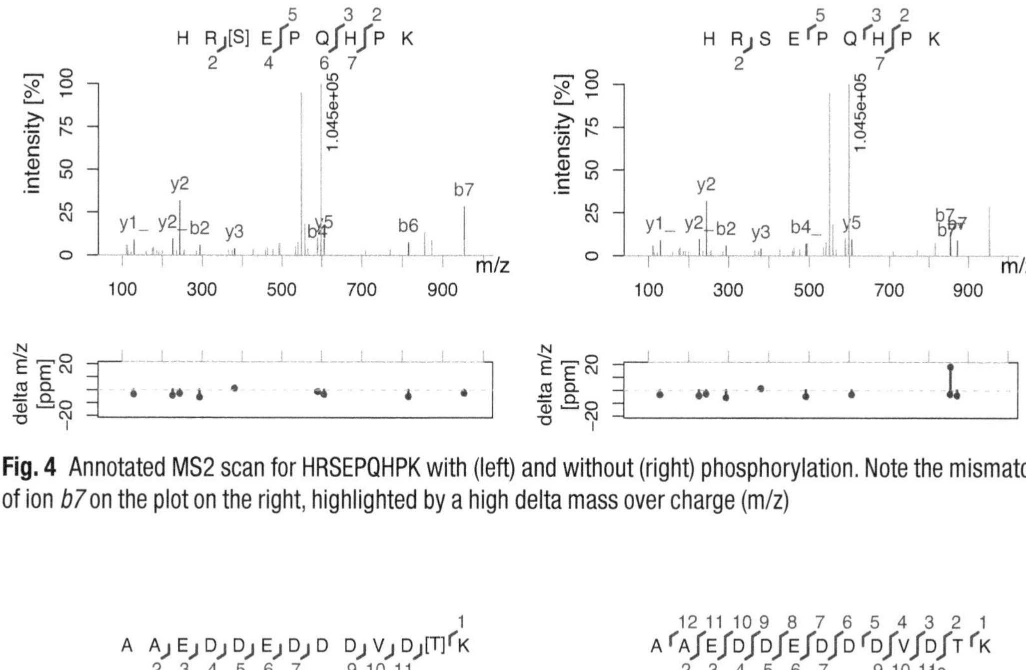

Fig. 4 Annotated MS2 scan for HRSEPQHPK with (left) and without (right) phosphorylation. Note the mismatch of ion *b7* on the plot on the right, highlighted by a high delta mass over charge (m/z)

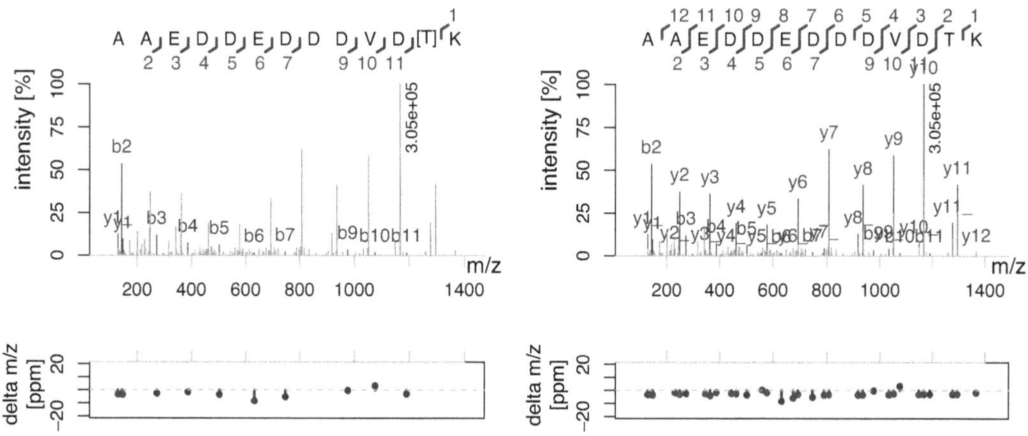

Fig. 5 Annotated MS2 scan for AAEDDEDDDVDTK with (left) and without (right) phosphorylation

4 Notes

1. The versions of R and the specific Bioconductor release are tight. It is thus necessary to use a recent version of R itself for `BiocManager::install()` to install the latest versions of Bioconductor packages.
2. The code that generates the Sage-derived `PSM` object can be simplified using the `sager` package [19], which provides functions customized for Sage identification and quantitation data.

Funding

GD acknowledges the support of a Special Research Funds (FSR) from the UCLouvain.Data and Software AvailabilityAll data needed to reproduce the code is available on `Zenodo (10.5281/zenodo.15854241)`. All software packages are open-source and freely available as part of the Bioconductor project as well as in their respective GitHub repositories (https://github.com/RforMassSpectrometry/PSMatch).

Competing Interests The authors report no competing interests.

References

1. Sticker A, Goeminne L, Martens L, Clement L (2020) Robust summarization and inference in proteome-wide label-free quantification. Mol Cell Proteomics 19(7):1209–1219. https://doi.org/10.1074/mcp.RA119.001624
2. Zhu Y, Orre LM, Zhou Tran Y, Mermelekas G, Johansson HJ, Malyutina A, Anders S, Lehti¨o J (2020) DEqMS: a method for accurate variance estimation in differential protein expression analysis. Mol Cell Proteomics 19(6):1047–1057. https://doi.org/10.1074/mcp.TIR119.001646
3. Hediyeh-Zadeh S, Webb AI, Davis MJ (2023) MsImpute: estimation of missing peptide intensity data in label-free quantitative mass spectrometry. Mol Cell Proteomics 22(8):100558. https://doi.org/10.1016/j.mcpro.2023.100558
4. Balliau T, Frambourg A, Langella O, Martin ML, Zivy M, Blein-Nicolas M (2025) MCQR: Enhancing the processing and analysis of quantitative proteomics data by incorporating chromatography and mass spectrometry information. J Proteome Res. 24(6):2861–2873. https://doi.org/10.1021/acs.jproteome.4c01119
5. Panse C, Grossmann J (2023) protViz: visualizing and analyzing mass spectrometry related data in proteomics. R package version 0.7.7. https://CRAN.R-project.org/package=protViz
6. Gabriels R, Declercq A, Bouwmeester R, Degroeve S, Martens L (2023) psm_utils: A high-level python API for parsing and handling peptide-spectrum matches and proteomics search results. J Proteome Res 22(2):557–560. https://doi.org/10.1021/acs.jproteome.2c00609
7. Deutsch EW, Mendoza L, Moritz RL (2025). Quetzal: comprehensive peptide fragmentation annotation and visualization. Journal of proteome research. 24(4):2196. https://doi.org/10.1021/acs.jproteome.5c00092
8. Ming L, Zou Y, Zhao Y, Zhang L, He N, Chen Z, Li SS-C, Li L (2020) MMS2plot: an R package for visualizing multiple MS/MS spectra for

groups of modified and non-modified peptides. Proteomics 20(15-16):2000061
9. Gentleman RC, Carey VJ, Bates DM, Bolstad B, Dettling M, Dudoit S, Ellis B, Gautier L, Ge Y, Gentry J, Hornik K, Hothorn T, Huber W, Iacus S, Irizarry R, Leisch F, Li C, Maechler M, Rossini AJ, Sawitzki G, Smith C, Smyth G, Tierney L, Yang JYH, Zhang J (2004) Bioconductor: open software development for computational biology and bioinformatics. Genome Biol 5(10):80. https://doi.org/10.1186/gb-2004-5-10-r80
10. Huber W, Carey VJ, Gentleman R, Anders S, Carlson M, Carvalho BS, Bravo HC, Davis S, Gatto L, Girke T, Gottardo R, Hahne F, Hansen KD, Irizarry RA, Lawrence M, Love MI, MacDonald J, Obenchain V, OleÅ› AK, PagÃ¨s H, Reyes A, Shannon P, Smyth GK, Tenenbaum D, Waldron L, Morgan M (2015) Orchestrating high-throughput genomic analysis with bioconductor. Nat Methods 12(2):115–121. https://doi.org/10.1038/nmeth.3252
11. Jones AR, Eisenacher M, Mayer G, Kohlbacher O, Siepen J, Hubbard SJ, Selley JN, Searle BC, Shofstahl J, Seymour SL, Julian R, Binz P-A, Deutsch EW, Hermjakob H, Reisinger F, Griss J, VizcaÃno JA, Chambers M, Pizarro A, Creasy D (2012) The mzIdentML data standard for mass spectrometry-based proteomics results. Mol Cell Proteomics 11(7):111–014381. https://doi.org/S1535-9476(20)33012-7
12. Chick JM, Kolippakkam D, Nusinow DP, Zhai B, Rad R, Huttlin EL, Gygi SP (2015) A mass-tolerant database search identifies a large proportion of unassigned spectra in shotgun proteomics as modified peptides. Nat Biotechnol 33(7):743–749. https://doi.org/10.1038/nbt.3267
13. Gholamizoj S, Ma B (2022) SPEQ: quality assessment of peptide tandem mass spectra with deep learning. Bioinformatics 38(6):1568–1574. https://doi.org/10.1093/bioinformatics/btab874
14. Codilupi T, Szybinski J, Arunasalam S, Jungius S, Dunbar AC, Stivala S, Brkic S, Albrecht C, Vokalova L, Yang JL, Buczak K, Ghosh N, Passweg JR, Rovo A, Angelillo-Scherrer A, Pankov D, Dirnhofer S, Levine RL, Koche R, Meyer SC (2024) Development of resistance to type II JAK2 inhibitors in MPN depends on AXL kinase and is targetable. Clin Cancer Res 30(3):586–599. https://doi.org/10.1158/1078-0432.CCR-23-0163
15. Lazear MR (2023) Sage: an open-source tool for fast proteomics searching and quantification at scale. J Proteome Res 22(11):3652–3659. https://doi.org/10.1021/acs.jproteome.3c00486
16. Perez-Riverol Y, Bandla C, Kundu DJ, Kamatchinathan S, Bai J, Hewapathirana S, John NS, Prakash A, Walzer M, Wang S, VizcaÃno JA (2025) The PRIDE database at 20 years: 2025 update. Nucleic Acids Res 53(D1):543–553. https://doi.org/7874848
17. Deflandre, G.& Gatto, L. (2025). Data for ‘PSMatch: an R/Bioconductor package to explore proteomics identification data.’ [Dataset]. Zenodo. https://doi.org/10.5281/zenodo.15854241
18. Rainer J, De Graeve M, Louail P, Gibb S, Gatto L (2025) An open infrastructure for mass spectrometry data in R. OSF Preprints. https://doi.org/10.31219/osf.io/cwt2v_v1
19. Gatto L sager: analysing sage results with R. R package version 0.3.3

Check for updates

Chapter 6

A Uniquome-Based Method for Proteins Identification by Mass Spectrometry

Vasileios Pierros, Evangelos Kontopodis, Dimitrios J. Stravopodis and George Th, Tsangaris

Abstract

Protein identification by mass spectrometry (MS) is a pivotal step in proteomics. Numerous methods have been developed to securely and effectively identify proteins derived from experimentally detected peptides by mass spectrometry. The dominant approach is based on the assumption that each experimentally identified peptide can be matched with a peptide included in a database of peptide sequences, generated by in silico digestion of proteins with a specific proteolytic enzyme. In this way, the protein containing the peptide can be identified. In a more advanced approach, the proteins and their in silico-digested peptides in the database are transformed into theoretical mass spectrometry spectra, and search engines match the experimentally obtained spectra to these theoretical spectra generated from protein and peptide sequences. We developed an alternative method for protein identification using Core Unique Peptides (CrUPs) and the Uniquome, termed as Uniquome-Based Protein Identification Method (UB-PIM). According to this method, instead of searching for peptides in the database of *in silico*-digested peptides, we search for CrUPs within the experimentally obtained peptides by mass spectrometry. If a peptide contains at least one CrUP, it can be directly correlated to the protein from which the CrUP is derived. Because of the unique nature of CrUPs, peptides obtained by MS can securely and uniquely identify the protein of origin. This provides a reference space in which even single-peptide identifications can achieve high specificity, reducing the ambiguity caused by shared or homologous sequences and improving the interpretability of MS data. Furthermore, UB-PIM can be applied to any type of peptide and is effective with both Data-Independent Acquisition (DIA) and Data-Dependent Acquisition (DDA) approaches, as well as with top-down and bottom-up proteomics. This allows confident protein identification from minimal evidence, expands the scope of detectable proteins, and remains computationally efficient, rapid, and universally applicable.

Keywords Protein identification method, Mass spectrometry, Core Unique Peptides, Uniquome, Uniquome-based method, UB-PIM

1 Introduction

Proteomics is a set of complex methods and technologies aimed at identifying, cataloging, and studying the total protein content of a biological material. The conventional approach includes the

Frédérique Lisacek (ed.), *Bioinformatics for Mass Spectrometry*, Methods in Molecular Biology, vol. 3071,
https://doi.org/10.1007/978-1-0716-5519-1_6,

separation of proteins in the biological sample, their digestion by proteolytic enzymes, the analysis of the resulting peptides by mass spectrometry, protein identification using bioinformatics tools, systematic entry of the results into databases, and finally their processing and evaluation. The introduction of mass spectrometry (MS) proved pivotal, and the development of MS-based proteomics has profoundly transformed the ability to identify and quantify proteins in complex biological samples [1].

At its core, proteomic workflows employ either "bottom-up" or "top-down" approaches—digesting proteins into peptides or analyzing intact proteins directly, respectively. MS methods can be broadly divided into Data-Independent Acquisition (DIA) and Data-Dependent Acquisition (DDA) techniques [2,3]. DIA is the MS technique primarily used in proteomics to identify and quantify proteins in complex biological samples. It systematically fragments all ions within predefined mass-to-charge (m/z) windows across the entire m/z range, unlike DDA, where only a subset of ions is selected for fragmentation. In both techniques, protein identification relies heavily on matching observed mass-to-charge (m/z) spectra to peptide/protein sequences through computational search engines and scoring systems, aided by statistical validation techniques such as false discovery rate (FDR) control [2–4].

Briefly, protein identification in a proteomics workflow is based on a FASTA-format file containing peptide sequences, generated by *n silico* digestion of proteins with a specific proteolytic enzyme (commonly trypsin, but also chymotrypsin, etc.). To generate the FASTA file, all reviewed and unreviewed proteins of a given species (e.g., *Homo sapiens*) from the UniProt database are digested *in silico* according to the rules of the selected enzyme, and the resulting theoretical peptides are compiled into the FASTA file [5]. For protein identification, each experimentally observed peptide by MS is matched with a peptide contained in the FASTA file, thereby identifying the protein in which the peptide occurs [6].

In a more advanced approach, proteins and their *in silico*-digested peptides included in the FASTA file are transformed into theoretical MS spectra, and search engines (such as Mascot, MaxQuant, MSFragger, or Comet) match the experimentally obtained spectra to these theoretical spectra generated from protein and peptide sequences [7,8].

Despite their widespread use, these methods have important limitations. For reliable protein identification, the analysis of at least two peptides per protein is typically required, and many peptides identified by the mass spectrometer do not ultimately lead to unambiguous protein characterization and are therefore rejected during bioinformatics processing. Increasing the number of search engines and integrating diverse algorithmic approaches can marginally improve protein identification coverage, but such gains are often limited by computational burden and diminishing returns [9]. These weaknesses of current methods have highlighted the need for a new approach to protein identification. Such an approach is based on the assumption that the amino acid sequence of each protein should contain at least one peptide whose sequence is absolutely unique to the proteome of the organism to which it belongs, thereby enabling differential and unambiguous characterization of the protein [10–12].

In this context, the concept of the Uniquome offers a groundbreaking strategy, originally introduced for constructing a proteomic atlas for each organism by cataloging peptides whose amino acid sequences are absolutely unique to that organism's proteome. Within the Uniquome, two novel peptide entities were defined: (a) the Core Unique Peptide (CrUP), which is a peptide sequence present specifically and exclusively in only one protein of a given proteome and which also represents the minimal amino acid length necessary for unique identification; and (b) the Composite Unique Peptide (CmUP), an extended sequence entity formed by the linear, sequential, or overlapping unification of two or more CrUPs [10,11]. Thus, a typical CrUP is the shortest amino acid sequence found exclusively in a single protein within a proteome, ensuring that the sequence does not occur in any other protein of the examined proteome. The term "core" emphasizes that this peptide is minimal in length yet sufficient to uniquely represent its source protein [13].

Analysis of the human proteome revealed that 99.2% of reviewed proteins contain CrUPs, suggesting that leveraging these peptide entities could dramatically enhance identification specificity—even in cases supported by only a single peptide. This challenges the conventional requirement for multipeptide evidence. By computing and indexing these unique sequences, the Uniquome provides a reference space in which single-peptide identifications can achieve high specificity, thereby reducing ambiguity caused by shared or homologous sequences and improving the interpretability of MS data [12,13].

Based on this framework, we developed an MS-based protein identification method in which the counterpart of the FASTA file used in conventional approaches is replaced with a file containing the CrUPs of the Uniquome of the studied organism. In the present study, this method was applied to the identification of *Homo sapiens* proteins included in the human proteome. The obtained results indicate that the Uniquome could transform MS-based protein identification in both DIA and DDA approaches by removing ambiguity at the sequence level, enabling confident identification from minimal evidence, and expanding the scope of detectable proteins—all while being computationally efficient, rapid, and universally applicable.

2 Materials

2.1 Databases Used

The following databases were used:

- UniProt (www.uniprot.org) (Release 2025_03);
- Unipept (https://unipept.ugent);
- InterPro (https://www.ebi.ac.uk/interpro) (Release 97.0, November 9, 2023);
- National Center for Biotechnology Information (https://www.ncbi.nlm.nih.gov);
- Kyoto Encyclopedia of Genes and Genomes (https://www.genome.jp/kegg/);
- Japan ProteOme STandard Repository (https://repository.jpostdb.org);
- Trans-Proteomic Pipeline (TPP) (http://www.tppms.org)

3 Methods

3.1 Uniquome and Uniquome (+) Databases

In the latest release of the human proteome in the UniProt database (Release 2025_3), 20,420 reviewed proteins and 184,785 unreviewed proteins were included [14]. The total set of 20,420 reviewed human proteins was analyzed for the presence of the novel CrUP and CmUP peptide entities. Following the procedure described previously [13], a total of 7,279,076 peptides ranging from 4 to 100 amino acids (aa) in length were characterized as CrUPs in 20,251 (99.16%) human proteins (Table 1), whereas 170 proteins (0.82%) did not contain any Unique Peptide (UP) species of the same length (4–100 aa). Further analysis revealed that the 7,279,076 CrUPs could generate 67,921 CmUPs. The overall density of the human proteome in CrUPs is 64%, indicating that 64 CrUPs are generated per 100 aa, while the coverage of the human proteome by UP species is 93% (Table 1).

For protein identification by CrUPs, the construction of the Human Uniquome database is crucial. In our previous studies, the Uniquome database was constructed exclusively from reviewed proteins in UniProt. To expand this database, we examined the possibility of including unreviewed proteins from UniProt. Analysis of these proteins revealed several cases: (a) proteins with identical sequences to reviewed proteins but with different accession numbers. For example, the reviewed protein SAMP_Human (P02743, Serum Amyloid P-component) has an identical amino acid sequence to the unreviewed protein V9HWP0_Human (V9HWP0, Pentraxin Family member) (Fig. 1). In this case, the CrUPs of SAMP_Human

Table 1
Uniquome and Uniquome (+) analysis

	Uniquome	Uniquome (+)
Proteins included	**20,420** (Only reviewed)	**20,859** (Reviewed + Unreviewed)
Core Unique Peptides	**7,279,076**	**7,279,514**
Composite Unique Peptides	67,921	67,932
Density of CrUP	64%	64%
Density of CmUP	0.6%	0.6%
Coverage	93%	93%
Proteins with Unique Peptides	**20,251**	**20,385**
Proteins without Unique Peptides	170	474

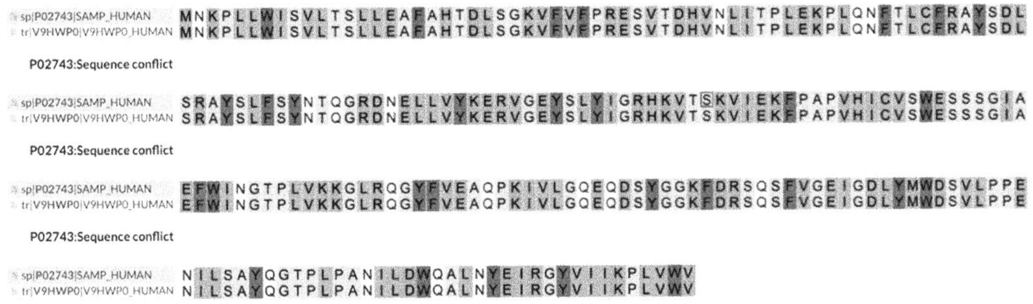

Fig. 1 Alignment of the SAMP_Human (P02743, Serum Amyloid P-component) and the V9HWP0_Human (V9HWP0, Pentraxin Family member) amino acid sequences. The sequence of these proteins is totally identical.

lose their uniqueness, as they also belong to V9HWP0_Human. (b) Fragments of reviewed proteins, which share identical sequence segments with the complete protein but are assigned different accession numbers. For instance, the reviewed protein VW5B1_Human (Q5TIE3, von Willebrand factor A domain-containing protein 5B1) has a length of 1220 amino acids, whereas the unreviewed protein E9PP07_Human (E9PP07, von Willebrand factor A domain-containing 5B1) is an 82-aa fragment corresponding to the N-terminal region of VW5B1_Human (Fig. 2). In this case, the CrUPs located within the first 82 amino acids of VW5B1_Human lose their uniqueness, as they also belong to E9PP07_Human.

To expand the Human Uniquome database while accounting for such cases, we constructed the Uniquome (+) database. This extended version was designed to include only unreviewed proteins that do not "overlap" with any reviewed proteins. Specifically,

Fig. 2 The alignment of the reviewed protein VW5B1_Human (Q5TIE3, von Willebrand factor A domain-containing protein 5B1) and the unreviewed protein E9PP07_Human (E9PP07, von Willebrand factor A domain-containing 5B1) indicated that E9PP07_Human is a fragment of the VW5B1_Human protein.

unreviewed proteins that did not contain any CrUPs of the reviewed proteins in their sequence were isolated. These proteins, lacking CrUPs derived from reviewed proteins, indicate that they share no amino acid sequence with reviewed proteins and are not fragments of them. We identified 439 such unreviewed proteins. These proteins were added to the 20,420 reviewed proteins, creating a new dataset of 20,859 proteins (Table 1). In this expanded set, CrUPs and CmUPs were recalculated, resulting in 7,279,514 CrUPs and 67,932 CmUPs, respectively, which together constitute the Uniquome (+) (Table 1).

3.2 The Uniquome Based-Protein Identification Method (UB-PIM)

We developed a protein identification algorithm based on the Uniquome and Uniquome (+) databases, which checks whether an MS-identified peptide contains at least one CrUP. If the check is positive, the peptide is directly associated with the corresponding protein from which the CrUP is derived. Since each CrUP is unique to its source protein, any MS-identified peptide containing it can unambiguously and reliably identify that protein. In other words, the presence of a CrUP within a peptide ensures that the peptide uniquely identifies a single protein (*see* **Note 1**).

The developed method, named as Uniquome-Based Protein Identification Method (UB-PIM), has been implemented on the Uniquome.com website (*see* **Note 2**). Figure 3 shows a screenshot of the Uniquome.com homepage, where protein identification

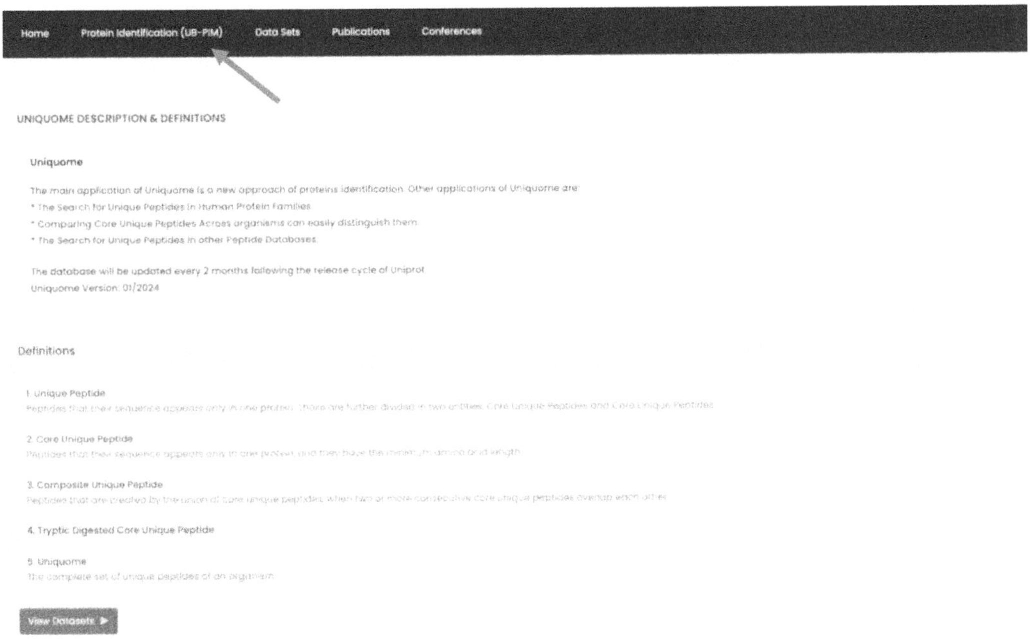

Fig. 3 The Uniquome.com homepage, where protein identification via the developed method is introduced as *"Protein identification (UB-PIM)."* The position in the page is marked by the red arrow

via the developed method is introduced as "*Protein identification (UB-PIM).*" Selecting this option leads the user to the main interface of the UB-PIM method (Fig. 4). On this screen, required and optional fields are clearly indicated by an asterisk (*).

In the first required field, the user must select which database will be used for protein identification. Two options are available: the latest version of the Human Uniquome [appearing as *Core Unique Peptides of Homo sapiens (Human) ver. 4_2025*] or the latest version of the Uniquome (+) [appearing as *Core Unique Peptides of Homo sapiens (Human) plus (+) ver. 4_2025*]. The second required field is the user's email address, to which the results will be sent. The third required field is the "Job Title," where the user provides a descriptive name for the task. Finally, under "Select File," the user uploads the file containing the MS-derived peptides to be analyzed. This file may be either a plain text file (with one peptide per line) or a FASTA-format file. In the case of a text file, peptides are automatically numbered (e.g., 000001), whereas in a FASTA file, peptides can be grouped under user-defined names. For demonstration purposes, a list of 30 peptides in text format was used to perform the method (Table 2). Optional fields allow the user to enter their name and a more detailed description of the submitted task, but these are not mandatory.

After all required fields are completed, clicking "Create Job" initiates the method. UB-PIM is executed asynchronously at scheduled intervals (every few hours). The process begins by retrieving all proteins and their corresponding Core Unique Peptides (CrUPs) from the selected database. Each user-provided peptide is then checked to determine whether it contains one or more CrUPs from a given

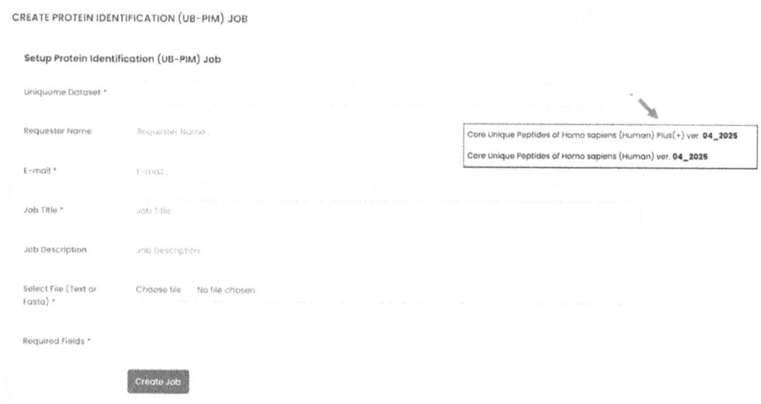

Fig. 4 The main interface of the UB-PIM method after the Uniquome.com homepage. Required and optional fields are indicated by an asterisk (*). Red arrow indicates that in the required position "Uniquome Dataset," the user could select the Core Unique Peptides of Homo sapiens (Human) Plus (+) or the Core Unique Peptides of *Homo Sapiens* (Human)

Table 2
Peptides

AVPPFRQVGHALKEK
RVIWLIGQWISVK
HTFILSAPAQLGLLR
IVERDVSHTLKLLR
LHYTDLADLQPSGLWK
LDADAVSPTVSQLSAVFEK
LILKGVILLIVDK
CPAGQQPPQHR
LQSLEQEK
ISELEGNLQTLR
LSLGYYRGFWPVLAR
RSALNNHC
KTVVPKVIVATPKVK
KDFDTLLYTC
KGKLKPVIKSQPVPK
SVVFIVLLILQWK
FMVQSMFAPTDTSDM
LKGIPVLVGLLDHPKKGQGEPLDDR
AKINKYVTILHLPK
LTINTTIC
VPEARLLGLALAWLRDSDFL
DYGFERSSSSESSTNK
YVPLLLLLLLR
VLEPLLLLLHPK
TLLAPLSRGLYLLK
RSALNNHCLQSLEQEK
AKINKYVTILLTINTTIC
LQSLEQEKCPAGQQ
TEDGDWEWSDDEMDEK
LIGFAQLSVS

protein (*see* **Note 3**). A peptide may match multiple CrUPs from the same protein (*see* **Note 4**). However, if a peptide does not contain any CrUP or contains CrUPs from more than one protein, it is flagged as *"Unable to identify protein"* and excluded from the results matrix (*see* **Note 5**). After processing all peptides, the method compiles the distinct set of proteins identified—while accounting for the fact that a single protein may be matched by multiple peptides (*see* **Note 3**).

3.3 UB-PIM Results Interpretation

The results are exported as an Excel file, which is automatically sent to the email address specified by the user. The resulting Excel file contains five worksheets (Fig. 5):

1. Job Details—Provides information about the submitted job.
2. Results—Lists all peptides included in the user file, along with the proteins they identify, as well as peptides unable to identify a protein.
3. Identified Proteins—Summarizes the proteins identified from the peptides in the input file.
4. CrUPs Matches—Provides detailed information on the Core Unique Peptides (CrUPs) matched to each user peptide.
5. Findings Matrix—Displays all peptides that identify a protein, with an "X" marking the intersection between peptides and their corresponding proteins.

In our example, the *Results* worksheet includes all peptides listed in Table 2, along with the corresponding protein identified by each peptide after applying the developed method. Four peptides were

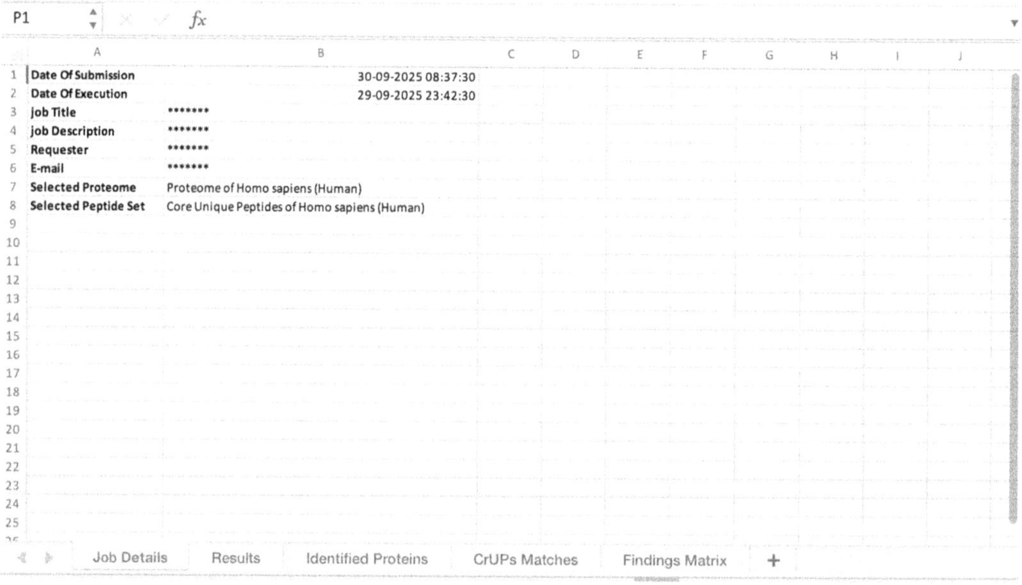

Fig. 5 The Excel file which the user receives automatically after applying the file containing the MS-derived peptides to be analyzed by UB-PIM. The Excel file includes five worksheets as shown in the bottom of the Excel file

classified as *"Unable to identify protein"* and are highlighted in red (Fig. 6).

The next worksheet, *Identified Proteins*, provides a summary only of the proteins identified by the uploaded peptides and their associated CrUPs (Fig. 7). In this example, 21 proteins were identified by 26 of the 30 uploaded peptides, while 4 peptides did not identify any protein (*see* **Notes 3** and **4**). The first column lists the

Peptide Name/Number	Peptide	Protein Accession Number	Protein Name
000001	AVPPFRQVGHALKEK	P56705	WNT4_HUMAN Protein Wnt-4
000002	RVIWLIGQWISVK	Q9UI26	IPO11_HUMAN Importin-11
000003	HTFILSAPAQLGLLR	Q8TDX9	PK1L1_HUMAN Polycystin-1-like protein 1
000004	IVERDVSHTLKLLR	Q96F81	DISP1_HUMAN Protein dispatched homolog 1
000005	LHYTDLADLQPSGLWK	Q6ZU45	CL20A_HUMAN Putative C-type lectin domain family 20 member A
000006	LDADAVSPTVSQLSAVFEK	Q96SB3	NEB2_HUMAN Neurabin-2
000007	LILKGVILLIVDK	P78344	IF4G2_HUMAN Eukaryotic translation initiation factor 4 gamma 2
000008	CPAGQQPPQHR	Q96SB3	NEB2_HUMAN Neurabin-2
000009	LQSLEQEK	Q96SB3	NEB2_HUMAN Neurabin-2
000010	ISELEGNLQTLR	Q96SB3	NEB2_HUMAN Neurabin-2
000011	LSLGYYRGFWPVLAR	Q5H9E4	S2553_HUMAN Solute carrier family 25 member 53
000012	RSALNNHC	Q14588	ZN234_HUMAN Zinc finger protein 234
000013	KTVVPKVIVATPKVK	Q8WZ42	TITIN_HUMAN Titin
000014	KDFDTLLYTC	Q2PZI1	D19L1_HUMAN Protein C-mannosyl-transferase DPY19L1
000015	KGKLKPVIKSQPVPK	P13667	PDIA4_HUMAN Protein disulfide-isomerase A4
000016	SVVFIVLLILQWK	Q86TL2	STIMA_HUMAN Store-operated calcium entry regulator STIMATE
000017	FMVQSMFAPTDTSDM	O95292	VAPB_HUMAN Vesicle-associated membrane protein-associated protein B/C
000018	LKGIPVLVGLLDHPKKGQGEPLDDR	Unable to Identify Protein	Unable to Identify Protein
000019	AKINKYVTILHLPK	P12544	GRAA_HUMAN Granzyme A
000020	LTINTTIC	P01222	TSHB_HUMAN Thyrotropin subunit beta
000021	VPEARLLGLALAWLRDSDFL	Unable to Identify Protein	Unable to Identify Protein
000022	DYGFERSSSSESSTNK	Q96PQ0	SORC2_HUMAN VPS10 domain-containing receptor SorCS2
000023	YVPLLLLLLLLR	E2RYF7	PBMU2_HUMAN Protein PBMUCL2
000024	VLEPLLLLLLHPK	Q5JWR5	DOP1A_HUMAN Protein DOP1A
000025	TLLAPLSRGLYLLK	Q8IWL3	HSC20_HUMAN Iron-sulfur cluster co-chaperone protein HscB
000026	RSALNNHCLQSLEQEK	Unable to Identify Protein	Unable to Identify Protein
000027	AKINKYVTILLTINTTIC	Unable to Identify Protein	Unable to Identify Protein
000028	LQSLEQEKCPAGQQ	Q96SB3	NEB2_HUMAN Neurabin-2
000029	TEDGDWEWSDDEMDEK	Q9UEW8	STK39_HUMAN STE20/SPS1-related proline-alanine-rich protein kinase
000030	LIGFAQLSVS	Q9UEW8	STK39_HUMAN STE20/SPS1-related proline-alanine-rich protein kinase

Job Details | Results | Identified Proteins | CrUPs Matches | Findings Matrix

Fig. 6 The "Results" worksheet. Lists all peptides included in the user file, along with the proteins they identify. Peptides unable to identify a protein appeared in red color marked by the "Unable to Identify Protein"

Protein Accession Number	Protein Name	# Peptides in Set Identifying Protein	# CrUPs identified	# CrUPs in Protein	CrUPs Identified
P56705	WNT4_HUMAN Protein Wnt-4	1	7	210	PPFRQ,QVGHA,AVPPFR,PFRQVG,FRQVGH,VGHALK,GHALKE
Q9UI26	IPO11_HUMAN Importin-11	1	6	727	VIWLI,IWLIG,WLIGQ,IGQWI,GQWIS,WISVK
Q8TDX9	PK1L1_HUMAN Polycystin-1-like protein 1	1	8	2133	HTFIL,FILSAP,AQLGLL,QLGLLR,ILSAPAQ,LSAPAQL,SAPAQLG,APAQLGL
Q96F81	DISP1_HUMAN Protein dispatched homolog 1	1	7	1118	VERDVS,ERDVSH,RDVSHT,VSHTLK,SHTLKL,HTLKLL,TLKLLR
Q6ZU45	CL20A_HUMAN Putative C-type lectin domain family 20 member A	1	9	300	LHYTDL,HYTDLA,YTDLAD,TDLADL,LADLQP,ADLQPS,DLQPSG,QPSGLW,PSGLWK
P78344	IF4G2_HUMAN Eukaryotic translation initiation factor 4 gamma 2	1	4	678	GVILLI,ILLIVD,LILKGVI,ILKGVILL
Q5H9E4	S2553_HUMAN Solute carrier family 25 member 53	1	10	237	LSLGYY,SLGYYR,LGYYRG,GYYRGF,YYRGFW,YRGFWP,RGFWPV,GFWPVL,FWPVLA,WPVLAR
Q14588	ZN234_HUMAN Zinc finger protein 234	1	3	262	RSALNN,SALNNH,ALNNHC
Q8WZ42	TITIN_HUMAN Titin	1	7	23707	TVVPKV,VVPKVI,VPKVIV,VIVATP,IVATPK,PKVIVAT,VATPKVK
Q2PZI1	D19L1_HUMAN Protein C-mannosyl-transferase DPY19L1	1	4	408	DTLLYT,TLLYTC,KDFDTLL,DFDTLLY
P13667	PDIA4_HUMAN Protein disulfide-isomerase A4	1	8	446	GKLKPV,KLKPVI,LKPVIK,KPVIKS,PVIKSQ,VIKSQP,KSQPVP,SQPVPK
Q86TL2	STIMA_HUMAN Store-operated calcium entry regulator STIMATE	1	6	206	LILQW,SVVFIV,IVLLIL,VLLILQ,VVFIVLL,VFIVLLI
O95292	VAPB_HUMAN Vesicle-associated membrane protein-associated protein B/C	1	8	137	QSMFA,FMVQSM,MVQSMF,SMFAPT,MFAPTD,FAPTDT,APTDTS,PTDTSD
P12544	GRAA_HUMAN Granzyme A	1	6	181	KINKY,NKYVT,KYVTI,YVTILH,VTILHL,ILHLPK
P01222	TSHB_HUMAN Thyrotropin subunit beta	1	2	98	LTINTT,INTTIC
Q96PQ0	SORC2_HUMAN VPS10 domain-containing receptor SorCS2	1	8	845	DYGFER,YGFERS,GFERSS,SESSTN,ESSTNK,FERSSSS,SSSESST,ERSSSSESS
E2RYF7	PBMU2_HUMAN Protein PBMUCL2	1	3	137	YVPLLL,VPLLLLLLLL,PLLLLLLLLR
Q5JWR5	DOP1A_HUMAN Protein DOP1A	1	6	1804	LLLHPK,VLEPLLL,LLLLLHP,LEPLLLL,EPLLLLL,PLLLLLLH
Q8IWL3	HSC20_HUMAN Iron-sulfur cluster co-chaperone protein HscB	1	8	166	LSRGLY,SRGLYL,RGLYLL,GLYLLK,TLLAPLS,LLAPLSR,LAPLSRG,APLSRGL
Q96SB3	NEB2_HUMAN Neurabin-2	5	17	535	DADAVS,ADAVSP,SAVFEK,DAVSPTV,CPAGQQ,PAGQQPP,AGQQPPQ,GQQPPQH,QQPPQHR, QSLEQEK,EGNLQT,ISELEGN,SELEGNL,ELEGNLQ,GNLQTLR,CPAGQQ,QSLEQEK
Q9UEW8	STK39_HUMAN STE20/SPS1-related proline-alanine-rich protein kinase	2	9	253	GDWEW,DWEWS,EDGDWE,WSDDEM,SDDEMD,DDEMDE,DEMDEK,FAQLSV,AQLSVS

Job Details | Results | Identified Proteins | CrUPs Matches | Findings Matrix

Fig. 7 The "Identified Proteins" worksheet. The identified proteins by the uploaded set of peptides, the number of CrUPs and a comma-separated list of all matched CrUPs for each protein are shown

UniProt accession number of each identified protein, and the second column provides its full protein name. The third column shows the number of peptides from the input set that contributed to the identification of each protein. For example, 16 proteins were identified by a single peptide, one protein by 2 peptides, and 1 protein by 5 peptides. The fourth column lists the number of identified CrUPs per protein, while the fifth column shows the total number of CrUPs present in each identified protein. The final column contains a comma-separated list of all matched CrUPs for each protein (*see* **Note 6**).

The *CrUPs Matches* worksheet presents the detailed CrUP information underlying protein identification (Fig. 8). Specifically, it shows the CrUP(s) included in each peptide, the protein of origin, the position of the CrUP within the peptide, and the position of the CrUP within the corresponding protein sequence (*see* **Note 7**).

D222 fx 4

	A	B	C	D	E	F	G
1	000001	AVPPFRQVGHALKEK					
2		CrUP	Protein	Location in Peptide	CrUP Location in Protein		
3		PPFRQ	P56705 / WNT4_HUMAN Protein Wnt-4	3	225		
4		QVGHA	P56705 / WNT4_HUMAN Protein Wnt-4	7	229		
5		AVPPFR	P56705 / WNT4_HUMAN Protein Wnt-4	1	223		
6		PFRQVG	P56705 / WNT4_HUMAN Protein Wnt-4	4	226		
7		FRQVGH	P56705 / WNT4_HUMAN Protein Wnt-4	5	227		
8		VGHALK	P56705 / WNT4_HUMAN Protein Wnt-4	8	230		
9		GHALKE	P56705 / WNT4_HUMAN Protein Wnt-4	9	231		
10							
11	000002	RVIWLIGQWISVK					
12		CrUP	Protein	Location in Peptide	CrUP Location in Protein		
13		VIWLI	Q9UI26 / IPO11_HUMAN Importin-11	2	496		
14		IWLIG	Q9UI26 / IPO11_HUMAN Importin-11	3	497		
15		WLIGQ	Q9UI26 / IPO11_HUMAN Importin-11	4	498		
16		IGQWI	Q9UI26 / IPO11_HUMAN Importin-11	6	500		
17		GQWIS	Q9UI26 / IPO11_HUMAN Importin-11	7	501		
18		WISVK	Q9UI26 / IPO11_HUMAN Importin-11	9	503		
19							
20	000003	HTFILSAPAQLGLLR					
21		CrUP	Protein	Location in Peptide	CrUP Location in Protein		
22		HTFIL	Q8TDX9 / PK1L1_HUMAN Polycystin-1-like protein 1	1	1846		
23		FILSAP	Q8TDX9 / PK1L1_HUMAN Polycystin-1-like protein 1	3	1848		
24		AQLGLL	Q8TDX9 / PK1L1_HUMAN Polycystin-1-like protein 1	9	1854		
25		QLGLLR	Q8TDX9 / PK1L1_HUMAN Polycystin-1-like protein 1	10	1855		
26		ILSAPAQ	Q8TDX9 / PK1L1_HUMAN Polycystin-1-like protein 1	4	1849		
27		LSAPAQL	Q8TDX9 / PK1L1_HUMAN Polycystin-1-like protein 1	5	1850		
28		SAPAQLG	Q8TDX9 / PK1L1_HUMAN Polycystin-1-like protein 1	6	1851		
29		APAQLGL	Q8TDX9 / PK1L1_HUMAN Polycystin-1-like protein 1	7	1852		
30							
31	000004	IVERDVSHTLKLLR					
32		CrUP	Protein	Location in Peptide	CrUP Location in Protein		
33		VERDVS	Q96F81 / DISP1_HUMAN Protein dispatched homolog 1	2	371		
34		ERDVSH	Q96F81 / DISP1_HUMAN Protein dispatched homolog 1	3	372		
35		RDVSHT	Q96F81 / DISP1_HUMAN Protein dispatched homolog 1	4	373		
36		VSHTLK	Q96F81 / DISP1_HUMAN Protein dispatched homolog 1	6	375		
37		SHTLKL	Q96F81 / DISP1_HUMAN Protein dispatched homolog 1	7	376		
38		HTLKLL	Q96F81 / DISP1_HUMAN Protein dispatched homolog 1	8	377		
39		TLKLLR	Q96F81 / DISP1_HUMAN Protein dispatched homolog 1	9	378		
40							
41	000005	LHYTDLADLQPSGLWK					
42		CrUP	Protein	Location in Peptide	CrUP Location in Protein		
43		LHYTDL	Q6ZU45 / CL20A_HUMAN Putative C-type lectin domain family 20 member A	1	42		
44		HYTDLA	Q6ZU45 / CL20A_HUMAN Putative C-type lectin domain family 20 member A	2	43		
45		YTDLAD	Q6ZU45 / CL20A_HUMAN Putative C-type lectin domain family 20 member A	3	44		
46		TDLADL	Q6ZU45 / CL20A_HUMAN Putative C-type lectin domain family 20 member A	4	45		
47		LADLQP	Q6ZU45 / CL20A_HUMAN Putative C-type lectin domain family 20 member A	6	47		
48		ADLQPS	Q6ZU45 / CL20A_HUMAN Putative C-type lectin domain family 20 member A	7	48		
49		DLQPSG	Q6ZU45 / CL20A_HUMAN Putative C-type lectin domain family 20 member A	8	49		
50		QPSGLW	Q6ZU45 / CL20A_HUMAN Putative C-type lectin domain family 20 member A	10	51		
51		PSGLWK	Q6ZU45 / CL20A_HUMAN Putative C-type lectin domain family 20 member A	11	52		
52							
53	000006	LDADAVSPTVSQLSAVFEK					
54		CrUP	Protein	Location in Peptide	CrUP Location in Protein		
55		DADAVS	Q96SB3 / NEB2_HUMAN Neurabin-2	2	200		
56		ADAVSP	Q96SB3 / NEB2_HUMAN Neurabin-2	3	201		
57		SAVFEK	Q96SB3 / NEB2_HUMAN Neurabin-2	14	212		
58		DAVSPTV	Q96SB3 / NEB2_HUMAN Neurabin-2	4	202		
59							
60	000007	LILKGVILLIVDK					
61		CrUP	Protein	Location in Peptide	CrUP Location in Protein		
62		GVILLI	P78344 / IF4G2_HUMAN Eukaryotic translation initiation factor 4 gamma 2	5	113		
63		ILLIVD	P78344 / IF4G2_HUMAN Eukaryotic translation initiation factor 4 gamma 2	7	115		
64		LILKGVI	P78344 / IF4G2_HUMAN Eukaryotic translation initiation factor 4 gamma 2	1	109		
65		ILKGVILL	P78344 / IF4G2_HUMAN Eukaryotic translation initiation factor 4 gamma 2	2	110		
66							
67	000008	CPAGQQPPQHR					
68		CrUP	Protein	Location in Peptide	CrUP Location in Protein		
69		CPAGQQ	Q96SB3 / NEB2_HUMAN Neurabin-2	1	272		
70		PAGQQPP	Q96SB3 / NEB2_HUMAN Neurabin-2	2	273		
71		AGQQPPQ	Q96SB3 / NEB2_HUMAN Neurabin-2	3	274		
72		GQQPPQH	Q96SB3 / NEB2_HUMAN Neurabin-2	4	275		
73		QQPPQHR	Q96SB3 / NEB2_HUMAN Neurabin-2	5	276		
74							

Job Details | Results | Identified Proteins | CrUPs Matches | Findings Matrix | +

Fig. 8 The "CrUPs Matches" worksheet. Detailed information on the CrUPs match to each uploaded peptide, together with the identified protein and the location of each CrUP in the peptide and in the identified protein is presented

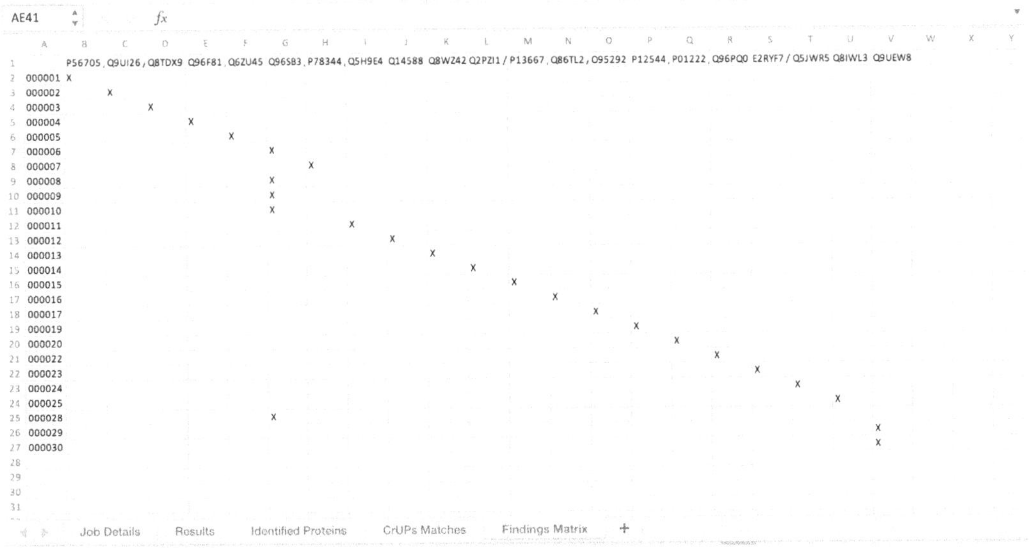

Fig. 9 The "Findings Matrix" worksheet displays all peptides that identify a protein, with an "X" marking the intersection between peptides and their corresponding proteins

Finally, the *Findings Matrix* worksheet provides a global view of the results by displaying all peptides that identify a protein, with an "X" marking the intersection between each peptide and its corresponding protein (Fig. 9) (*see* **Note 8**).

4 Notes

1. The UB-PIM approach contrasts with traditional protein identification methods, which search peptides within a FASTA format file containing theoretical peptides generated from protein digestion by a specified proteolytic enzyme. In conventional methods, if an identified peptide is present in the FASTA file, it is assigned to the protein from which it is theoretically derived [15].
2. According to the UB-PIM's development concept, the method is independent of peptide type (e.g., tryptic digest peptides, chymotryptic digest peptides) and can be applied to any peptide as well as to any MS identified peptide. Based on this principle, UB-PIM is effective for protein identification in DIA procedures but is especially suitable for DDA procedures, where peptides are not predefined and only a subset of ions is selected for fragmentation during MS/MS. This indicates that UB-PIM enables secure protein identification in both bottom-up and top-down proteomic approaches.

3. Conventional proteomics guidelines require two unique peptides per protein to confirm an identification [16]. While this rule minimizes false positives, it also excludes many biologically important but hard-to-detect proteins, such as small proteins, low-abundance proteins, and proteins with few proteotypic tryptic sites. These include biomarkers, tissue-specific proteins, hormones, and pathogen proteins, which are rare but crucial and often difficult to detect and quantify. By applying UB-PIM, even a single match to a CrUP can represent a high-confidence identification. This expands the detectable proteomic space and improves sensitivity without increasing false-positive identifications. In this way, many biologically significant proteins (e.g., biomarkers, tissue-specific proteins, pathogen proteins) that yield only a few peptides under standard digestion—or are present at low abundance (e.g., signaling peptides, cytokines, membrane proteins)—can now be detected and quantified, overcoming the limitations of the two-peptide rule.
4. It is not entire peptides but rather CrUPs within those peptides that are identified and quantified. This leads to rapid and secure identifications that accurately reflect protein abundance, shortens the pipeline from discovery to clinical or translational assays, and improves reproducibility. Instead of searching across an entire proteome database, smaller targeted databases containing only CrUP sequences (and their parent proteins) can be constructed—for example, databases restricted to cancer-related proteins. Such reduced databases minimize search space, memory usage, and false discovery rates, making them ideal for high-throughput clinical proteomics.
5. Regarding peptides for which identification is not became possible through UB-PIM; it may be due to the fact that either the MS-identified peptide does not contain any CrUP or it contains CrUPs from more than one protein included in the searched database. In the first case, the MS-identified peptide may be a family peptide or belong to the common region of homologous proteins, which, by definition, does not contain CrUPs [13]. The second case is more complicated. The identification of CrUPs from more than one protein in the MS-identified peptide is possible to indicate that the MS-identified peptide could include a single-amino-acid polymorphisms (SAPs) as a result of a single-nucleotide polymorphism (SNP) or an amino acid mismatch. Preliminary results about the role of SAPs on CrUPs alterations indicated that the difference of even one amino acid in a protein/peptide can cause extensive changes in the CrUPs (manuscript in preparation). In addition to the above, the identification of more than one CrUPs in the MS-identified peptide could suggest that the peptide belongs to a chimeric protein or is located at the point of creation of a

fusion protein, or more generally, it derives from a protein produced from a new coding sequence. All of the above are under consideration and will be included in a subsequent version of the developed method.

6. Because CrUPs are derived from exact sequences, UB-PIM can distinguish isoforms, sequence variants, and posttranslationally processed forms—an advantage particularly valuable for splice variant analysis or in differentiating highly similar paralogous proteins, which traditional peptide sets often cannot achieve. A future advancement of our method, currently in preparation, would involve transforming CrUPs into mass spectra (using the SEQUEST HT algorithm) and directly correlating CrUP-derived spectra with experimental mass spectra, thereby leveraging all the advantages outlined above [18].
7. In peptide identification, quantification, and targeted proteomics, shared peptides often cause signal contamination between proteins. The UB-PIM method avoids this entirely because of the uniqueness of CrUPs ensuring that quantitation reflects true protein abundance rather than mixed contributions from homologues [17]. Since CrUPs are highly specific and stable, they are also excellent candidates for targeted MS methods (SRM, PRM, DIA). This advantage is particularly evident in label-related methods, where UB-PIM could also support the design of ready-made peptide panels for validation and biomarker quantitation.
8. At the informatics level, the precomputed uniqueness of CrUPs simplifies postprocessing of search results and enables instantaneous checking during peptide searches, resulting in faster and more secure identifications [17]. Thus, another advantage of the developed approach is its straightforward integration into widely used search engines such as Proteome Discoverer, MaxQuant, MSFragger, and ProFound.

Acknowledgments

This study has not been funded by any source whatsoever.

Author Contributions VP: Software and Methodology. EK: Validation, Software, Methodology, and Formal analysis. DJS: Writing—review & editing, Writing—original draft, and Validation. GThT: Conceptualization, Methodology, Writing—original draft, Writing—review & editing, Visualization, Validation, and Supervision.

Conflicts of Interest The authors declare financial interests in the form of issued and pending patent applications.

References

1. Shuken SR (2023) An Introduction to mass spectrometry-based proteomics. J Proteome Res 22(7):2151–2171. https://doi.org/10.1021/acs.jproteome.2c00838
2. Meyer JG (2021) Qualitative and quantitative shotgun proteomics data analysis from data-dependent acquisition mass spectrometry. Methods Mol Biol 2259:297–308. https://doi.org/10.1007/978-1-0716-1178-4_19
3. Bichmann L, Gupta S, Röst H (2024) Data-independent acquisition peptidomics. Methods Mol Biol 2758:77–88. https://doi.org/10.1007/978-1-0716-3646-6_4
4. Noor Z, Ahn SB, Baker MS, Ranganathan S, Mohamedali A (2021) Mass spectrometry-based protein identification in proteomics—a review. Brief Bioinform 22(2):1620–1638. https://doi.org/10.1093/bib/bbz163
5. Mansuri MS, Bathla S, Lam TT, Nairn AC, Williams KR (2024) Optimal conditions for carrying out trypsin digestions on complex proteomes: from bulk samples to single cells. J Proteomics 297:105109. https://doi.org/10.1016/j.jprot.2024.105109
6. Aslam B, Basit M, Nisar MA, Khurshid M, Rasool MH (2017) Proteomics: technologies and their applications. J Chromatogr Sci 55(2):182–196. https://doi.org/10.1093/chromsci/bmw167
7. Matthiesen R (2020) LC-MS spectra processing. Methods Mol Biol 2051:59–77. https://doi.org/10.1007/978-1-4939-9744-2_2
8. Tyanova S, Temu T, Cox J (2016) The MaxQuant computational platform for mass spectrometry-based shotgun proteomics. Nat Protoc 11(12):2301–2319. https://doi.org/10.1038/nprot.2016.136
9. Castaño JD, Beaudry F (2025) Optimization of protein identifications through the use of different chromatographic approaches and bioinformatic pipelines. Rapid Commun Mass Spectrom 39(1):e9937. https://doi.org/10.1002/rcm.9937
10. Pierros V, Kontopodis E, Stravopodis DJ, Tsangaris GT (2022) Unique peptide signatures of SARS-CoV-2 virus against human proteome reveal variants' immune escape and infectiveness. Heliyon 8(4):e09222. https://doi.org/10.1016/j.heliyon.2022.e09222
11. Kontopodis E, Pierros V, Stravopodis DJ, Tsangaris GT (2022) Prediction of SARS-CoV-2 Omicron variant immunogenicity, immune escape and pathogenicity, through the analysis of spike protein-specific core unique peptides. Vaccines (Basel) 10(3):357. https://doi.org/10.3390/vaccines10030357
12. Moeckel C, Mareboina M, Konnaris MA, Chan CSY, Mouratidis I, Montgomery A, Chantzi N, Pavlopoulos GA, Georgakopoulos-Soares I (2024) A survey of k-mer methods and applications in bioinformatics. Comput Struct Biotechnol J 23:2289–2303. https://doi.org/10.1016/j.csbj.2024.05.025
13. Kontopodis E, Pierros V, Vorgias CE, Papassideri IS, Stravopodis DJ, Tsangaris GT (2025) Uniquome: construction and decoding of a novel proteomic atlas that contains new peptide entities. Comput Struct Biotechnol J 27:2123–2138. https://doi.org/10.1016/j.csbj.2025.05.027
14. The UniProt Consortium (2025) UniProt: the universal protein knowledgebase in 2025. Nucleic Acids Res 53(D1):D609–D617. https://doi.org/10.1093/nar/gkae1010
15. Sigloch FC, Grüning B (2025) Protein FASTA Database Handling (Galaxy Training Materials). https://training.galaxyproject.org/training-material/topics/proteomics/tutorials/database-handling/tutorial.html. Accessed 1 Oct2025
16. Käll L, Canterbury JD, Weston J, Noble WS, MacCoss MJ (2007) Semi-supervised learning for peptide identification from shotgun proteomics datasets. Nature Methods 4(11):923–925. https://doi.org/10.1038/nmeth1113
17. Batut B, Hiltemann S, Bagnacani A, Baker D, Bhardwaj V, Blank C, Bretaudeau A, Brillet-Guéguen L et al. (2018) Community-driven data analysis training for biology. Cell Syst 6(6):752–758.e1. https://doi.org/10.1016/j.cels.2018.05.012
18. Yates JR, Eng 3rd JK, McCormack AL, Schieltz D (1995) Method to correlate tandem mass spectra of modified peptides to amino acid sequences in the protein database. Anal Chem 67(8):1426–1436. https://doi.org/10.1021/ac00104a020

Chapter 7

ProtyQuant: A Step-by-Step Protocol for Label-Free Protein Quantification in Data-Dependent Acquisition (DDA) Shotgun Proteomics—From Raw Spectra to Comparative Analysis

Robert Winkler

Abstract

This chapter explains how to use ProtyQuant software to do label-free protein quantification in data-dependent acquisition (DDA) shotgun proteomics. Using the innovative "accumulated peptide probabilities" (APP) metric, it unifies protein inference and quantification, addressing significant issues in conventional workflows. The protocol gives a step-by-step guide that starts with preparing raw mass spectrometry data and searching databases. It then moves on to validating peptides and ends with using ProtyQuant to infer proteins and estimate their amounts. It further outlines the optional postprocessing analysis using R. Emphasizing reproducibility and accessibility, the entire workflow is integrated within MASSyPupX, a free and open-source Linux distribution, enabling researchers to perform high-throughput proteomics studies efficiently from a portable USB drive. This guide is designed to empower researchers in academic, clinical, and industrial settings to generate statistically robust and interpretable protein abundance data.

MASSyPupX can be accessed via Zenodo (https://doi.org/10.5281/zenodo.14618430) and its project page (https://codeberg.org/LabABI/MASSyPupX).

Keywords Shotgun proteomics, Data-dependent acquisition, ProtyQuant, Comparative analysis, Label-free quantification, Mass spectrometry, Protein inference

1 Introduction

Shotgun proteomics has become an indispensable tool for large-scale protein identification and quantification in complex biological samples. One method is data-dependent acquisition (DDA), in which the most prevalent precursor ions are dynamically chosen for fragmentation by mass spectrometers, producing complex MS/MS spectra. Although DDA provides comprehensive proteome coverage, there are substantial challenges due to its dependence on computational workflows for peptide identification, protein inference, and quantification. Label-free quantification (LFQ) workflows involve

Frédérique Lisacek (ed.), *Bioinformatics for Mass Spectrometry*, Methods in Molecular Biology, vol. 3071,
https://doi.org/10.1007/978-1-0716-5519-1_7,

intricate steps: **peptide-spectrum matching (PSM)** to assign spectra to peptides, **statistical validation** of identifications (e.g., via PeptideProphet), **protein inference** to resolve ambiguities from shared peptides, and **quantification** using spectral counting or intensity-based metrics. Traditional methods often treat these steps as discrete processes, leading to fragmented workflows and potential inaccuracies in protein abundance estimation.

ProtyQuant [1] addresses these challenges by unifying protein inference and quantification into a single, streamlined framework. Building on the foundational work of Huang et al. [2] and He et al. [3], ProtyQuant replaces simplistic spectral counting with **accumulated peptide probabilities (APP)**—summing validated peptide probabilities to derive a holistic measure of "protein presence." This metric integrates both the likelihood of a protein's existence (*Pr*) and its relative abundance (*APP*), enhancing sensitivity and accuracy. By automating critical steps in LFQ workflows, ProtyQuant simplifies the transformation of raw mass spectrometry data into interpretable, statistically robust results.

Key innovations of ProtyQuant include:

1. **Advanced Protein Inference**: Modified implementations of Huang and He's algorithms (Multiple Counting, Equal Division, Linear Programming) resolve peptide–protein degeneracy while optimizing quantification.
2. **Compatibility and Accessibility**: Support for standardized inputs (mzML, pepXML, FASTA) and integration with major pipelines (e.g., Trans-Proteomic Pipeline, Comet, Mascot) ensure interoperability. Outputs are lightweight, human-readable TSV files, ideal for downstream statistical analysis.
3. **Unified Probabilistic Scoring**: By combining peptide validation (via tools like PeptideProphet) with probabilistic quantification, ProtyQuant minimizes false positives and improves reproducibility.
4. **Cross-Platform Usability**: A Python-based GUI and Docker support cater to novice users, while command-line scripts and high-performance computing compatibility meet expert needs.

This protocol provides a step-by-step guide for analyzing DDA-based shotgun proteomics data using ProtyQuant, from raw data processing to comparative protein quantification. The chapter covers:

1. **Using the MASSyPupX Live USB** for DDA proteomics data processing.
2. **Raw Data Preparation**: Conversion of mass spectrometry outputs (e.g., raw files) to open formats (e.g., mzML).
3. **Database Search**:Peptide identification using search engines (e.g., Comet, Mascot) and generation of pepXML files.

4. **Peptide Validation**: Statistical filtering of PSMs with PeptideProphet to control false discovery rates.
5. **Protein Inference and Quantification**:Execution of ProtyQuant to resolve protein groups, calculate *APP* scores, and generate abundance tables.
6. **Data Integration**:Merging results across samples, normalization, and statistical comparison (e.g., differential expression analysis).
7. **Output Interpretation**: Guidelines for analyzing ProtyQuant's TSV reports and visualizing results.

All programs for reproducing the steps of the protocol and the example data are compiled in MASSyPupX, a free and open-source (FOSS) Linux distribution for mass spectrometry data processing [4, 5]. The complete workflow can be run from a USB flash drive without installation.

This protocol uses published DDA data from a public repository for demonstrating the complete pipeline. However, users can process their own DDA proteomics project datasets or choose different datasets for research or teaching.

By following this protocol, researchers can efficiently leverage ProtyQuant's unified framework to advance reproducible, high-throughput proteomics studies in academic, clinical, and industrial settings.

2 Materials

- Two USB flash storage drives with at least 32 GB.
- Laptop, personal computer, or server with x86-64 architecture (AMD or Intel) and at least 16 GB RAM.
- MASSyPupX ISO file. The project page is https://codeberg.org/LabABI/MASSyPupX and links to the torrent download. Alternatively, the ISO is available from https://doi.org/10.5281/zenodo.14618430.
- Balena Etcher (https://etcher.balena.io/), or Rufus (https://rufus.ie/).

3 Methods

The main workflow consists of three steps (see Fig. 1):

1. **Peptide-Spectrum Matching**
2. **Peptide Validation**
3. **Protein Inference Calculations and Quantification**

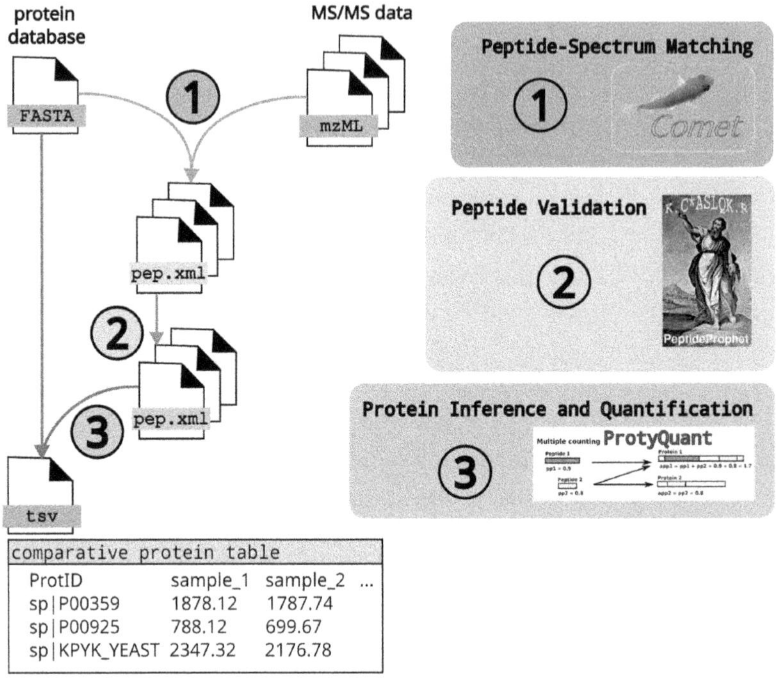

Fig. 1 Workflow for comparing data-dependent acquisition (DDA) shotgun proteomics data with ProtyQuant

3.1 Prepare and Boot MASSyPupX

1. Flash the MASSyPupX ISO file to one of the USB drives using Balena Etcher or Rufus.
2. Boot the computer from the MASSyPupX USB drive (*see* **Notes 1–3** and Fig. 2).
3. On the boot screen, accept the default option.
4. When the system is completely booted up, the Connman System Tray (CMST) pops up, a graphical interface for connecting to wired and wireless networks (*see* **Note 4**).
5. Format the second USB drive with FAT32 and give it a label, e.g., USB-DATA. You can use the antiX formatting tool: Applications (left bottom corner) => antiX => Format USB. Plug out and plug in again to ensure that the USB drive is recognized and mounted.

3.2 Download and Convert Data from ProteomeXchange

The conversion is computationally demanding. See **Note 5** for recommendations.

1. Open a web browser: Click on the navigator symbol on the bottom panel, or choose Applications => Internet => Firefox ESR.
2. Go to the ProteomeCentral of ProteomeXchange https://proteomecentral.proteomexchange.org/.

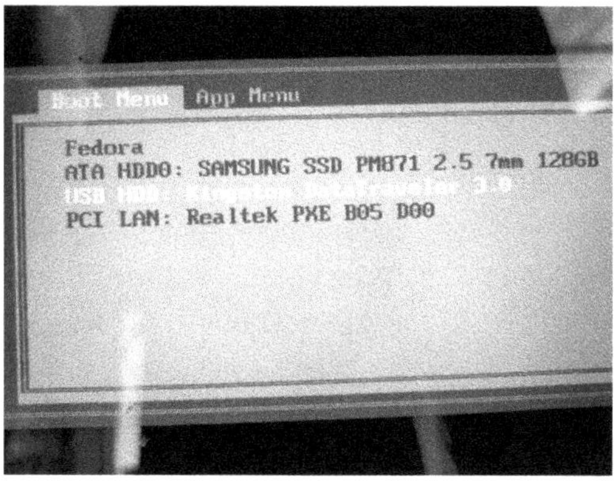

Fig. 2 Booting MASSyPupX from a USB drive

3. Find the data set by identifier PXD001819.
4. Open the data set page, and find the data set FTP location (at the bottom).
5. Click on the link of the data FTP location. When asked which application should be used for opening it, select FileZilla.
6. On the left side of the FileZilla interface ('Local') select the directory /media/antix/USB-DATA; on the right side ("Remote") choose the data files of interest and drag them to the left. For the example workflow, we use the files USP1_50000amol_R1.raw, USP1_50000amol_R2.raw, USP1_50000amol_R3.raw, USP1_5000amol_R1.raw, USP1_5000amol_R2.raw, and USP1_5000amol_R3.raw.
7. Open a Terminal from the Application Menu. Change to superuser mode with sudo su (password: antix). Then, start the Docker daemon and the ProteoWizard [6] container with:

```
service docker start
docker run -it -v /media/antix/USB-DATA/:/data
chambm/pwiz-skyline-i-agree-to-the-vendor-licenses
bash
```

   See **Note 5**.
8. Convert all .raw files to the HUPO standard format mzML [7] using:

```
wine msconvert *.raw
```

9. Using the default parameters, the resulting .mzML files have a size of about 3.6 GB each, which can be confirmed by ls -al. Since the .raw files are about 1.6 GB each, keeping both versions is not possible on a 32 GB flash drive, and the .raw files need to be removed during the conversion. Removing files from the terminal is possible with the rm command (*see* **Note 5**).

3.3 Prepare the FASTA Database

1. Go to the webpage https://www.sigmaaldrich.com/MX/es/technical-documents/technical-article/protein-biology/protein-mass-spectrometry/ups1-and-ups2-proteomic and click Download the UPS FASTA for downloading the sequences of the spiked-in standard proteins, and save the file ups1-ups2-sequences .fasta in the directory /media/antix/USB-DATA.
2. Go to the Uniprot (https://www.uniprot.org) site; choose "Table View" and search for "yeast." In the left menu, select "S. cerevisiae," then "Reviewed (Swiss-Prot)," and download all sequences in canonical FASTA format. Choose "Compressed No."
3. Open the File Manager (Panel at the bottom), and move the downloaded Uniprot FASTA file to media/antix/USB-DATA.
4. Merge the two FASTA databases: Open a Terminal from the Applications menu and type:

```
cd /media/antix/USB-DATA
cat *.fasta > combined.fasta
wc -l *.fasta
```

Which should give the output:

```
60597 combined.fasta
60502 uniprotkb_yeast_AND_model_
organism_5592_2025_07_01.fasta
   95 ups1-ups2-sequences.fasta
121194 total
```

The combined FASTA database contains the sequences of human proteins of the UPS database and the Swiss-Prot FASTA sequences of Baker's yeast (*Saccharomyces cerevisiae*).

3.4 Run the Peptide-Spectrum Matching with Comet Spectral Matching is Performed by Comet That is Monitored as Shown in Fig. 3

1. Create a new Comet parameters file.

```
cd /media/antix/USB-DATA
comet -p
mv comet.params.new comet.params
```

2. Open the parameter file with geany comet.params & and edit the following lines:

```
database_name = combined.fasta
decoy_search = 1
spectrum_batch_size = 1500
```

3. Then process all .mzML files with comet:

```
time comet *.mzML
```

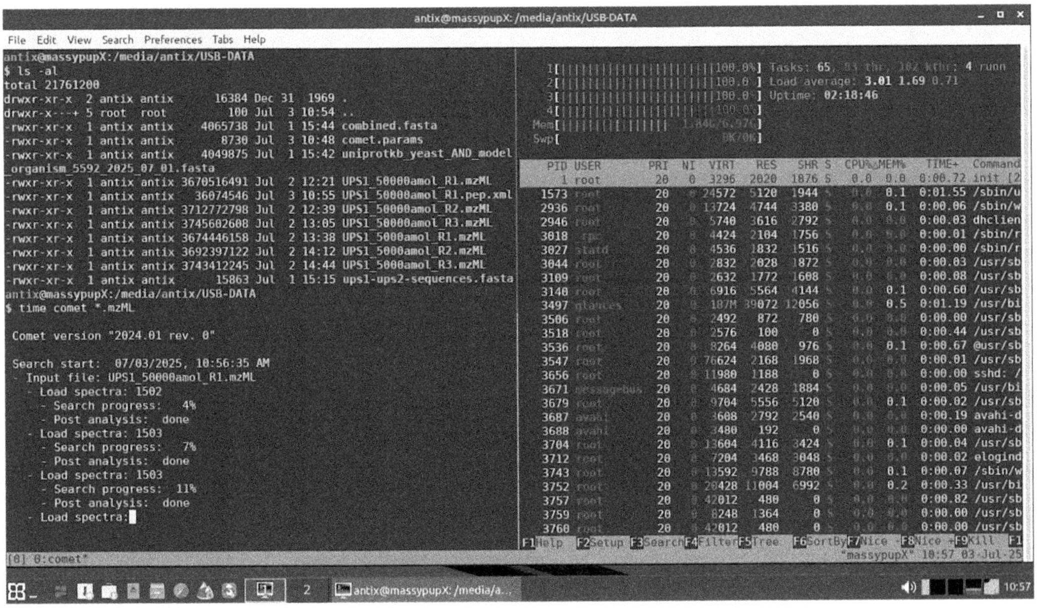

Fig. 3 Running Comet from the Linux terminal of MASSyPupX. The console window was split using tmux. On the left side, the files were listed with ls -al. Following this, the Peptide-Spectrum Matching was started with time comet *.mzML. On the right side, the CPU and RAM use as well as running processes are visualized with htop

About the processing time, *see* **Note 6**.

4. As a result, six pep.xml files with about 130 MB each are created, which can be verified with ls -al.

3.5 Peptide Validation with PeptideProphet

1. The next steps require validation of the identified peptides. PeptideProphet [8] adds probabilities to the hits in the .pep.xml files:

```
for i in *.pep.xml; do PeptideProphetParser $i;
done
```

3.6 Protein Inference with ProteinProphet (Optional)

1. With ProteinProphet [9], the peptide hits can be assembled into proteins:

```
for i in *.pep.xml; do ProteinProphet $i $i.prot.
xml NOGROUPS; done
```

2. A Python script converts the XML file with the protein identifications to a .tsv text table:

```
for i in *.prot.xml; do python3 protxml:to_tsv.py
$i; done
```

3. The results can be reviewed with LibreOffice (see Fig. 4):

```
libreoffice UPS..protXML.tsv &
```

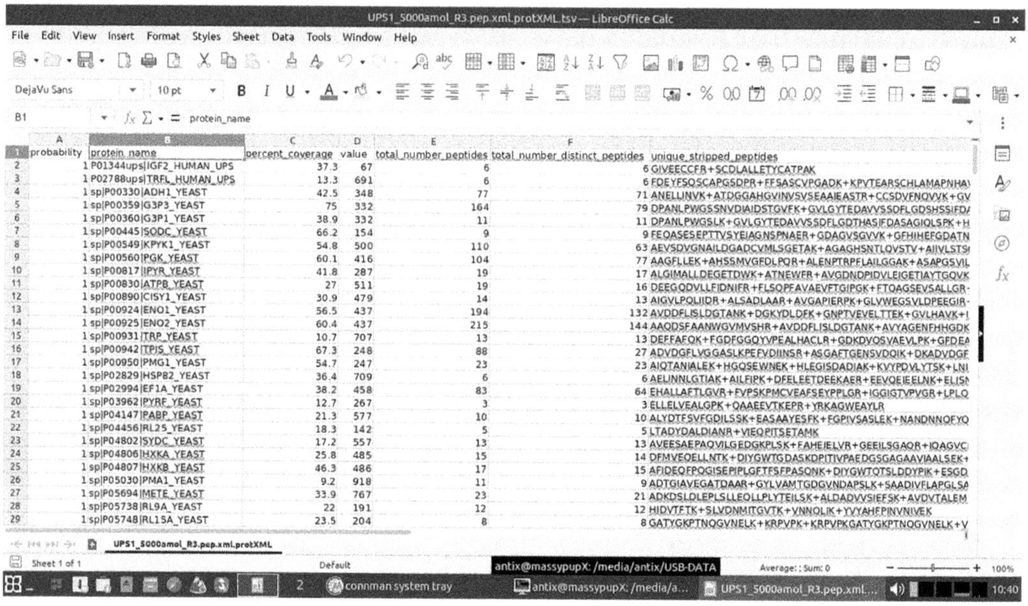

Fig. 4 Review of protein identifications obtained after PeptideProphet/ProteinProphet validation in LibreOffice

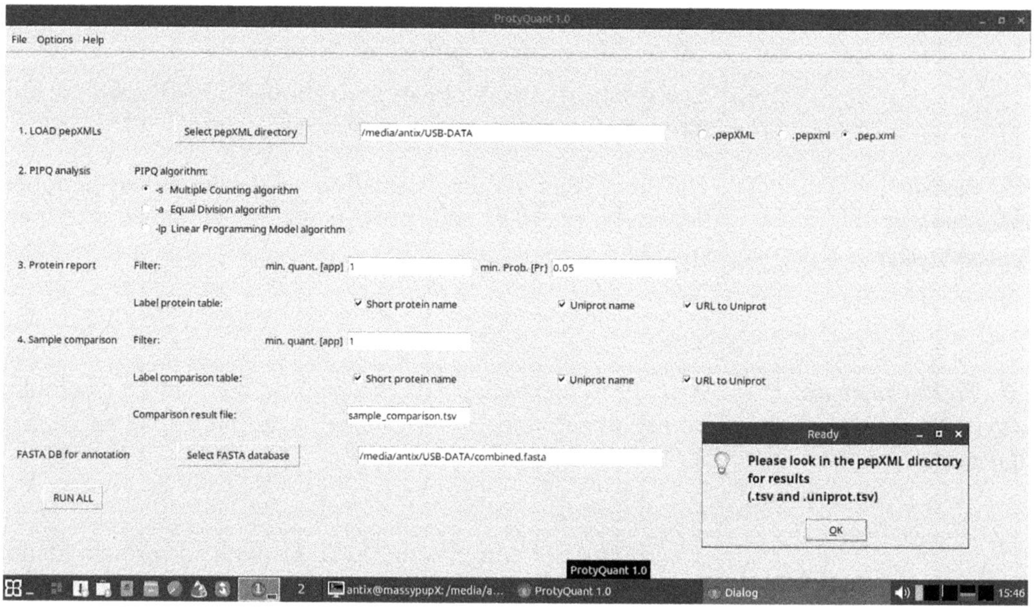

Fig. 5 Graphical user interface of ProtyQuant

3.7 Protein Inference and Quantification with ProtyQuant

1. Start ProtyQuant from the terminal with protyquant &.
2. Choose the directory containing the validated.pep.xml peptide hits and the .FASTA file. The other parameters can be left on their default settings (see Fig. 5).

3. Push RUN ALL. When the analysis is finished, an information window pops up (see Fig. 5).
4. The results can be reviewed by libreoffice sample_comparison.uniprot.tsv (see Fig. 6).

3.8 Analysis with R (Optional)

1. Create a file with the column headers and proteins matching an identifier pattern in a new file:

```
grep -e ProtID -e ALBU_HUMAN sample_comparison.
uniprot.tsv > extracted-hits.tsv
```

2. Start rkward & and create an R-script:

```
# Read the TSV file
data <- read.delim("extracted-hits.tsv",
stringsAsFactors = FALSE)

# Debug: Print column names to verify
cat("Column names in the data:\n")
print(names(data))

# Debug: Print the first row of data
cat("\nFirst row of data:\n")
print(data[1,])

# Extract intensity values from the first row
intensity_values <- as.numeric(data[1,
c("X50000amol_R1", "X50000amol_R2", "X50000amol_R3",
```

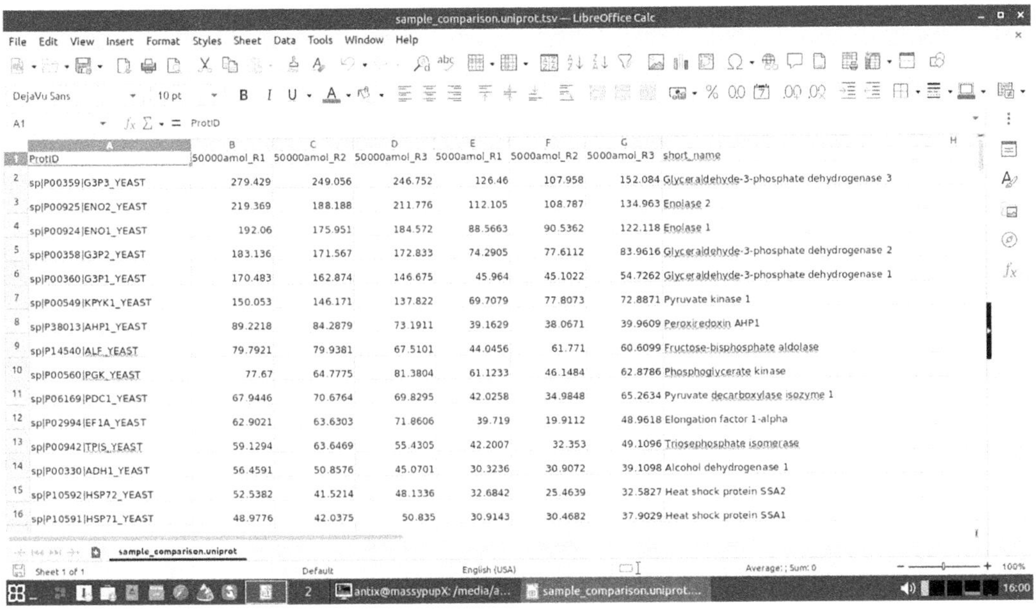

	ProtID	50000amol_R1	50000amol_R2	50000amol_R3	5000amol_R1	5000amol_R2	5000amol_R3	short_name
2	sp|P00359|G3P3_YEAST	279.429	249.056	246.752	126.46	107.958	152.084	Glyceraldehyde-3-phosphate dehydrogenase 3
3	sp|P00925|ENO2_YEAST	219.369	188.188	211.776	112.105	108.787	134.963	Enolase 2
4	sp|P00924|ENO1_YEAST	192.06	175.951	184.572	88.5663	90.5362	122.118	Enolase 1
5	sp|P00358|G3P2_YEAST	183.136	171.567	172.833	74.2905	77.6112	83.9616	Glyceraldehyde-3-phosphate dehydrogenase 2
6	sp|P00360|G3P1_YEAST	170.483	162.874	146.675	45.964	45.1022	54.7262	Glyceraldehyde-3-phosphate dehydrogenase 1
7	sp|P00549|KPYK1_YEAST	150.053	146.171	137.822	69.7079	77.8073	72.8871	Pyruvate kinase 1
8	sp|P38013|AHP1_YEAST	89.2218	84.2879	73.1911	39.1629	38.0671	39.9609	Peroxiredoxin AHP1
9	sp|P14540|ALF_YEAST	79.7921	79.9381	67.5101	44.0456	61.771	60.6099	Fructose-bisphosphate aldolase
10	sp|P00560|PGK_YEAST	77.67	64.7775	81.3804	61.1233	46.1484	62.8786	Phosphoglycerate kinase
11	sp|P06169|PDC1_YEAST	67.9446	70.6764	69.8295	42.0258	34.9848	65.2634	Pyruvate decarboxylase isozyme 1
12	sp|P02994|EF1A_YEAST	62.9021	63.6303	71.8606	39.719	19.9112	48.9618	Elongation factor 1-alpha
13	sp|P00942|TPIS_YEAST	59.1294	63.6469	55.4305	42.2007	32.353	49.1096	Triosephosphate isomerase
14	sp|P00330|ADH1_YEAST	56.4591	50.8576	45.0701	30.3236	30.9072	39.1098	Alcohol dehydrogenase 1
15	sp|P10592|HSP72_YEAST	52.5382	41.5214	48.1336	32.6842	25.4639	32.5827	Heat shock protein SSA2
16	sp|P10591|HSP71_YEAST	48.9776	42.0375	50.835	30.9143	30.4682	37.9029	Heat shock protein SSA1

Fig. 6 Review of ProtyQuant protein identification and quantification in LibreOffice. ProtyQuant creates a comparative table with Uniprot links; the file names are used to identify the columns. In the shown figure, the sample names were manually edited

```
                              "X5000amol_R1", "X5000amol_
R2", "X5000amol_R3")])

# Verify we have numeric values
cat("\nIntensity values:\n")
print(intensity_values)

# Prepare condition groups
conditions <- factor(c(rep("50000amol", 3),
rep("5000amol", 3))
               levels = c("50000amol", "5000amol"))

# Calculate statistics
means <- tapply(intensity_values, conditions, mean)
sds <- tapply(intensity_values, conditions, sd)

# Set plot parameters
y_min <- min(intensity_values) * 0.8
y_max <- max(intensity_values) * 1.2
colors <- c("#4e79a7", "#f28e2b")

# Create the plot
pdf("protein_intensity_comparison.pdf")

par(mar = c(5, 5, 4, 2)) # Set margins
plot(1, type = "n"
   xlim = c(0.5, 2.5), ylim = c(y_min, y_max)
   log = "y", xaxt = "n", xlab = "",
   ylab = "accumulated peptide probability (APP)",
   main = data[1, "short_name"],
   cex.lab = 1.2, cex.main = 0.9)

# Add group labels
axis(1, at = 1:2, labels = c("50000 amol", "5000
amol"), cex.axis = 1)
mtext("spiked-in protein concentration", side = 1,
line = 3, cex = 1.2)

# Plot bars
bar_width <- 0.6
rect(1 - bar_width/2, 1e-100, 1 + bar_width/2,
means[1]
   col = colors[1], border = NA)
rect(2 - bar_width/2, 1e-100, 2 + bar_width/2,
means[2],
   col = colors[2], border = NA)

# Add error bars
arrows(1, means[1] - sds[1], 1, means[1] + sds[1],
    angle = 90, code = 3, length = 0.05, lwd =
1.5)
arrows(2, means[2] - sds[2], 2, means[2] + sds[2],
    angle = 90, code = 3, length = 0.05, lwd =
1.5)
```

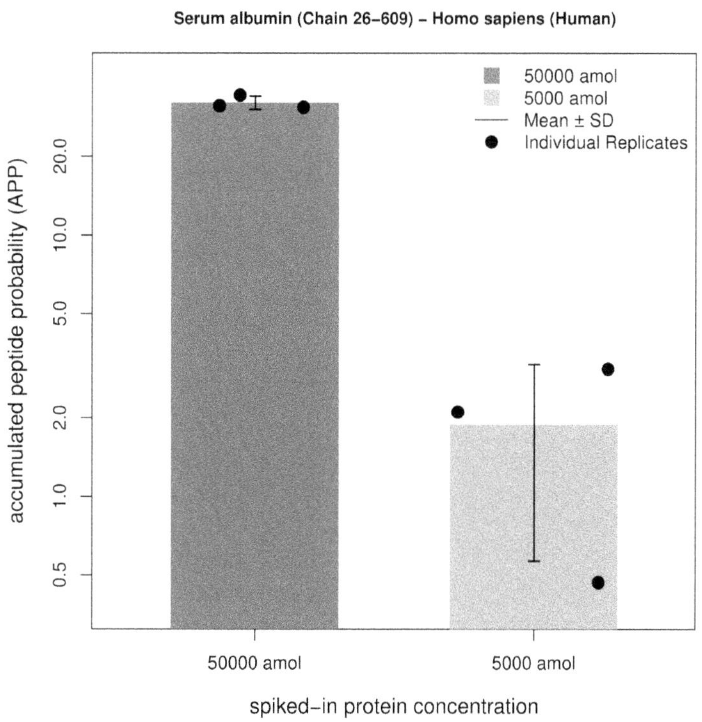

Fig. 7 Comparison of estimated concentrations of spiked-in human serum albumin for two sample groups, using R. The graph is log-scaled and indicates higher variability of APP-based quantification for low concentrations. The concentration difference is overestimated compared to the true value, but in the correct order of magnitude

```
# Add individual points
set.seed(123)
x_jitter <- jitter(rep(1:2, each = 3), factor =
1.5)
points(x_jitter, intensity_values, pch = 19, cex =
1.5, col = "black")

# Add legend
legend("topright"
    legend = c("50000 amol", "5000 amol", "Mean ±
    SD", "Individual Replicates")
    pch = c(15, 15, NA, 19)
    lty = c(NA, NA, 1, NA)
    col = c(colors, "black", "black"),
    pt.cex = c(2, 2, NA, 1.5),
    bty = "n")

dev.off()
cat("Plot successfully created as 'protein_
intensity_comparison.pdf'\n")
```

3. Select CD to script directory and Run all.
4. The script generates Fig. 7. The graph indicates higher variability of label-free quantifications using the accumulated peptide probability (APP) for low protein concentrations. The 5000 amol samples were close to the detection limit for human serum albumin. Therefore, the APP-based analysis slightly overestimated the concentration difference. Nevertheless, the difference of about one order of magnitude was correctly detected.

4 Notes

1. MASSyPupX is not a program but a complete Linux operating system. It can be booted and used directly from a USB flash drive and can be installed on personal computers and servers. Running and installing MASSyPupX as a program from a host system is not possible.
2. Booting from an external USB storage device usually requires pressing and holding a predefined key just after starting the computer. Which key allows choosing the boot device depends on the computer model. Sometimes the instructions for entering boot options are prompted during start-up. Instructions for particular computer models can be found in internet searches. Often, Function keys such as F12 enable choosing the boot options, and special keys such as ESC and DEL give access to the BIOS settings. It might be necessary to modify the boot parameters in the BIOS settings, for example, to allow for Legacy Boot.
3. Running MASSyPupX in "live" mode means that any modifications, such as system settings and file operations, realized during a session are lost after shutting down the system. There are several options to preserve settings and work files between boots:
   - Installing MASSyPupX on a hard drive.
   - Mounting internal or external storage media with the data and result files; in the example, we are using a second USB drive as a working directory.
   - Using "*Persistence*" files (homefs or rootfs). Persistence files could be even stored on the same USB flash drive as the MASSyPupX distribution if the USB drive was created with the included live-usb-maker. See the antiX documentation (https://antixlinux.com/) for details.

4. MASSyPupX has all programs installed for processing DDA proteomics data, and a network connection is only necessary for downloading additional data. For working completely offline, course materials may be distributed on a remastered version of MASSyPupX or on a separate USB drive.
5. The Docker daemon needs to be started with dockerd --storage-driver=vfs on live installations. Downloading the ProteoWizard container requires sufficient storage, and the conversion is relatively slow. Therefore, it is recommended to perform the data file conversion on a fully installed version of MASSyPupX or to use the ProteoWizard converter (command line or Windows GUI version) on a separate computer.
6. The processing time is about 1 h for the six example files, using a laptop with four AMD CPUs @ 1600 MHz and 8 GB RAM. Thus, in a course, one could program a coffee or lunch break after starting the job.
7. The directory /usr/local/ProtyQuant/scripts/ contains scripts for running ProtyQuant from the command line.
8. The described workflow does not work for data-independent acquisition (DIA) shotgun proteomics data. For DIA data, alternative software, for example, DIA-NN [10], can be run on MASSyPupX.

Acknowledgments

I thank Nancy Shyrley García-Rojas for reviewing the text and testing the protocol. Her feedback was invaluable in ensuring the accuracy and clarity of this chapter.

Conflict of Interest The authors declare no conflicts of interest that are relevant to the content of this chapter.

References

1. Winkler R (2021) ProtyQuant: comparing label-free shotgun proteomics datasets using accumulated peptide probabilities. J Proteomics 230:1–12. https://doi.org/10.1016/j.jprot.2020.103985 103985
2. Huang T, Zhu P, He Z (2012) Protein inference and protein quantification: two sides of the same coin. arXiv:12102515 [cs, q-bio]. http://arxiv.org/abs/1210.2515
3. He Z, Huang T, Liu X, Zhu P, Teng B, Deng S (2016) Protein inference: a protein quantification perspective. Comput Biol Chem 63:21–29. https://doi.org/10.1016/j.compbiolchem.2016.02.006
4. García-Chávez JN, Winkler R (2025) Off-site processing of data-dependent and data-independent acquisition shotgun proteomics data with MASSyPupX. J Proteomics 321:1–9. https://doi.org/10.1016/j.jprot.2025.105494 105494
5. Winkler R (2025) MASSyPupX USB (ISO). https://doi.org/10.5281/zenodo.14618430

6. Kessner D, Chambers M, Burke R, Agus D, Mallick P (2008) ProteoWizard: open source software for rapid proteomics tools development. Bioinformatics 24:2534–2536. https://doi.org/10.1093/bioinformatics/btn323
7. Martens L, Chambers M, Sturm M, Kessner D, Levander F, Shofstahl J, Tang WH, Rompp A, Neumann S, Pizarro AD, Montecchi-Palazzi L, Tasman N, Coleman M, Reisinger F, Souda P, Hermjakob H, Binz P-A, Deutsch EW (2011) mzML—a community standard for mass spectrometry data. Mol Cell Proteomics 10:1–7. https://doi.org/10.1074/mcp.R110.000133 R110.000133
8. Keller A, Nesvizhskii AI, Kolker E, Aebersold R (2002) Empirical statistical model to estimate the accuracy of peptide identifications made by MS/MS and database search. Anal Chem 74:5383–5392. https://doi.org/10.1021/ac025747h
9. Nesvizhskii AI, Keller A, Kolker E, Aebersold R (2003) A statistical model for identifying proteins by tandem mass spectrometry. Anal Chem 75:4646–4658. https://doi.org/10.1021/ac0341261
10. Demichev V, Messner CB, Vernardis SI, Lilley KS, Ralser M (2020) DIA-NN: neural networks and interference correction enable deep proteome coverage in high throughput. Nature Methods 17:41–44. https://doi.org/10.1038/s41592-019-0638-x

Check for updates

Chapter 8

Integrated Analysis of Posttranslational Modifications and Total Proteome: Methods for Distinguishing Abundance from Usage Changes

Witold Wolski, Antje Dittmann, Christian Panse, Laura Kunz and Jonas Grossmann

Abstract

Posttranslational modifications (PTMs), particularly phosphorylation, regulate protein function at substoichiometric levels, making their quantitative analysis technically challenging. A critical limitation in PTM studies is distinguishing between changes in modification abundance due to altered protein expression versus genuine changes in modification stoichiometry (the fraction of protein molecules that are modified). This chapter presents a comprehensive workflow integrating Tandem-Mass-Tag (TMT)-based quantitative proteomics with specialized computational analysis to distinguish differential PTM abundance (DPA) from differential PTM usage (DPU) using the `prolfqua`, `prolfquapp`, and prophosqua *R* packages. The workflow introduces a robust statistical framework for integrated PTM-protein analysis; comprehensive visualization tools, including N-terminus to C-terminus (N-to-C) plots for protein-centric PTM mapping and sequence motif analysis for kinase prediction. This integrated approach enables researchers to identify actual signaling changes in PTM studies, prioritize functionally relevant modification sites, understand pathway-level regulation in disease contexts, and generate robust datasets suitable for publication in high-impact journals.

Keywords Posttranslational modifications, Phosphoproteomics, TMT labeling, Differential PTM abundance analysis, Differential PTM usage analysis, Mass spectrometry, Data integration, Kinase activity

1 Introduction

1.1 Biological Significance of PostTranslational Modifications

Protein phosphorylation is a reversible PTM that modulates a protein's conformation, enzymatic activity, subcellular localization, or binding interactions. Because this modification can be reversed rapidly, it acts as a molecular switch in cellular signaling networks. As a result, cells respond to environmental cues without requiring new protein synthesis. Dysregulation of phosphorylation is implicated in

Frédérique Lisacek (ed.), *Bioinformatics for Mass Spectrometry*, Methods in Molecular Biology, vol. 3071,
https://doi.org/10.1007/978-1-0716-5519-1_8,

cancer, neurodegeneration, and metabolic disorders [1]. Large-scale phosphoproteomic profiling by high-throughput mass spectrometry (MS) (e.g., TMT-based workflows) is therefore crucial for dissecting disease-associated signaling alterations and identifying candidate therapeutic targets.

1.2 Technical Challenges in PTM Analysis

The analysis of PTMs, such as phosphorylation, presents significant analytical challenges due to the low abundance of modified peptides. Typically, phosphorylated peptides represent less than 1%–5% of the total peptide population [2]. Additionally, phosphopeptides generally exhibit poorer ionization efficiency in MS-based proteomics compared to nonmodified peptides, further complicating their detection and quantification. This low abundance and challenging ionization behavior necessitate specialized experimental workflows and computational strategies to achieve accurate quantification [3].

A key analytical challenge is distinguishing genuine changes in PTM levels, due to altered kinase and phosphatase activity, from changes caused by altered protein abundance [4,5]. Without proper correction, this confounding factor can lead researchers to misinterpret PTM data, incorrectly attributing higher phosphorylation signals solely to increased modification efficiency or pathway activation, rather than to elevated protein expression [6]. Controlling for this confounder is critical to discern rapid, signaling-driven phosphorylation events from slower, transcriptionally or translationally regulated changes in protein abundance.

Because phosphorylation is substoichiometric, only a small fraction of protein molecules carry the modification; therefore, enrichment is required to isolate phosphopeptides. Enrichment steps (such as Immobilized Metal Affinity Chromatography (IMAC) or Metal Oxide Affinity Chromatography (MOAC)) introduce additional sample handling and technical variability [6–8]. Therefore, each stage, from digestion to cleanup to MS acquisition, must be carefully optimized to improve reproducibility [9]. Effective experimental design combined with tailored normalization strategies are therefore essential to recover reliable, biologically meaningful PTM signals.

Isobaric TMT labeling enables multiplexed PTM analysis of up to 35 samples in a single run, reducing inter-sample variability and missing values to improve statistical power [10,11]. Phosphopeptides are typically enriched by IMAC or MOAC, which provide complementary selectivity and broaden coverage [12,13]. Adding offline fractionation before liquid chromatography–tandem mass spectrometry (LC–MS/MS) further increases phosphosite depth and distributes peptide load, improving quantification consistency across fractions and conditions [14].

In TMT-based proteomics, a *plex* refers to the complete set of isobarically tagged samples analyzed together in a single LC–MS/MS run. When the total number of samples exceeds the available TMT channels, each plex should include at least one biological replicate from every experimental group and a common reference or pooled channel for normalization [15]. Replicates of the same group are then distributed across different plexes rather than clustered within a single run, preventing any group's replicates from being confined to one batch. If all samples fit within the channel capacity of a single plex, these balancing measures are inherently satisfied in that run.

1.3 Computational Tools and Workflows

The computational landscape for quantitative protein PTM analysis has undergone significant evolution, with numerous specialized software platforms supporting bottom-up proteomics workflows. Following MS data acquisition, spectra undergo database searching to identify peptide sequences, which is subsequently followed by protein inference to determine protein identities based on these identified peptides. The ecosystem of PTM analysis includes several well-established data-dependent acquisition (DDA)-TMT compatible software suites, both free and commercial, such as *Andromeda* integrated within *MaxQuant* [16], *Proteome Discoverer* (Thermo Fisher Scientific), *FragPipe* [17], and *PeptideShaker* [18].

Among the analytical steps following peptide identification, PTM site localization scoring represents a particularly crucial computational challenge. Site localization determines the confidence with which a modification can be assigned to a specific amino acid residue within a peptide sequence. This process is especially challenging when peptides contain multiple potential modification sites, necessitating algorithms to evaluate the quality and specificity of site-determining fragment ions. Specialized tools have been developed explicitly for robust PTM site localization, including *PTMProphet* [19], *PhosphoRS* [20], and *Ascore* [21]. Accurate localization is essential, as incorrect site assignments can lead to biological misinterpretations, particularly in signaling pathway analyses and kinase–substrate predictions.

In DDA workflows, *FragPipe* [17] represents a widely adopted and integrated platform that comprehensively addresses PTM analysis, including site localization challenges. *FragPipe* integrates the *MSFragger* search engine [22], enabling highly sensitive peptide identification and employs *PTMProphet* [23] functionality for site localization scoring. It also generates detailed quantitative outputs, including site-level quantification reports and multisite feature reports that group peptides sharing identical modification patterns. A detailed description of the *FragPipe* workflow is provided in the experimental methods section (see Sect. 3.1).

While the above discussion has focused primarily on DDA workflows, data-independent acquisition (DIA)-based approaches represent alternative strategies with distinct computational considerations for PTM analysis. Specialized DIA platforms such as *Spectronaut* [24,25] and *DIA-NN* [26] offer PTM support through dedicated site-localization scoring algorithms and site-level quantification capabilities. Unlike DDA, DIA workflows require specialized computational approaches to extract and quantify modification-specific information from highly multiplexed fragmentation spectra. Both *Spectronaut* and *DIA-NN* incorporate advanced statistical frameworks designed explicitly for PTM identification and quantification in DIA workflows.

1.4 Data Analysis Frameworks

Proteomic data analysis can be conducted at multiple levels of granularity, each providing distinct analytical perspectives. A peptidoform represents a specific peptide sequence with a particular set of modifications, while site-level analysis aggregates signals for each residue position across multiple peptides. Differential analysis can therefore target either peptidoforms or individual modification sites, with each approach offering unique advantages for different research questions.

FragPipe introduces an additional analytical concept through multisite features, which refers to the set of all identified peptide forms sharing the same set of modification sites under investigation, regardless of sequence derivatives, cleavage state, or whether the modifications are unambiguously localized. This approach groups peptidoforms with different sequences, which can arise from missed cleavages or alternative cleavage patterns, but identical possible modification sites into the same multisite feature, providing a balanced approach between peptidoform specificity and site-level aggregation. Discarding all peptidoforms with ambiguous localization would result in the loss of a large fraction of PTM information. Furthermore, multisite features preserve information about phosphorylation sites that occur together on the same peptide molecule, revealing comodification patterns that indicate coordinated regulation by kinases or functional relationships between sites. Traditional site-level analysis treats each phosphorylation site independently, losing this valuable information about which specific combinations of sites are modified together in biological samples.

Site-level reports collapse signals from all peptidoforms mapping to the same PTM at an amino acid residue into a single quantitative value, ensuring each site has exactly one intensity measurement per sample. In downstream analysis, site-level intensities integrate seamlessly with site-centric enrichment tools and kinase-activity inference platforms (PhosR [27], PTM-SEA [28], RoKAI [29], etc.), streamlining biological interpretation and network reconstruction analyses [30].

1.5 Statistical Analysis of PTM Data

Statistical analysis of PTM data requires specialized methods due to confounding effects between PTM levels and protein abundances. Several statistical packages have emerged to address the critical issue of integrating PTM-feature quantifications with protein abundance data. Notable examples include `MSstatsPTM` [4] and `msqrob2PTM` [5], which explicitly model both modified peptide and protein-level changes to distinguish genuine PTM regulation from protein abundance effects accurately.

The `msqrob2PTM` framework exemplifies the statistical approaches now available for PTM analysis, defining two complementary analytical strategies: DPA and DPU [5]. DPA directly models the log^2 intensities of each modified peptide (peptidoform), detecting absolute changes in PTM levels between experimental conditions. In contrast, DPU adjusts the PTM intensities by the corresponding protein-level changes, effectively testing for changes in the relative usage or stoichiometry of a modification site. This dual approach enables researchers to distinguish between PTM changes driven by protein abundance differences (DPA) versus those reflecting genuine changes in modification efficiency or regulatory activity (DPU).

These two frameworks differ in their analytical approaches for addressing confounding between PTM and changes in protein abundance. `MSstatsPTM` employs a "model then correct" strategy, separately fitting linear models to modified and unmodified peptide data and then combining statistical inferences postmodeling. In contrast, `msqrob2PTM` follows a "correct then model" approach, first adjusting PTM intensities by their corresponding protein abundances at the data level and subsequently performing statistical testing on these normalized values.

A further option, discussed in [31], is to model the PTM and total proteome data as a factorial design that jointly includes a biologically relevant explanatory variable, e.g., genotype, and a "component" indicator, i.e., modified peptide and its total proteome parent protein, and fits a linear model with their interaction. It then tests the statistical significance of the interaction term to flag modified peptides whose dynamics deviate from the corresponding total proteome protein abundance. Here, the package limma is used to fit the model and perform the testing.

While all three methodologies ultimately estimate PTM abundance changes corrected for protein-level effects, each method can yield different test results due to varying assumptions about the relationship between PTMs and protein abundances within a sample, as well as differences in how variance and degrees of freedom are estimated.

- MSstatsPTM fits the PTM and the protein in separate models, then computes an adjusted effect as a difference of estimated contrasts: $\Delta_{adj} = \Delta_{PTM} - \Delta_{protein}$ with standard error combined in quadrature for the test statistic, and Welch's approximation to estimate the degrees of freedom, which yields p-values for the adjusted fold change directly on the contrast scale. MSstatsPTM assumes that there is no correlation between PTM and the protein abundances within the samples.
- The single-model approach [31] uses limma moderated (empirical-Bayes) variances for the interaction contrast within one fit, which shrinks feature-wise variances toward a common prior; *df* are moderated accordingly. The single-model formulation can capture shared covariance by adding a sample blocking/random effect. Still, as illustrated [31], the baseline model omits a

sample term, effectively treating PTM and protein measurements as independent.

- msqrob2PTM assumes that the PTM and the protein abundances within a sample are correlated and therefore corrects PTM abundances with the matched protein abundances. Then, for modeling, only the corrected PTM intensities are used.

The three pipelines target the PTM-vs-protein effect through different modeling paths and make different assumptions about within-sample covariance and differ in variance moderation. Therefore, even with the same raw data and contrasts, we can expect differences in estimated effects, standard errors, and p-values.

To offer researchers additional flexibility, we developed `prophosqua` [32], an *R* package designed to streamline statistical analysis of phosphoproteomics and other PTM datasets. Using `prophosqua` and the `prolfqua` package [33], the `MStatsPTM model-first and msqrob2PTM` *correct first* approaches can be executed to obtain protein-corrected PTM data and provide intuitive visualizations for comprehensive exploratory analysis.

1.6 The Protein Assignment Problem in Differential PTM-Feature Usage

Identification and quantification tools need to assign peptides to proteins, i.e., find the leading or representative protein identifier (ID). This assignment may be dataset-specific and may differ between the PTM and the total proteome dataset. Therefore, using protein IDs to match PTM features with proteins from the total run may lead to a loss of information due to differing representative IDs [34]. For example, a phosphopeptide might be assigned, by protein inference, to protein isoform A in the PTM dataset. In contrast, in the total proteome dataset, protein abundance is quantified under isoform B, which prevents proper integration and DPU analysis.

A robust solution involves matching stripped peptide sequences from the PTM dataset to peptide sequences of proteins quantified in the total proteome dataset, bypassing protein ID dependency entirely. When peptides are shared among multiple protein isoforms, the averaged protein abundance across all matching isoforms provides a more stable reference for normalization. While this sequence-based matching approach requires additional computational implementation, it reduces the loss of valuable information due to protein assignment differences between datasets.

1.7 Chapter Overview and Learning Objectives

This chapter introduces an analytical workflow that addresses the biological challenge of accurately distinguishing DPA from DPU. Distinguishing between protein-level effects and genuine modification-specific changes is critical for meaningful interpretation of PTM data. The workflow emphasizes reproducibility, automation, and biological interpretation.

DPA tests raw PTM signal changes between experimental conditions, identifying any modification abundance changes regardless of underlying protein-level effects. This analysis captures the total impact of experimental perturbations on PTM levels, including both direct modification effects and indirect effects mediated through changes in protein abundance.

DPU evaluates protein-normalized PTM changes, explicitly identifying modification sites where stoichiometry is genuinely altered independent of protein abundance. This analysis pinpoints sites where experimental conditions directly impact modification efficiency, kinase activity, or phosphatase activity, thereby highlighting specific regulatory events.

Learning objectives include: (1) Conducting integrated statistical analysis of PTM and total proteome data using `prolfquapp` and prophosqua *R* packages. (2) Interpreting DPA vs. DPU results for biological insights and generating publication-quality visualizations. (3) Performing sequence motif analysis for kinase prediction and pathway reconstruction.

2 Materials

2.1 Demonstration Dataset

This protocol utilizes the Atg16l1 macrophage dataset from Maculins et al. [35] to demonstrate the bioinformatics workflow and integrated statistical analysis. The dataset comprises two TMT-11-plex sets measuring six conditions(WT/KO × uninfected/early/late infection) with both phospho-enriched and total proteome samples, making it ideal for illustrating the principles of integrated PTM analysis. In the original publication, the authors also performed ubiquitin remnant enrichment, although we focus on the phosphoproteome and total proteome datasets for demonstration purposes. This same dataset was previously used in the MSstatsPTM publication [4], providing additional validation of the analytical approaches.

As discussed below, TMT enables high-precision quantification across many conditions, particularly when combined with fractionation. In contrast, DIA is well suited when avoiding fractionation or scaling to larger sample numbers (*see* **Note 1**).

2.2 System Requirements

1. Search and quantification software: *FragPipe* 22.0 (free download for academic use from fragpipe.nesvilab.org; system requirements: 32GB+ RAM, 50GB+ storage)
2. Statistical platform: *R* version ≥ 4.0.0 (https://www.r-project.org/) and RStudio (https://posit.co/)

   - **System requirements**: 16GB+ RAM, 20 GB+ storage.
   - **Expected runtime**: 10 min to 1 h depending on dataset size and number of fractions.

2.3 R Packages

Most relevant *R* packages used for the analysis are:

- tidyverse [36]—Data manipulation and visualization tools.
- ggseqlogo [37]—Sequence motif visualization.
- `prolfqua` [33]—Core statistical framework for proteomics data analysis.
- `prolfquapp` [38]—Streamlined differential expression analysis pipeline.
- `prolfquappPTMreaders` [39]—Specialized readers for PTM data formats.
- prophosqua [32]—Integration and visualization of PTM and protein data.

For detailed installation instructions, see the package documentation on

- github.com/fgcz/`prolfqua` .
- github.com/`prolfqua` /`prolfquapp`.
- github.com/`prolfqua` /prophosqua.

2.4 Mass Spectrometry Data Processing

Raw MS data acquired using DDA combined with TMT labeling are processed using specialized software to:

1. Search spectra against a protein sequence database, including species-specific sequences, common contaminants, and decoy sequences.
2. To extract reporter ion intensities.
3. To compute site localization probabilities

The section (see Sect. 3.1) describes in detail how we parameterized the *FragPipe* software [17].

3 Methods

A key challenge, when analyzing PTMs, is to distinguish changes in PTM abundance that are due to altered protein expression from those that reflect a change in the modification stoichiometry (i.e., the fraction of the protein pool that is modified). This protocol provides a step-by-step computational workflow to analyze PTM data in the context of total protein abundance changes. We leverage the *R* packages `prolfquapp` [38] for streamlined differential expression analysis [33] and `prophosqua` [32] for the integration, analysis, and visualization of PTM and total proteome data. The workflow comprises three parts: (1) analyzing the raw mass spectrometric data with *FragPipe*, (2) performing differential expression analysis for the PTM-enriched and total proteome datasets, and (3) integrating the PTM and total proteome data.

3.1 Analyzing TMT Raw Data with FragPipe

We used the TMT 11-plex phospho workflow in *FragPipe* 22.0 [17] with *MSFragger* (version 4.1) [22]. Database searching was performed against a species-specific protein sequence database supplemented with common contaminants and reversed decoy sequences. The search parameters included:

- Fixed modifications:
- Carbamidomethylation of cysteine (+57.0215 Da).
- TMT labeling of lysine residues and peptide N-termini (+ 229.1629 Da).
- Variable modifications:

- Phosphorylation on serine, threonine, and tyrosine residues (+ 79.9663 Da).
- Oxidation of methionine (+15.9949 Da).
- Acetylation at protein N-termini (+42.0106 Da).

Reporter ions are quantified by *IonQuant* (version 1.10.27), with downstream processing in *TMTIntegrator (5.0.9)* . *TMTIntegrator* [17] performs quantification and normalization specifically tailored for multiplexed TMT experiments.

Selected *TMTIntegrator* parameters are:

- Best PSM selection: OFF.
- Outlier removal enabled: OFF.
- Allow overlabel/underlabel: OFF.
- Use MS1 intensity: OFF.
- Aggregation method: Median.

For phospho-enriched samples, PTM sites are scored using *PTMProphet* [23]; only sites withlocalization scores ≥ 0.75 are retained. *FragPipe* workflow parameters are stored in configuration files (`fragpipe.workflow`). The workflow file used to parametrize *FragPipe* can be downloaded from [40].

FragPipe provides two types of PTM-features reports for the phospho-enriched samples:

- Multisite features: groups peptidoforms sharing identical modification sites, as described in Sect. 1.4.
- Site-level reports collapse signals from all peptidoforms mapping to the same amino acid position into a single quantitative value per PTM site, ensuring each site has exactly one intensity per sample. In downstream analysis, site-level intensities can be fed into site-centric enrichment or kinase-activity inference tools (`PhosR`, `PTM-SEA`, `RoKAI`, etc.), streamlining biological interpretation and network reconstruction.

An essential step in the data preprocessing is the quality control (QC) of the quantified samples, which involves estimating the missed cleavage rates or labeling efficiency (*see* **Note 2**).

3.2 Data Setup for Statistical Analysis

The following sections demonstrate the computational workflow through *R* code snippets that implement each step of the integrated PTM analysis. All code examples are also available within the prophosqua package [32] and can be easily adapted to different datasets by modifying file paths and experimental parameters.

The first step is to download the example PTM dataset, which serves as a comprehensive demonstration of integrated PTM analysis. This dataset is derived from the Atg16l1 macrophage study by Maculins et al. [35], which investigates the role of autophagy in bacterial infection responses. The dataset contains both total proteome and PTM-enriched measurements from a well-designed factorial experiment with the following characteristics:

- Experimental design: 2×3 factorial (Genotype: WT/KO; Timepoint: Uninfected/Early/Late infection).
- Biological context: Macrophage responses to Shigella infection in wild-type vs Atg16l1-knockout cells.
- Technical approach: TMT 11-plex labeling for precise quantification across conditions.
- Data types: Total proteome protein-level and phospho-enriched phospho-site quantification results.

This dataset is particularly valuable for demonstrating PTM analysis concepts because the Atg16l1 knockout creates both direct protein abundance changes (loss of Atg16l1 itself) and secondary effects on autophagy-related signaling pathways.

The first step involves obtaining the *FragPipe* 22 mass spectrometry output files [40] and setting up the analysis environment in *R*. Depending on the sample type, different *FragPipe* output files are required:

- total proteome samples: psm.tsv files, containing peptide-spectrum match (PSM) information.
- phospho-enriched samples: `abundance_single-site.tsv` files for individual modification site analysis.

This dataset is available from Zenodo [41] https://doi.org/10.5281/zenodo.15879865. The code below downloads the data, extracts it into a folder, and removes the zip archive.

```
need_download <-!dir.exists("PTM_example")
zenodo_url <- "https://zenodo.org/records/15879865/
files/PTM_experiment_FP_22_Maculins_and_QC.zip"
destfile <- basename(zenodo_url)
# Download, unzip and delete
download.file(zenodo_url, destfile, mode = "wb")
unzip(destfile, exdir = ".")
unlink(destfile)
```

3.3 Setting Up prolfquapp Scripts

This section demonstrates how to perform **differential expression analysis (DEA)** on both total proteome and PTM-enriched proteome data using the `prolfquapp` package. The analysis provides the foundation for subsequent integration by preprocessing, normalizing the data, and establishing which proteins and PTM-features show significant changes between experimental conditions.

The `prolfquapp` [38] package provides a set of shell scripts that automate the analysis workflow. These scripts are copied into the working directory by running the command in the Linux shell and are used in subsequent steps.

```
R --vanilla -e "library(prolfquapp); copy_shell_
script(workdir = '.')"
```

3.4 Creating Sample Annotation

Sample annotation is the foundation of any meaningful DEA, serving as the bridge between raw mass spectrometry files and biological interpretation. The annotation file maps each raw data file to its corresponding experimental conditions (WT or KO) and additional explanatory variables, such as Time.

We start by reading the sample names from the *FragPipe* output files using the `prolfqua_dataset.sh` script to create a dataset annotation file compatible with `prolfquapp`, that maps sample names to their experimental conditions (see Table 1).

```
./prolfqua_dataset.sh --indir PTM_example/data_
total/FP_22/ \
--dataset "PTM_example/data_total/dataset.tsv" \
--software prolfquapp.FP_TMT
```

We use R to extract experimental factors, the Genotype and Timepoint, from the channel labels using the tidyr::separate function. If the channel names do not encode the experimental conditions, the explanatory variables must be provided by adding columns to the annotation file using a text editor.

```
annot <- readr::read_tsv("PTM_example/data_total/
dataset.tsv")
```

Table 1
Annotation table created from FragPipe outputs, with a set of columns required by prolfquapp

Channel	Name	Group	Subject	CONTROL
KO_Early_1	KO_Early_1	NA	NA	NA
KO_Early_2	KO_Early_2	NA	NA	NA
KO_Late_1	KO_Late_1	NA	NA	NA
KO_Late_2	KO_Late_2	NA	NA	NA
KO_Uninfect_1	KO_Uninfect_1	NA	NA	NA
KO_Uninfect_2	KO_Uninfect_2	NA	NA	NA

```
annot <- annot |> tidyr::separate("name",
c("Genotype", "Timepoint", NA),
                                  sep = "_", remove = FALSE)
annot$group <- NULL
annot$subject <- NULL
annot$CONTROL <- NULL
```

We then call the annotation_add_contrasts function in the `prolfqua` [33] to generate factorial contrasts automatically. The annotation_add_contrasts function adds two key columns to the sample annotation:

- ContrastName: A descriptive name for each statistical contrast.
- Contrast: The mathematical formula defining the contrast in terms of group means.

```
annot2 <- prolfqua::annotation_add_contrasts(
annot, primary_col = "Genotype", secondary_col =
"Timepoint",
decreasing = TRUE, interactions = FALSE)$annot
write_tsv(annot2, "dataset_with_contrasts.tsv")
```

Note that setting the argument interactions = FALSE in the annotation_add_contrasts function disables the generation of interaction contrasts for factor level combinations. Interaction contrasts assess whether the effect of one factor depends on the level of another, thereby capturing non-additive interplay between factors. The file dataset_with_contrasts.tsv is the final dataset file that will be used for the DEA (see Table 2).

For the experimental design in this protocol (2 × 3 factorial: Genotype [KO/WT] × Time [Uninfected/Early/Late]), the function generated four distinct contrasts:

1. KO_vs_WT (Main effect): ((G_KO_Uninfect + G_KO_Late + G_KO_Early)/3 – (G_WT_Uninfect + G_WT_Late + G_WT_Early)/3) – Compares overall genotype effects across all timepoints.
2. KO_vs_WT_at_Uninfect: G_KO_Uninfect – G_WT_Uninfect – Compares genotypes specifically in the *uninfected* condition.
3. KO_vs_WT_at_Late: G_KO_Late – G_WT_Late – Compares genotypes specifically in the *late* infection condition.
4. KO_vs_WT_at_Early: G_KO_Early – G_WT_Early – Compares genotypes specifically in the *early* infection condition.

Table 2
Annotation table with explanatory variables and contrasts

Channel	Name	Group	ContrastName	Contrast
KO_Early_1	KO_Early_1	KO_Early	KO_vs_WT	((G_KO_Uninfect + G_KO_Late + G_KO_Early)/3 – (G_WT_Uninfect + G_WT_Late + G_WT_Early)/3)
KO_Early_2	KO_Early_2	KO_Early	KO_vs_WT_at_Uninfect	G_KO_Uninfect – G_WT_Uninfect
KO_Late_1	KO_Late_1	KO_Late	KO_vs_WT_at_Late	G_KO_Late – G_WT_Late
KO_Late_2	KO_Late_2	KO_Late	KO_vs_WT_at_Early	G_KO_Early – G_WT_Early
KO_Uninfect_1	KO_Uninfect_1	KO_Uninfect		
KO_Uninfect_2	KO_Uninfect_2	KO_Uninfect		

If we want to compare timepoints instead of genotypes, we set: `primary_col = "Timepoint"` and `secondary_col = "Genotype"`.

3.5 Execution of Differential Abundance Analysis

Before performing DEA, we generate a YAML configuration file using the `prolfqua_yaml.sh` script. The configuration file (`config.yaml`) defines key analysis parameters, including the normalization method (variance stabilization normalization (VSN) by default), protein summarization strategy (Tukey's median polish by default), and quality-filtering thresholds such as q-value cutoffs for peptide or protein-level evidence.

```
./prolfqua_yaml.sh -y config.yaml
```

Once the sample annotation and configuration files are prepared, DEA is executed via the `prolfqua` _dea.sh script. This script reads the experimental design (sample IDs, group labels, and contrasts) alongside the configuration file. Single-site or protein intensities are log_2-transformed and variance-stabilized before modeling, improving homoscedasticity across conditions.

Linear models are fitted, and contrasts are tested using the `prolfqua` *R*-package's contrast API. Rather than imputing missing values, the model adjusts the degrees of freedom based on the number of actual observations per feature, which naturally down-weights features with high missingness and improves the reliability of the fold-change and p-value estimates. Empirical Bayes variance moderation is applied to stabilize variance estimates, particularly when sample size is limited.

First, we run the DEA analysis of the total proteome quantification results. The script uses *FragPipe* `psm.tsv` file as input, aggregates reporter ions abundances to the peptide level, and then estimates the protein abundances. The code below demonstrates how the `prolfqua_dea.sh` shell script can be executed on the Linux command line.

```
./prolfqua_dea.sh -i PTM_example/data_total/
FP_22 \
   -d dataset_with_contrasts.tsv \
   -y config.yaml \
   -w total_proteome \

   -s prolfquapp.FP_TMT
```

The `prolfqua _dea` application expects the following parameters:

- -i: Specifies input directory with *FragPipe* results.
- -d: Uses the annotated dataset file.
- -y: Applies configuration settings in the config.yaml file.
- -w: Sets the name of the work unit.
- -s: Specifies software format (*FragPipe* TMT).

After running the DEA, the results are stored in a time-stamped directory. This code constructs the path to the results directory using the `prolfquapp::zipdir_name() function, which follows the standard naming convention for DEA results.`

```
prolfquapp::zipdir_name("DEA", workunit_id =
"total_proteome")
## [1] "DEA_20260123_WUtotal_proteome_vsn"
```

The DEA of the single-site output of enriched phosphoproteome examines changes at individual modification sites and is performed by running the code below. The analysis provides site-specific information about which particular amino acids are modified under different conditions. The abundance_single-site_None.tsv file generated by *FragPipe* is used as input.

```
./prolfqua_dea.sh -i PTM_example/data_ptm/FP_22 \
-d dataset_with_contrasts.tsv \
-y config.yaml \
-w singlesite_PTM \
-s prolfquappPTMreaders.FP_singlesite
```

Please note that using the -s argument, we point to the correct reader function in the prolfquappPTMreaders R package to read the *FragPipe* single-site output.

For each data type, prolfquapp generates a comprehensive suite of outputs:

- Dynamic HTML reports containing interactive PCA plots, boxplots, volcano plots, and heatmaps for quality control and exploratory analysis.
- XLSX tables (DE_*.xlsx) summarizing $\log_2$ fold changes, moderated t-statistics, raw p-values, and Benjamini–Hochberg–adjusted p-values.
- GSEA rank files (GSEA_*.rnk) for downstream gene-set enrichment analysis.
- SummarizedExperiment objects (.rds) for seamless import into Shiny-based visualization tools.

These results can be integrated using the prophosqua package for combined analysis of total proteome and PTM data. However, it is essential to note that there are scenarios where performing differential PTM-feature abundance analysis is the only practical option (*see* **Note 3**).

3.6 Integrated PTM Analysis Using prophosqua

The prophosqua package [32] provides tools for integrating and analyzing PTM data with total proteome measurements. It enables researchers to distinguish between changes in protein abundance and changes in the usage of modification sites. This section explains how to load the results of the differential abundance analysis, integrate the PTM and total proteome data, and determine differential PTM-feature usage.

Differential expression results from prolfquapp, provided in Excel format, for the total proteome and the multi- or single-site PTM analyses are imported into *R*. These datasets are then integrated by performing a left join on protein identifiers, merging PTM-level statistics with the corresponding protein-level values for each condition.

We start by defining the project name and analysis type.

```
library(prophosqua)
wu_id <- "CustomPhosphoAnalysis"
fgcz_project <- "PTM_analysis"
oid_fgcz <- "fgcz_project"
datetoday <- format(Sys.Date(), "%Y%m%d")
project_dir <- "."
```

We also create a results directory to store the PTMPost-translational modification (PTM)/total proteome integration results:

```
descri <- wu_id
res_dir <- file.path(
  project_dir,
  paste0(fgcz_project, "_", datetoday, "_
singlesite_", descri))if (!dir.exists(res_dir)) {
  dir.create(res_dir, recursive = TRUE)
}
```

In the code below, we define the paths to the DEA results in the XLSX files stored by `prolfquapp`. We need to specify the workunit names and the date when the DEA analysis was run.

```
# Use today's date if running DEA, otherwise use
existing resuls datedea <- if(do_dea) datetoday
else "20251214"
tot_file <- prophosqua::dea_xlsx_path(project_dir,
"total_proteome", datedea)
ptm_file <- prophosqua::dea_xlsx_path(project_dir,
"singlesite_PTM", datedea)
# Directory paths
stopifnot(file.exists(tot_file))
stopifnot(file.exists(ptm_file))
```

Finally, we load the differential expression results into the *R* session.

```
required_cols <- c("protein_Id", "protein_length",
"contrast")
tot_res <- prophosqua::load_and_preprocess_
data(tot_file, required_cols)
tot_res <- prophosqua::filter_contaminants(tot_res)
phospho_res <- prophosqua::load_and_preprocess_
data(ptm_file, required_cols)
phospho_res <
- prophosqua::filter_contaminants(phospho_res)
```

The data integration step is critical for enabling DPU analysis and represents one of the most important technical aspects of the workflow. This step joins phosphorylation site data with total protein data to create a unified dataset where each PTM-feature is linked to its corresponding protein-level measurements.

The left_join operation used here ensures that all PTM features are retained in the analysis, even if their parent proteins were not detected in the total proteome experiment. This approach maximizes the information content by allowing protein normalization where possible, while also reporting PTM-features that cannot be mapped to total proteome measurements.

```
join_column <- c(
  "fasta.id" = "protein_Id",
  "contrast",
  "description",
```

```
    "protein_length",
  )
  suffix_a <- ".site"
  suffix_b <- ".protein"
  combined_site_prot <- dplyr::left_join(
    phospho_res,
    tot_res,
    by = join_column,
    suffix = c(suffix_a, suffix_b))
```

We also need to check the match rate between PTM sites and proteins, that is, the number of PTM sites that have a protein match divided by the total number of PTM sites. Currently, prolfquapp matches PTM and protein features using exact protein IDs, which introduces a limitation: discrepancies in representative protein identifiers between the PTM-enriched and total proteome datasets may result in the loss of mapping information (see Note 4).

```
  # Calculate match rate by contrast
  match_rates <- combined_site_prot |>
    group_by(contrast) |>
    summarize(
      total_sites = n(),
      matched_sites = sum(!is.na(diff.protein)),
      match_rate = round(matched_sites / total_sites
  * 100, 1)
  )
```

Table 3 shows the match rates for each contrast.

3.7 Visualizing PTM-Feature Abundance (DPA)

DPA tests the raw PTM-feature signal change between conditions, without any correction for its parent protein's expression level. This analysis is used to flag any PTM-feature whose abundance changes, even if the parent protein itself is also up- or down-regulated.

DPA results are visualized using N-to-C plots, generated by the `prophosqua` [32] `n_to_c_expression_multicontrast` function. These multi-panel plots map both the $\log_2$ fold changes of individual modification sites along the primary sequence of the protein and the $\log_2$ fold change of the protein abundance from the

Table 3
Match rates between PTM sites and proteins

Contrast	Total sites	Matched sites	Match rate
KO_vs_WT	21,733	21,000	96.6
KO_vs_WT_at_Early	21,733	21,000	96.6
KO_vs_WT_at_Late	21,733	21,000	96.6
KO_vs_WT_at_Uninfect	21,733	21,000	96.6

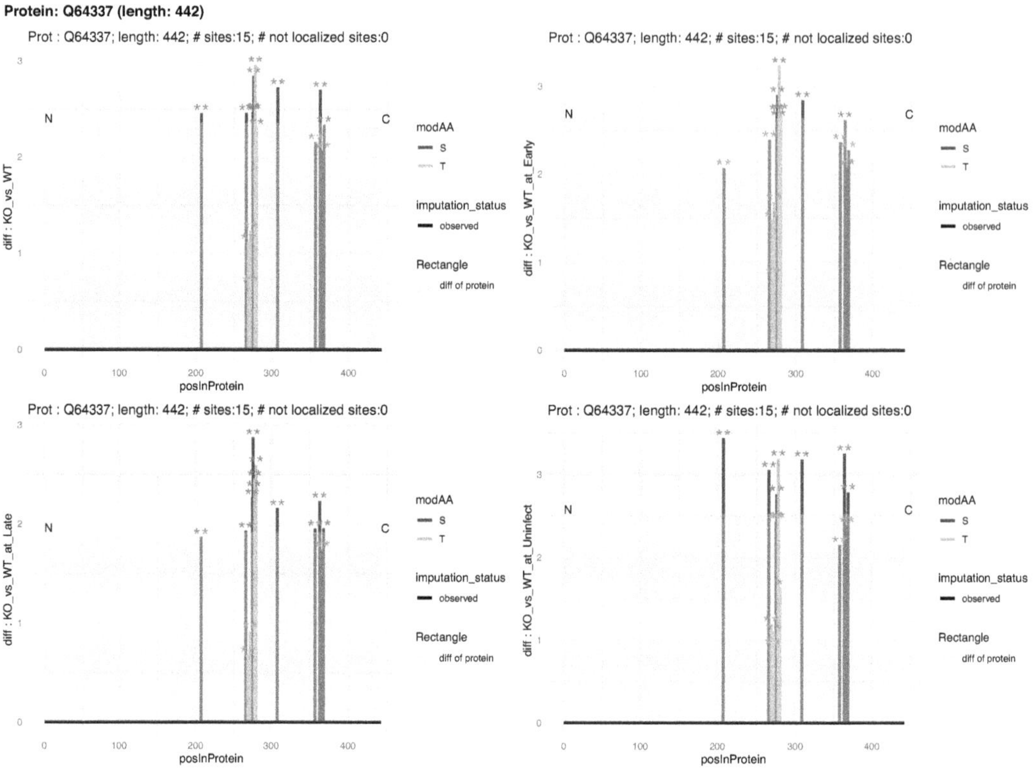

Fig. 1 Multi-panel N-to-C expression plot showing differential expression of total protein abundance (light yellow rectangle) and phosphorylation sites (vertical lines, color coded by type of residue) along protein sequences across all contrasts. Each panel represents a different experimental contrast, enabling direct comparison of protein and PTM expression changes across conditions. The vertical lines are either dashed or solid, depending on the site's imputation status. The x-axis represents amino acid position from N to C terminus, while the y-axis shows $\log_2$ fold changes. Significantly regulated sites (FDR $<$ 0.05 and FDR $<$ 0.01) are highlighted with one or two red asterisks, respectively

total proteome experiment across all contrasts simultaneously. This representation provides a clear visual summary of both site-specific and overall protein expression changes across multiple experimental comparisons. The N-to-C expression plot (see Fig. 1) shows that the protein SQSTM1 (UniProt id Q64337) and the phosphorylation sites are upregulated in the KO vs WT comparison across different timepoints.

Because generating N-to-C plots is computationally intensive, we generate plots only for proteins with at least one significant phosphorylation site in any contrast. The False Discovery Rate (FDR) threshold for determining significance can be adjusted in the `n_to_c_expression_multicontrast` or `n_to_c_usage_multicontrast` functions, allowing researchers to balance computational efficiency with comprehensive visualization based on their specific considerations. When no significant PTM sites are detected despite biological expectations, one possible reason is insufficient statistical power, which can be mitigated by increasing the sample size (*see* **Note 5**).

```
plot_data <- n_to_c_expression_multicontrast(
  combined_site_prot, FDR_threshold = 0.01,
  max_plots = 100, include_proteins = c("Q64337"))
x1 <- which(plot_data$protein_Id == "Q64337")
fig1_plot <- plot_data$plot[[x1]]
```

The next code snippet exports 100 N-to-C expression plots to a PDF file for further analysis and presentation (see also **Note 6** why we limit the number of N-to-C plots written into a file).

```
pdf(file.path(res_dir,
"Site_differential_Expression_multicontrast.pdf"),
    width = 14, height = 10)
for (i in seq_len(min(100, nrow(plot_data)))){
  print(plot_data$plot[[i]])
}
dev.off()
```

3.8 Analysis of Differential PTM-Feature Usage (DPU)—Model-First

DPU tests the protein-normalized changes of PTM-features. This analysis is essential for determining whether the fraction of a protein that is modified at a specific site changes between conditions. It isolates changes in modification stoichiometry from changes in overall protein abundance.

In the model-first approach, DPU is calculated by subtracting the total proteome protein $\log_2$ fold change from the PTM-feature's $\log_2$ fold change. The associated p-values are recalculated using a method that combines the variance from both the PTM and protein-level models, as implemented in `prophosqua` [32] `test_diff` function, or in the `MSstatsPTM` *R*-package [4].

The analysis involves three key mathematical components:

1. Effect Size Calculation (Δ_I).
2. Standard Error Propagation (SE_I).
3. Degrees of Freedom Calculation (df_I).

The differential PTM usage effect size is calculated as the difference between PTM and total proteome protein $\log_2$ fold changes for each comparison:

$$\Delta_I = \log_2 FC_{PTM} - \log_2 FC_{Protein}$$

This difference represents the protein-normalized change in PTM abundance, where:

- Positive values—indicate increased modification stoichiometry (higher fraction of protein is modified).
- Negative values—indicate decreased modification stoichiometry (lower fraction of protein is modified).
- Zero values—indicate no change in modification stoichiometry relative to protein expression.

The combined standard error accounts for uncertainty in both PTM and protein measurements:

$$SE_I = \sqrt{SE_{PTM}^2 + SE_{Protein}^2}$$

The error propagation ensures that the statistical significance of usage changes reflects uncertainty from both datasets, providing a more robust assessment of the reliability of the observed changes.

The effective degrees of freedom for the combined test are calculated using Welch's approximation:

$$df_I = \frac{\left(SE_{PTM}^2 + SE_{Protein}^2\right)^2}{\frac{SE_{PTM}^4}{DF_{PTM}} + \frac{SE_{Protein}^4}{DF_{Protein}}},$$

which provides appropriate degrees of freedom. The Welch approximation is essential when the two datasets have different experimental designs or sample sizes.

Using these calculated parameters, the t-statistic is computed:

$$t = \frac{\Delta_I}{SE_I}.$$

The p-value is determined using the t-distribution with df_I degrees of freedom. Multiple testing correction (e.g., Benjamini–Hochberg FDR) is applied across all tested sites.

The test_diff function calculates the difference of differences between PTM and total proteome data, revealing site-specific usage changes.

```
combined_test_diff <- prophosqua::test_
diff(phospho_res, tot_res, join_column =
join_column)
```

The PTM-feature usage fold changes, p-values, and FDR estimates are added to the dataframe combined_test_diff.

The DPU results are visualized using N-to-C plots generated by prophosqua::n_to_c_usage_multicontrast, which display the protein-normalized PTM-feature abundance changes across all contrasts simultaneously and highlight sites with significant changes in usage. These multi-panel plots reveal which sites are being used more or less actively, independent of total protein changes, enabling comparison across all experimental conditions.

```
plot_data_usage <- n_to_c_usage_multicontrast(
  combined_test_diff, FDR_threshold = 0.05,
  max_plots = 100, include_proteins = c("Q64337"))
```

In the *Atg16l1* macrophage dataset [35], SQSTM1 (UniProt id Q64337) protein levels are markedly increased in Atg16l1-knockout (KO) versus wild-type (WT), reflecting loss of autophagy-mediated turnover. Phosphosite analysis reveals that the raw phosphopeptide intensities at Ser_{28}, Ser_{308}, and Ser_{368} are similarly elevated in KO compared to WT (DPA), consistent with higher protein abundance (see Fig. 1). However, when applying DPU normalization, which subtracts the total proteome protein $\log_2$ fold change from each site's phospho $\log_2$ fold change, these apparent phosphorylation increases are substantially reduced, indicating that most of the phosphosite

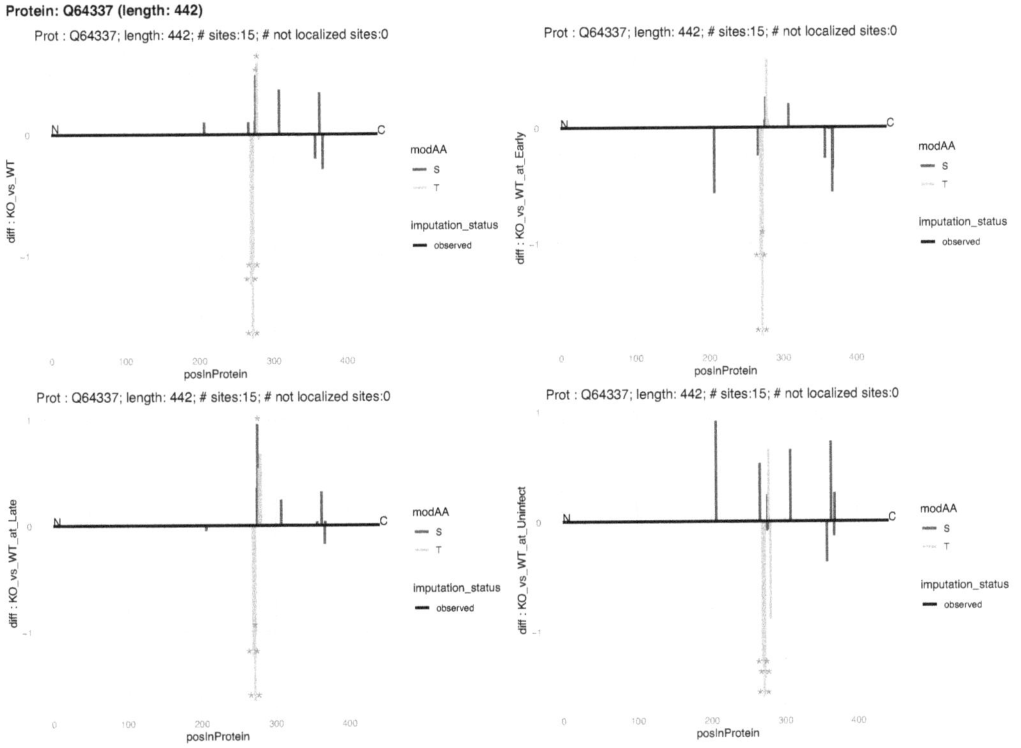

Fig. 2 Multi-panel N-to-C usage plot showing differential usage of phosphorylation sites (vertical lines, color-coded by type of residue) along the protein sequences across all contrasts. Each panel represents a different experimental contrast, enabling direct comparison of protein-normalized PTM changes across conditions. The vertical lines are either dashed or solid, depending on the site's imputation status. The x-axis represents amino acid position from N to C terminus, while the y-axis shows $\log_2$ fold changes. Significantly regulated sites (FDR < 0.05 and FDR < 0.01) are highlighted with one or two red asterisks, respectively

abundance changes arise from protein accumulation rather than genuine changes in modification stoichiometry (see Fig. 2). This example highlights how DPU effectively isolates regulatory events on SQSTM1 from confounding protein-level effects.

Additionally, individual high-confidence DPA/DPU sites should be carefully inspected for technical quality, including verification of protein coverage and assessment of missing values across treatment groups (*see* **Note 7**).

```
x1_usage <- which(plot_data_usage$protein_Id ==
"Q64337")
fig2_plot <- plot_data_usage$plot[[x1_usage]]
```

When evaluating statistical significance in DPU analysis, it is essential to avoid overinterpreting results (*see* **Note 8**). Statistical significance does not guarantee biological relevance; effect sizes should be considered alongside FDR values when prioritizing sites for further investigation. Key findings should be validated through orthogonal methods to confirm biological relevance.

3.9 Protein PTM Integration Report Generation

The integrated analysis results are compiled into a comprehensive HTML report using an *R* markdown template `_Overview_PhosphoAndIntegration_site.Rmd`. The report includes:

- Overview statistics on identified phosphorylation sites.
- Interactive scatter plots comparing protein vs PTM-feature changes.
- Volcano plots highlighting differential PTM-feature usage.

The report includes data tables that allow searching for a specific protein and phosphorylation feature, and highlighting it in the scatter plot showing the total proteome protein $log_2(FC)$ as a function of the PTM-site $log_2(FC)$, as well as in volcano plots. When analyzing the interactive reports, different combinations of DPA and DPU results provide distinct biological insights that guide interpretation. For instance, **DPA+ only** indicates changes driven primarily by protein abundance alterations, where PTM increases reflect higher protein levels rather than enhanced modification efficiency (*see* **Note 9**).

To generate the HTML report, we begin by creating a report configuration that includes the project ID, order ID, and a work unit description to be displayed in the interactive HTML report.

```
drumm <- prolfquapp::make_DEA_config_R6(
  PROJECTID = fgcz_project,
  ORDERID = oid_fgcz,
  WORKUNITID = descri
)
# Copy integration files
prophosqua::copy_phospho_integration()
```

Next, we can render the interactive HTML report with comprehensive analysis results, including tables, plots, and statistical summaries. This report is available for download from Zenodo [41].

```
# Render HTML report
rmarkdown::render(
  "_Overview_PhosphoAndIntegration_site.Rmd",
  params = list(
    data = combined_test_diff,
    grp = drumm
),
output_format = bookdown::html_document2(
  toc = TRUE,
  toc_float = TRUE
),
envir = new.env(parent = globalenv()))
# Copy HTML report to results directory (uses
existing if not re-rendered)
if (file.exists("_Overview_PhosphoAndIntegration_
site.html")) {
  file.copy(
    from = "_Overview_PhosphoAndIntegration_site.
html",
    to = file.path(res_dir,
"Result_phosphoAndTotalIntegration.html"),
    overwrite = TRUE
)
}
## [1] TRUE
```

In addition, the analysis results are exported to an Excel file for further analysis and sharing. The Excel file contains two worksheets:

- combinedSiteProteinData: Contains the merged PTM and protein-level data used for the DPA analysis.
- combinedStats: Contains the differential PTM-feature usage statistics (DPU), including $\log_2$ fold changes, p-values, and FDR estimates for each phosphorylation site.

This Excel format facilitates downstream analysis, data sharing with collaborators, and integration with other bioinformatics tools. The file is automatically saved with a timestamped filename in the results directory.

```
excel_result_list <- list(
  combinedStats = combined_test_diff,
  combinedSiteProteinData = combined_site_prot
)
writexl::write_xlsx(
  excel_result_list,
  path = file.path(
    res_dir,
```

```
    "Result_phosphoAndTotalIntegration.xlsx"
)
)
# Save example data for package (as.rda for data()
loading)
combined_test_diff_example <- combined_test_diff
save(combined_test_diff_example,
    file = here::here("data", "combined_test_diff_
example.rda"),
    compress = "xz")
```

The produced XLSX file can be downloaded from [41]. This comprehensive integrated analysis framework provides key advantages for PTM research, including the ability to distinguish genuine signaling changes from protein abundance effects (*see* **Note 10**).

3.10 Estimating PTM-Feature Usage by the Correct First Approach

Here we show how to estimate the DPU using the correct first approach. We first correct the PTM site abundances, using the total proteome, and then model the differential usage changes. This analysis starts by loading the prolfquapp outputs. This time, we read the transformed and normalized total proteome protein abundances.

```
tot_res_dir <- prophosqua::dea_res_dir(pro-
ject_dir,

"total_proteome", datedea)
ldata <-
arrow::read_parquet(file.path(tot_res_dir,"lfqdata_
normalized.parquet"))
ldata <- ldata |>
filter(!grepl("^rev_",protein_Id))
ldata <- prolfqua::get_UniprotID_from_fasta_
header(ldata)
tot_d <- ldata |>
  dplyr::select(channel, UniprotID,
normalized_abundance)
```

We are also loading normalized single-site abundances. In addition to loading the data, we create a prolfqua::LFQData object using the LFQData$new constructor. The LFQData object provides numerous methods for analyzing and visualizing quantitative proteomics data [33].

```
ptm_res_dir <- prophosqua::dea_res_dir(project_dir,
"singlesite_PTM", datedea)
ldata <-
arrow::read_parquet(file.path(ptm_res_dir,"lfqdata_
normalized.parquet"))
list <- yaml::read_yaml(file.path(ptm_res_
dir,"lfqdata.yaml"))
config <
- prolfqua::list_to_AnalysisConfiguration(list)
ptm_data <- prolfqua::LFQData$new(ldata,config)
```

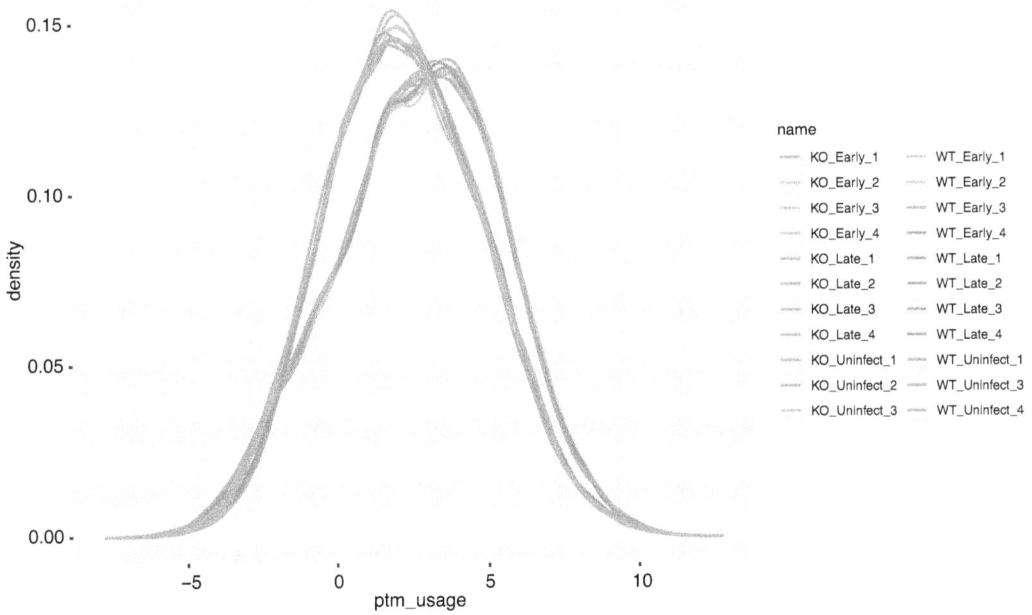

Fig. 3 Density plot of the total proteome corrected PTM site abundances

We merge the data for the single-site and total proteome protein abundances by protein identifier and sample. Then we subtract the total proteome protein abundances from the single-site abundances. The Density plot (see Fig. 3) shows the distribution of the corrected single-site abundances.

```
ptm_data$data <- dplyr::inner_join(
  ptm_data$data, tot_d,
  by = c("channel", protein_Id = "UniprotID"),
suffix =
c(".site",".total"))
ptm_data$data <- ptm_data$data |>
  dplyr::mutate(ptm_usage = normalized_abundance.
site - normalized_abundance.total)
ptm_data$config$table$set_response("ptm_usage")
pl <- ptm_data$get_Plotter()
fig3_plot <- pl$intensity_distribution_density()
```

Next, we fit a linear model for each PTM, which explains the corrected PTM site abundances using the grouping of the samples. The model is fitted for each site using the prolfqua::build_model function.

```
strategy_lm <- prolfqua::strategy_lm("ptm_usage ~
G_")
models <- prolfqua::build_model(data = ptm_data,
model_strategy = strategy_lm)
```

To compute the contrasts, we construct an array of contrasts from the annotation table, which we built in Sect. 3.4.

```
contrasts <- annot2$Contrast
names(contrasts) <- annot2$ContrastName
contrasts <- contrasts[!is.na(contrasts)]
```

The following code fragment computes contrasts from the linear models. In the presence of missing values, some PTMs cannot be fitted by the model. Therefore, we also estimate contrast using the ContrastsMissing method, which estimates missing values based on the inferred limit of detection (for more details, see [33]).

```
ctr <- prolfqua::Contrasts$new(models, contrasts)
x <- ctr$get_contrasts()
ctr_m <- prolfqua::ContrastsMissing$new(ptm_data,
contrasts)
ctr_all <- prolfqua::merge_contrasts_
results(ctr, ctr_m)$merged ctr_mod <
- prolfqua::ContrastsModerated$new(ctr_all)
ctr_df <- ctr_mod$get_contrasts()
```

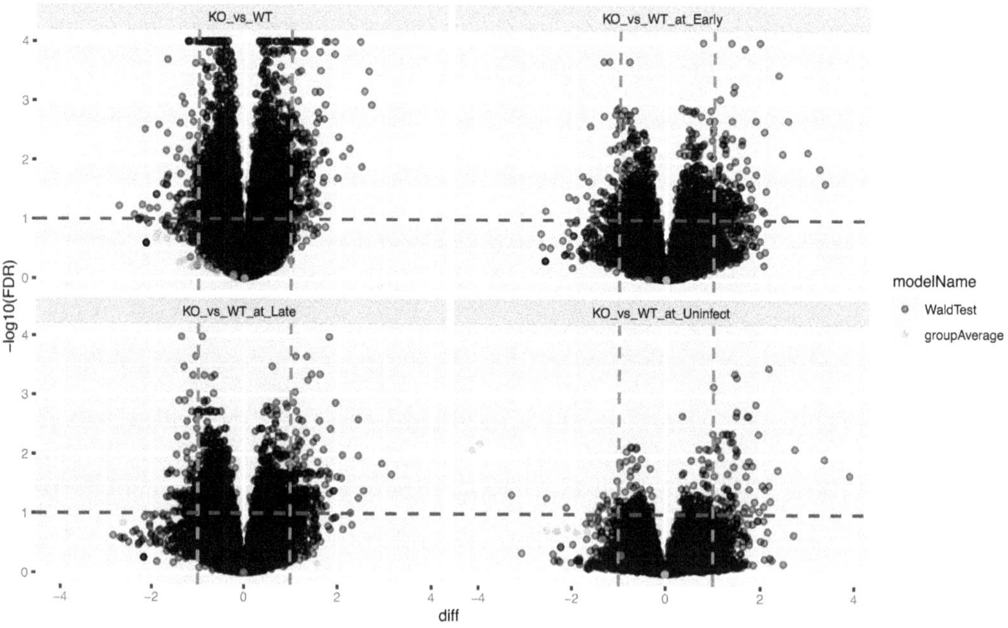

Fig. 4 Volcano plot showing the $-\log_{10}(FDR)$ as a function of the PTM site $\log_2$ fold change

Finally, we can create a volcano plot of the DPU results using the ctr_plotter variable, which is an instance of the prolfqua::ContrastsPlotter class (see Fig. 4).

```
ctr_plotter <- ctr_all$get_Plotter()
fig4_plot <- ctr_plotter$volcano()$FDR
```

To enable downstream analyses such as sequence motif analysis and PTM-SEA, we need to enrich the contrast results with sequence window information. The phospho_res dataframe loaded earlier contains all the annotation data we need.

```
# Select relevant columns from phospho_res for
joining
site_annotations <- phospho_res |>
  dplyr::select(
    site_Id = protein_Id,
    SequenceWindow,
    posInProtein,
    modAA,
    fasta.id,
    description,
    protein_length
) |>
dplyr::distinct()
# Join to contrast results
ctr_df_annotated <- ctr_df |>
  dplyr::left_join(site_annotations, by =
c("modelName" = "site_Id"))
# Rename columns to match the model-first output
format
ctr_df_annotated <- ctr_df_annotated |>
  dplyr::rename(
    diff_diff = diff,
    FDR_I = FDR
)
# Save correct-first results alongside model-first
output
saveRDS(ctr_df_annotated,
        file.path(res_dir, "ctr_df_annotated_
correctfirst.rds"))
```

The ctr_df_annotated dataframe now contains both the DPU statistics ($\log_2$ fold changes, p-values, FDR) and the sequence annotations needed for motif analysis and kinase activity inference.

3.11 Sequence Logo Analysis of Modified Sites

This section guides the biological interpretation of PTM changes by analyzing sequence motifs surrounding significantly modified sites to identify potential kinase recognition patterns. This analysis bridges the gap between statistical significance and biological mechanism (*see* **Note 8**) by leveraging the well-established principle that kinases recognize specific amino acid sequence patterns around their target sites.

Table 4
Consensus phosphorylation motifs of representative serine/threonine kinases

Kinase	Consensus_motif	Positional_notes
PKA/PKG	R-R-X-S/T	Arg at −3 and −2
PKC	R-X-S/T-X-R	Arg at −3 and +2
CK2	S/T-X-X-E/D	Acidic residue at +3
MAPK	P-X-S/T-P	Pro at −2 and +1

Motifs are defined relative to the phosphorylated residue (S/T at position 0). X = any residue; positions are relative to the phospho-acceptor serine/threonine (S/T)

Protein kinases exhibit sequence specificity, preferentially phosphorylating sites that match their recognition motifs. These motifs typically encompass positions −5 to +4 relative to the phosphorylation site, with specific positions being more critical than others. By analyzing the sequence patterns of significantly regulated sites, we can:

- Identify active kinases: Sites that share similar motifs likely represent targets of the same kinase or kinase family. Table 4 shows examples of kinase-specific sequence motifs.
- Understand pathway activation: Different kinase families are associated with different signaling pathways.
- Predict functional consequences: Kinase-specific phosphorylation often has predictable effects on protein function.
- Generate testable hypotheses: Motif predictions can guide follow-up experiments with kinase inhibitors or knockdowns.

To identify potential kinases responsible for the observed phosphorylation changes, sequence motif analysis is performed on significantly regulated PTM sites identified in the DPU analysis. For each contrast, amino acid sequences flanking the significantly regulated sites (i.e., $FDR \leq 0.05$ and $|\log_2(FC)| > 0.58$) are extracted and grouped by regulation status (i.e., upregulated or downregulated). The ggseqlogo *R* package is then used to generate sequence logo plots for each group. These plots visualize conserved amino acid patterns around the modification site, facilitating comparison to known kinase recognition motifs and enabling inference of upstream regulatory kinases (see Table 4).

The analysis focuses on sites with strong statistical evidence to ensure that motif patterns reflect genuine biological changes rather than technical artifacts. For this dataset, this stringent filtering yields many sites (see Table 5). However, for other datasets, the number of sites might be insufficient for creating sequence logos (*see* **Note 5**).

We start by filtering the DPU data by $|\log_2(FC)| > 0.58$ and $FDR \leq 0.05$. Table 5 summarizes the number of PTM-features for

Table 5
Number of significantly regulated sites, for each contrast, amino acid residue, and up- or down-regulation

Contrast	Regulation	modAA	Freq
KO_vs_WT	downregulated	S	309
KO_vs_WT_at_Early	downregulated	S	170
KO_vs_WT_at_Late	downregulated	S	221
KO_vs_WT_at_Uninfect	downregulated	S	72
KO_vs_WT	upregulated	S	499
KO_vs_WT_at_Early	upregulated	S	236
KO_vs_WT_at_Late	upregulated	S	315
KO_vs_WT_at_Uninfect	upregulated	S	140
KO_vs_WT	downregulated	T	85
KO_vs_WT_at_Early	downregulated	T	32
KO_vs_WT_at_Late	downregulated	T	79
KO_vs_WT_at_Uninfect	downregulated	T	12
KO_vs_WT	upregulated	T	97
KO_vs_WT_at_Early	upregulated	T	52
KO_vs_WT_at_Late	upregulated	T	65
KO_vs_WT_at_Uninfect	upregulated	T	14
KO_vs_WT	downregulated	Y	10
KO_vs_WT_at_Early	downregulated	Y	7
KO_vs_WT_at_Late	downregulated	Y	5
KO_vs_WT_at_Uninfect	downregulated	Y	1
KO_vs_WT	upregulated	Y	42
KO_vs_WT_at_Early	upregulated	Y	43
KO_vs_WT_at_Late	upregulated	Y	41
KO_vs_WT_at_Uninfect	upregulated	Y	1

each contrast, modification site, and up- or down-regulation. We observe that for serine (S) and threonine (T), there are many up- and downregulated sites, while for tyrosine (Y), there are only a few.

```
log_2FC_threshold <- 0.58
significant_sites <- combined_test_diff |>
  dplyr::filter(
    FDR_I < 0.05,
    abs(diff_diff) > log_2FC_threshold,
    !is.na(posInProtein),
    !grepl("^_", SequenceWindow),
```

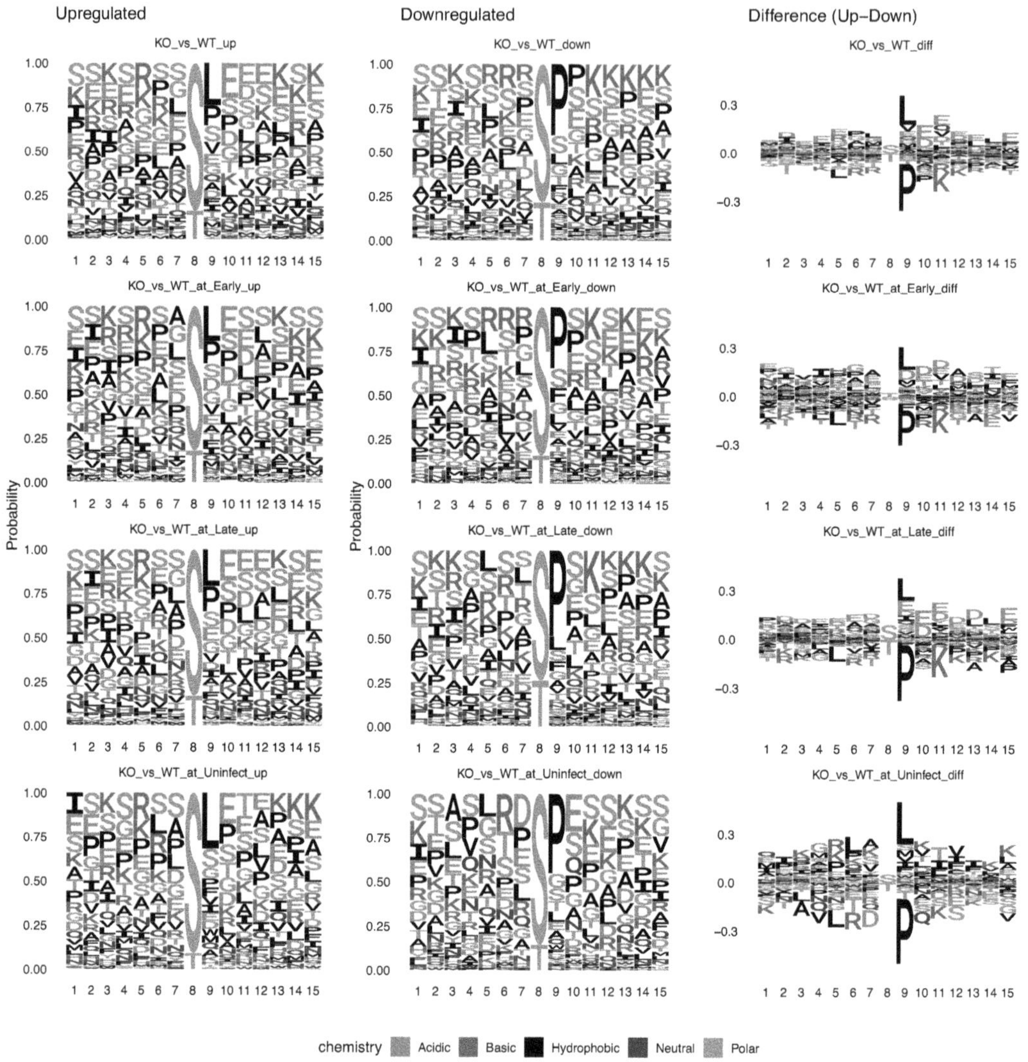

Fig. 5 Sequence logo plots showing amino acid motifs surrounding significantly regulated phosphorylation sites at serine and threonine residues. Each row represents a contrast with three columns: upregulated sites (left), downregulated sites (center), and the difference (right). In the difference column, letters above the baseline indicate amino acids enriched in upregulated sites, while letters below the baseline indicate enrichment in downregulated sites. Letter height corresponds to relative frequency at each position surrounding the phosphorylation site (position 8)

```
    !grepl("_$", SequenceWindow)
) |>
  dplyr::mutate(
    regulation = dplyr::case_when(
      diff_diff > 0 ~ "upregulated",
      diff_diff < 0 ~ "downregulated",
      TRUE ~ "no_change"
    )
  )
```

Before generating sequence logos, we must ensure that the sequence windows are properly centered on the phosphorylation site. This code validates that the eighth character (center position) of each sequence window matches the reported modified amino acid (*see* **Note 11**).

```
significant_sites$seventh_chars <
-toupper(substr(significant_sites$SequenceWindow,
8, 8))
significant_sites <- significant_sites |>
filter(seventh_chars == modAA)
```

We generate sequence logos for sites of significantly regulated serine and threonine phosphorylation. The plots are organized by contrast in a 3-column layout showing upregulated sites, downregulated sites, and their difference (Up - Down). The difference column highlights amino acids that are enriched in upregulated sites (positive, above baseline) versus downregulated sites (negative, below baseline), helping to identify kinase-specific motifs (see Fig. 5).

```
sig_sites_ST <- significant_sites |>
  dplyr::filter(modAA == "S" | modAA == "T")
fig5_plot <- plot_seqlogo_with_diff(sig_sites_ST)
```

This analysis helps identify the signaling pathways and kinases responsible for the observed changes in phosphorylation under your experimental conditions. When motif analysis yields weak or absent sequence patterns, several factors may require optimization, including the number of significant sites, the sequence window size used for motif extraction (*see* **Note 12**).

3.12 Kinase Activity Inference Using ptm-pipeline

Kinase activity can be inferred from phosphoproteomics data by testing whether known or predicted substrates of a kinase are coordinately shifted in a ranked list of regulated phosphosites. Two complementary resources for defining kinase–substrate sets are (i) curated signatures from PTMsigDB (distributed via MSigDB's PTMsig collection) and (ii) motif specificity models from The Kinase Library.

PTMsigDB/MSigDB PTMsig collection. The PTMsig collection (PTMsigDB) provides literature-curated, PTM site-specific signature sets for site-centric enrichment analysis, including kinase–substrate signatures, perturbation signatures (e.g., growth factor or inhibitor treatment), and pathway-related site signatures. Here, a "signature" refers to a named set of many phosphosites (analogous to a gene set in classic GSEA), not to a single sequence window. To maximize compatibility across search engines and protein database versions, PTMsigDB distributes signatures using multiple site identifier schemes, most commonly phosphosite-centered flanking sequences (typically ± 7 amino acids around the modified residue; 15-mers) as well as UniProt-centric residue identifiers (and optionally PhosphoSitePlus site-group IDs). PTMsigDB provides organism-specific releases for human, mouse, and rat; in the original PTMsigDB publication, the total number of available signature sets was on the order of hundreds (reported as ≈490), and counts vary by database version and organism.

The Kinase Library. A more comprehensive, model-based resource is The Kinase Library [42], which profiled substrate sequence preferences for 303 human Ser/Thr kinases (covering >84% of those predicted to be active) and provides position-specific scoring matrices (PSSMs). Unlike curated substrate lists that require exact matches to previously observed sites, PSSMs assign a quantitative compatibility score to any observed sequence window, enabling kinase–substrate relationships to be inferred without perfect matching. This makes it straightforward to construct "kinase substrate sets" by applying a score threshold or selecting the top-ranked sites per kinase, which can then be used as input signatures for enrichment testing.

PTM Signature Enrichment Analysis (PTM-SEA) using GSEA. Once kinase–substrate sets are defined (from PTMsigDB, motif/PSSM scoring, or both), kinase activity can be inferred quantitatively using PTM-SEA [28], which adapts the Gene Set Enrichment Analysis (GSEA) framework [43] to site-centric data. Phosphosites (or their sequence-window identifiers) are ranked by a differential signal (e.g., $\log_2$ fold change or a t-statistic). GSEA then tests whether members of a kinase's substrate set tend to occur near the top or bottom of this ranked list more than expected under a null model of uniform distribution. The enrichment score is computed by walking down the ranked list, increasing a running sum when a set member is encountered and decreasing it otherwise. Positive enrichment is consistent with increased kinase activity (substrates preferentially upregulated), while negative enrichment is consistent with reduced activity.

Practical considerations and coverage. PTMsigDB signatures provide high-confidence, literature-supported kinase assignments, but they are constrained by database coverage and identifier matching (including the need for consistent site definitions and sequence context). These constraints become more pronounced outside human (and to a lesser extent mouse/rat). In contrast, motif-based scanning (e.g., basophilic motifs such as R-X-X-S/T for PKA/CaMKII-like kinases) and Kinase Library PSSMs generalize to any observed sequence window, enabling broader application across datasets where curated substrate evidence is sparse—while trading off direct experimental validation at the individual site level.

In ptm-pipeline [44], kinase activity inference is implemented as an automated workflow that uses prolfquapp/prophosqua to generate differential phosphorylation abundance (DPA) and differential phosphorylation usage (DPU) results under both the "model-first" and "correct-first" strategies. It then performs GSEA-based PTM-SEA using kinase substrate signatures from PTMsigDB as well as Kinase Library-derived sets generated by PSSM scoring of phosphosite sequence windows. The workflow is provided as a command-line tool and executed via a *Snakemake* pipeline. The source code is available at github.com/wolski/ptm-pipeline, with documentation and example reports at wolski.github.io/ptm-pipeline.

4 Notes

1. **Label-free quantification vs. TMT-based approach**
   TMT-based quantification offers several advantages, including higher throughput, reduced missing values, and improved quantitative precision, while also enabling efficient use of fractionation since all samples are represented within each fraction. In contrast, label-free approaches avoid labeling-related artifacts, can handle an unlimited number of samples, and typically have lower per-sample costs. Fractionation strategies benefit particularly from TMT, as all samples can be compared directly within each fraction with reliable cross-fraction integration. In contrast, data-independent acquisition (DIA) is often preferred when avoiding fractionation altogether. Overall, we recommend TMT for studies requiring high precision across multiple conditions, particularly when fractionation can be leveraged to increase depth. In contrast, label-free or DIA workflows are well-suited for discovery-phase studies or experiments where the number of samples exceeds the multiplexing capacity.
2. **Automated QC reporting with *FragPipe* TMT QC script**
   The prophosqua [32] package vignette QCReport.qmd can be used to generate a comprehensive TMT labeling efficiency report, directly from *FragPipe* PSM psm.tsv output file. The script evaluates labeling completeness at both peptide and PSM levels for N-terminal and lysine modifications, calculates missed cleavage rates for tryptic digestion efficiency, and provides quantitative balance assessment across all TMT channels. Key metrics include the percentage of modified N-termini and lysine residues (expected to be $\geq$95%), missed cleavage rates for lysine and arginine residues, and relative abundance distributions across channels to detect loading imbalances. To assess labeling efficiency, the TMT labeling of lysine residues and peptide N-termini must be specified as a variable modification in the search.

3. **Use of the integrated DPA/DPU approach**
   The integrated DPA/DPU approach is most suitable when the biological system under study exhibits substantial changes in protein expression. In such cases, distinguishing between signaling effects and expression-driven effects is essential, and including protein abundance measurements as part of the study design allows these two sources of variation to be effectively separated.
   In contrast, enriched phosphoproteome measurements without corresponding protein data can be considered when protein expression is expected to remain stable across conditions, for example, in experiments with very short timescales where transcriptional or translational changes are unlikely. This strategy may also be justified when budget limitations make comprehensive protein quantification infeasible.
4. **Match Rates**
   When evaluating integration between PTM and protein datasets, match rates above 80% generally indicate good consistency. In contrast, lower rates may reflect differences in protein assignment between the PTM-enriched and total proteome analyses. In cases where match rates fall below 60% , the underlying cause is often inconsistent protein identifiers between the two datasets. To address this, it is essential to verify that both analyses utilized the same FASTA database and identical protein inference settings, with particular attention to database versions and the inclusion of contaminant sequences. As a preventative measure, identical search parameters and databases should always be applied to both total proteome and PTM analyses, while sequence-based matching can further improve alignment across datasets.
5. **No significant PTM sites detected despite biological expectation**
   The absence of significant PTM sites, even when biological regulation is expected, can arise from several factors. Error propagation from both PTM and protein measurements may amplify overall uncertainty, masking accurate signals. Insufficient statistical power is another common cause, emphasizing the need for at least three biological replicates per group and, where possible, increasing sample size or reducing technical variability. In some cases, stringent statistical thresholds may obscure true positives, and adopting a more permissive cutoff (e.g., FDR=0.25 instead of 0.05) during exploratory analysis can be informative. Normalization problems, such as batch effects or inconsistent sample loading, may also contribute and should be checked carefully. Outliers in the data can further distort inference, making it useful to inspect density or scatter plots, for example, in prolfquapp reports. Finally, examining the distribution of p-values in quality-control summaries can

help assess whether the statistical model adequately explains the variance in the data.
When no significant differential peptidoform usage (DPU) sites are detected despite clear changes in PTMs, one possible explanation is that the PTM changes are proportional to changes in the parent protein, indicating no alteration in modification stoichiometry. In many cases, this represents a valid biological result rather than a technical issue, and it is advisable to verify it by inspecting individual examples where protein and PTM changes are closely correlated. Alternatively, such an outcome may suggest that the experimental perturbation primarily influences protein expression rather than signaling activity.

6. **Computational Performance Issues**
   Long processing times for N-to-C plots often occur because generating individual plots for hundreds of proteins is computationally intensive. A practical solution is to limit plot generation to proteins of particular interest, such as those with small FDR values and substantial $\log_2$ fold changes, rather than producing plots for all detected proteins.

7. **Quality Control Recommendations**
   When focusing on specific phosphorylation sites, it is essential to validate key findings by carefully inspecting high-confidence DPU sites to ensure both technical and biological plausibility. For example, this can be achieved by verifying the number of missing values in each treatment group. In addition, protein coverage should be verified, since meaningful interpretation of phosphorylation changes requires that the corresponding proteins have sufficient coverage in the total proteome. The number of peptides identified per protein, as reported in the prolfquapp output, can serve as a filtering criterion.

8. **Overinterpretation of statistical significance**
   Overinterpretation of statistical significance can occur when FDR estimates are reported without considering the design of the experiment or its biological context. A statistically significant finding does not necessarily imply biological relevance; therefore, it is essential to evaluate effect sizes alongside p-values to assess the magnitude and potential impact of a change. Key findings should also be validated with orthogonal methods to ensure that they reflect genuine biological regulation rather than statistical or technical artifacts.

9. **DPA vs DPU Interpretation**
   Conflicting results between DPA and DPU analyses can be due to:

   - DPA+ only: Changes driven by protein abundance alterations.

- DPU+ only: True signaling changes with stable protein levels.
- DPA+ and DPU+: Amplified signaling (both protein and modification efficiency increase).
- DPA− and DPU+: Compensatory regulation (modification increases despite protein decrease).

Such discrepancies between DPA and DPU are often informative rather than problematic. Sites that are significant in DPA but not in DPU typically reflect protein-driven changes, whereas DPU-only sites indicate genuine signaling regulation independent of protein abundance. To assess their biological plausibility, these results should be cross-referenced with known biology and supporting literature.

10. **Key insights from this analysis**
This integrated analysis distinguishes PTM feature abundance from usage changes, enabling site-specific insights into how phosphorylation and other PTMs are regulated. By linking DPA and DPU results, it clarifies whether observed changes reflect protein-level expression or signaling pathway activity, and it highlights key regulatory sites for functional validation. Visualization in sequence context and rigorous statistical modeling further support interpretation, allowing researchers to generate mechanistic hypotheses, identify pathway-level regulation, and prioritize targeted follow-up experiments.

11. **Extracting sequence windows with prophosqua**
Although *FragPipe* reports the sequence windows in the single-site report, we observed that some of the extracted windows are not correctly aligned, making the filtering step necessary. Some upstream quantification applications only report the modified amino acid and the position in the protein sequence. For those cases, the prophosqua function get_sequence_windows can be used. The function expects a dataframe with two columns: the protein sequences and the position of the modified amino acid. For details, see the prophosqua documentation [32].

12. **Motif analysis limitations**

Weak or absent sequence motifs in significantly regulated sites can arise for several reasons. One possibility is that too few significant sites were detected, which limits the statistical power of motif discovery. In such cases, it can be helpful to lower the threshold specifically for motif analysis to increase the number of sites included. Mixed kinase activities may also obscure motifs, and separating upregulated from downregulated sites often clarifies distinct regulatory patterns. Additionally, regulation may occur through noncanonical mechanisms, such as phosphatases or protein–protein interactions, rather than through specific kinase recognition motifs. Other technical considerations, such as the choice of sequence window size (e.g., ±5 versus ±7 amino acids), and biological factors like PTM crosstalk can also influence motif detection and interpretation. Overall, weak motif signals can result from an insufficient number of significant sites, the presence of multiple overlapping regulatory activities, or regulation by mechanisms that do not rely on canonical sequence motifs. To address these issues, one can lower the significance threshold for motif analysis to capture more sites. If no clear kinase sequence pattern emerges, keep in mind that regulation may involve nonkinase pathways, which underscores the importance of considering both statistical and biological explanations when interpreting motif analysis results.

Conflict of Interest

The authors declare no conflicts of interest that are relevant to the content of this chapter.

References

1. Gerritsen JS, White FM (2021) Phosphoproteomics: a valuable tool for uncovering molecular signaling in cancer cells. Expert Rev Proteomics 18:661–674. https://doi.org/10.1080/14789450.2021.1976152
2. Delom F, Chevet E (2006) Phosphoprotein analysis: from proteins to proteomes. Proteome Sci 4:15. https://doi.org/10.1186/1477-5956-4-15
3. Bortel P, Piga I, Koenig C et al. (2024) Systematic optimization of automated phosphopeptide enrichment for high-sensitivity phosphoproteomics. Mol Cell Proteomics 23:100754. https://doi.org/10.1016/j.mcpro.2024.100754
4. Kohler D, Tsai T-H, Verschueren E et al. (2023) MSstatsPTM: statistical relative quantification of posttranslational modifications in bottom-up mass spectrometry-based proteomics. Mol Cell Proteomics 22:100477. https://doi.org/10.1016/j.mcpro.2022.100477
5. Demeulemeester N, Gbelin M, Caldi Gomes L et al. (2024) msqrob2PTM: Differential abundance and differential usage analysis of MS-based proteomics data at the posttranslational modification and peptidoform level. Mol Cell Proteomics 23:100708. https://doi.org/10.1016/j.mcpro.2023.100708
6. Solari FA, Dell'Aica M, Sickmann A, Zahedi RP (2015) Why phosphoproteomics is still a challenge. Mol BioSyst 11:1487–1493. https://doi.org/10.1039/c5mb00024f
7. Wentzell PD, Gonalves TR, Matsushita M, Valderrama P (2021) Combinatorial projection pursuit analysis for exploring multivariate chemical data. Anal Chim Acta 1174:338716. https://doi.org/10.1016/j.aca.2021.338716
8. Humphrey SJ, Azimifar SB, Mann M (2015) High-throughput phosphoproteomics reveals in vivo insulin signaling dynamics. Nat Biotechnol 33:990–995. https://doi.org/10.1038/nbt.3327
9. Hogrebe A, von Stechow L, Bekker-Jensen DB, et al. (2018) Benchmarking common quantification strategies for large-scale phosphoproteomics. Nat Commun 9, 1045. https://doi.org/10.1038/s41467-018-03309-6
10. Zuniga NR, Frost DC, Kuhn K et al. (2024) Achieving a 35-plex tandem mass tag reagent set through deuterium incorporation. J Proteome Res 23:5153–5165. https://doi.org/10.1021/acs.jproteome.4c00668
11. Thompson A, Schaefer J, Kuhn K et al. (2003) Tandem mass tags: a novel quantification strategy for comparative analysis of complex protein mixtures by MS/MS. Anal Chem 75:1895–1904. https://doi.org/10.1021/ac0262560
12. Rauniyar N, Subramanian K, Lavalle-Adam M et al. (2015) Quantitative proteomics of human

fibroblasts with I1061T mutation in Niemann–Pick C1 (NPC1) protein provides insights into the disease pathogenesis. Mol Cell Proteomics 14:1734–1749. https://doi.org/10.1074/mcp.m114.045609
13. Diez IA, Govender I, Naicker P, Stoychev S, Jordaan J, Jensen ON (2021) Zirconium(IV)-IMAC revisited: improved performance and phosphoproteome coverage by magnetic microparticles. J Proteome Res 20:453–462. https://doi.org/10.1021/acs.jproteome.0c00508
14. Koenig C, Martinez-Val A, Franciosa G, Olsen JV (2022) Optimal analytical strategies for sensitive and quantitative phosphoproteomics using TMT-based multiplexing. Proteomics 22, e2100245. https://doi.org/10.1002/pmic.202100245
15. Burger B, Vaudel M, Barsnes H (2020) Importance of block randomization when designing proteomics experiments. J Proteome Res 20:122–128. https://doi.org/10.1021/acs.jproteome.0c00536
16. Ferretti D, Kyriakidou P, Xiao J et al. (2025) Isobaric labeling update in MaxQuant. J Proteome Res 24:1219–1229. https://doi.org/10.1021/acs.jproteome.4c00869
17. Chang H-Y, Deng Y, Li R, et al. (2026) Analysis of isobaric quantitative proteomic data using TMT-Integrator and FragPipe computational platform. Nat Commun 17:4010. https://doi.org/10.1038/s41467-026-70118-7
18. Farag YM, Horro C, Vaudel M, Barsnes H (2021) PeptideShaker online: a user-friendly web-based framework for the identification of mass spectrometry-based proteomics data. J Proteome Res 20:5419–5423. https://doi.org/10.1021/acs.jproteome.1c00678
19. Im JH, Buzzelli JN, Jones K et al (2020) FGF2 alters macrophage polarization, tumour immunity and growth and can be targeted during radiotherapy. Nat Commun 11. https://doi.org/10.1038/s41467-020-17914-x
20. Taus T, Kcher T, Pichler P et al. (2011) Universal and confident phosphorylation site localization using phosphoRS. J Proteome Res 10:5354–5362. https://doi.org/10.1021/pr200611n
21. Beausoleil SA, Villen J, Gerber SA et al. (2006) A probability-based approach for high-throughput protein phosphorylation analysis and site localization. Nature Biotechnol 24:1285–1292. https://doi.org/10.1038/nbt1240
22. Kong AT, Leprevost FV, Avtonomov DM et al. (2017) MSFragger: ultrafast and comprehensive peptide identification in mass spectrometry–based proteomics. Nature Methods 14:513–520. https://doi.org/10.1038/nmeth.4256
23. Shteynberg DD, Deutsch EW, Campbell DS et al. (2019) PTMProphet: fast and accurate mass modification localization for the trans-proteomic pipeline. J Proteome Res 18:4262–4272. https://doi.org/10.1021/acs.jproteome.9b00205
24. Joyce AW, Searle BC (2024) Computational approaches to identify sites of phosphorylation. Proteomics 24, e2300088. https://doi.org/10.1002/pmic.202300088
25. Bruderer R, Bernhardt OM, Gandhi T et al. (2015) Extending the limits of quantitative proteome profiling with data-independent acquisition and application to acetaminophen-treated three-dimensional liver microtissues. Mol Cell Proteomics 14:1400–1410. https://doi.org/10.1074/mcp.m114.044305
26. Demichev V, Messner CB, Vernardis SI et al. (2019) DIA-NN: neural networks and interference correction enable deep proteome coverage in high throughput. Nat Methods 17:41–44. https://doi.org/10.1038/s41592-019-0638-x
27. Kim HJ, Kim T, Hoffman NJ et al. (2021) PhosR enables processing and functional analysis of phosphoproteomic data. Cell Rep 34:108771. https://doi.org/10.1016/j.celrep.2021.108771
28. Krug K, Mertins P, Zhang B et al. (2019) A curated resource for phosphosite-specific signature analysis. Mol Cell Proteomics 18:576–593. https://doi.org/10.1074/mcp.TIR118.000943
29. Yilmaz S, Ayati M, Schlatzer D, et al. (2021) Robust inference of kinase activity using functional networks. Nat Commun 12:1177. https://doi.org/10.1038/s41467-021-21211-6
30. Savage SR, Zhang B (2020) Using phosphoproteomics data to understand cellular signaling: a comprehensive guide to bioinformatics resources. Clinical Proteomics 17:27. https://doi.org/10.1186/s12014-020-09290-x
31. Giai Gianetto Q (2021) Statistical analysis of post-translational modifications quantified by label-free proteomics across multiple biological conditions with r: illustration from SARS-CoV-2 infected cells. Statistical analysis of proteomic data. Springer, US, 267–302
32. Grossmann J, Wolski WE (2025) prolfqua/prophosqua. Zenodo. https://doi.org/10.5281/zenodo.15845272
33. Wolski WE, Nanni P, Grossmann J et al. (2023) Prolfqua: a comprehensive R-package for proteomics differential expression analysis. J Proteome Res 22:1092–1104. https://doi.org/10.1021/acs.jproteome.2c00441
34. Schlatter R, Philippi N, Wangorsch G et al. (2011) Integration of Boolean models

exemplified on hepatocyte signal transduction. Brief Bioinform 13:365–376. https://doi.org/10.1093/bib/bbr065
35. Maculins T, Verschueren E, Hinkle T, et al. (2021) Multiplexed proteomics of autophagy-deficient murine macrophages reveals enhanced antimicrobial immunity via the oxidative stress response. eLife 10, e62320. https://doi.org/10.7554/elife.62320
36. Wickham H, Averick M, Bryan J et al. (2019) Welcome to the tidyverse. J Open Source Softw 4:1686. https://doi.org/10.21105/joss.01686
37. Wagih O (2017) ggseqlogo: a versatile R package for drawing sequence logos. Bioinformatics 33:3645–3647. https://doi.org/10.1093/bioinformatics/btx469
38. Wolski WE, Grossmann J, Schwarz L et al. (2025) Prolfquapp a user-friendly command-line tool simplifying differential expression analysis in quantitative proteomics. J Proteome Res 24:955–965. https://doi.org/10.1021/acs.jproteome.4c00911
39. Wolski WE (2025) prolfqua/prolfquappPTMreaders. Zenodo. https://doi.org/10.5281/zenodo.15845243
40. Dittmann A, Grossmann J, Wolski WE (2025) FragPipe v22.0 analysis outputs for phosphoproteomics & total proteome of ATG16L1-deficient murine macrophages (Maculins et al., eLife 2021). Zenodo. https://doi.org/10.5281/zenodo.15850770
41. Wolski WE, Grossmann J, Dittmann A (2025) Complete computational workflow for integrated PTM and total proteome analysis: distinguishing expression from usage changes. Zenodo. https://doi.org/10.5281/zenodo.15830988
42. Johnson JL, Yaron TM, Huntsman EM et al. (2023) An atlas of substrate specificities for the human serine/threonine kinome. Nature 613:759–766. https://doi.org/10.1038/s41586-022-05575-3
43. Subramanian A, Tamayo P, Mootha VK et al. (2005) Gene set enrichment analysis: a knowledge-based approach for interpreting genome-wide expression profiles. Proc Natl Acad Sci 102:15545–15550. https://doi.org/10.1073/pnas.0506580102
44. Wolski WE (2025) wolski/ptm-pipeline. Zenodo. https://doi.org/10.5281/zenodo.18349420

Check for updates

Chapter 9

Pathway and Kinase Analysis in Proteomics and Phosphoproteomics with rROMA

Stéphane Liva, Anne-Claire Kröger, Michael Richard, Patrick Poullet and Loredana Martignetti

Abstract

Advances in data acquisition tools, particularly mass spectrometry-based techniques, have made high-throughput methods like proteomics and phosphoproteomics increasingly accessible. However, the volume and complexity of the data require specialized computational tools for analysis and interpretation to extract meaningful insights. Here, we describe the use of rROMA software to analyze pathway and kinase activity in proteomic and phosphoproteomic datasets. rROMA is an open-source package available on Github at: www.github.com/sysbio-curie/rROMA. As a case study, we applied rROMA to breast cancer samples from the Clinical Proteomic Tumor Analysis Consortium (CPTAC), which provides precise relative quantification of proteins and phosphorylation sites of a large cohort of breast cancer samples through Tandem mass tag (TMT)-based mass spectrometry. These analyses illustrate how to identify active pathways and kinase-driven mechanisms potentially involved in disease subtypes, along with their associated proteins and phosphosites.

Keywords Proteomics, Phosphoproteomics, Pathway analysis, Kinase activity

1 Introduction

High-throughput molecular techniques like proteomics and phosphoproteomics are becoming more and more accessible due to advancements in data acquisition tools, notably mass-spectrometry-based techniques. This progress enables precise measurement of thousands of peptides and proteins in complex biological specimens. However, the vast amount of data generated necessitates specialized software and algorithms for analysis. Additionally, interpreting these data in the context of biological processes is crucial, requiring dedicated analyses to convert them into meaningful biological information [1].

A widely used strategy in proteomics involves analyzing measurements at the level of individual proteins to identify biomarkers

Frédérique Lisacek (ed.), *Bioinformatics for Mass Spectrometry*, Methods in Molecular Biology, vol. 3071,
https://doi.org/10.1007/978-1-0716-5519-1_9,

or to pinpoint proteins with a causal role in disease development. However, in many complex or systemic diseases, disruptions in signaling pathways can result from alterations in different proteins or proteoform within the same pathway, with these alterations varying between patients [2]. For instance, in cancer, it has become increasingly evident that while different proteins may be dysregulated across patients, they often affect the same pathways [3]. As a result, molecular profiles tend to show greater similarity at the pathway level than at the level of individual proteins.

Gene-set analysis methods are commonly used in cancer genomics studies to identify dysregulated pathways. These approaches typically estimate pathway activity based on the expression levels of associated genes, despite the fact that it is the proteins, not the transcripts, that ultimately drive pathway function. In this chapter, we present how to use a gene set-based approach, called rROMA, for the analysis of proteomic and phosphoproteomic data.

The rROMA software was designed to quantify the activity of sets of genes characterized by their involvement in a common functional role [4,5]. Gene sets can thus correspond to genes with the same functional activities, genes regulated by the same motifs, genes belonging to the same signaling pathway, target genes of a transcription factor, or genes forming a group of frequently co-expressed genes.

The rROMA algorithm is a PCA-based approach for quantifying pathway activity. Using a gene expression data matrix, this algorithm applies a linear model of gene regulation to quantify the activity of a pathway represented by a given set of annotated proteins. It derives the pathway activity score by calculating the first principal component (PC1) of the corresponding submatrix (i.e., the expression matrix restricted to proteins in that set). The unifactor model underlying rROMA assumes the presence of a latent unobserved factor influencing the expression variables within a given protein set. This influence is represented by calculated weights, which reflect both the strength and direction of the factor's effect. These weights can have opposite signs, for example, when a latent factor activates certain proteins in the set while repressing others.

Notably, the algorithm estimates pathway activity for each individual sample without the need for predefined labels or group classifications. These activity scores can then be compared across samples, revealing underlying heterogeneity within the dataset related to the gene sets of interest. This approach can be especially valuable for identifying distinct sample or patient subtypes in cases where no prior stratification exists.

Furthermore, the strength of the rROMA method lies in its ability to calculate a *p*-value, a statistical measure that serves as a crucial indicator of the pathway activity's significance. This *p*-value enables the prioritization of biological processes that are pertinent to the underlying conditions in question. The rRoma workflow is illustrated in Fig. 1. First, the algorithm extracts the expression submatrix corresponding to a given gene set and performs PCA on this submatrix in the sample space using the SVD algorithm [6]. Two key

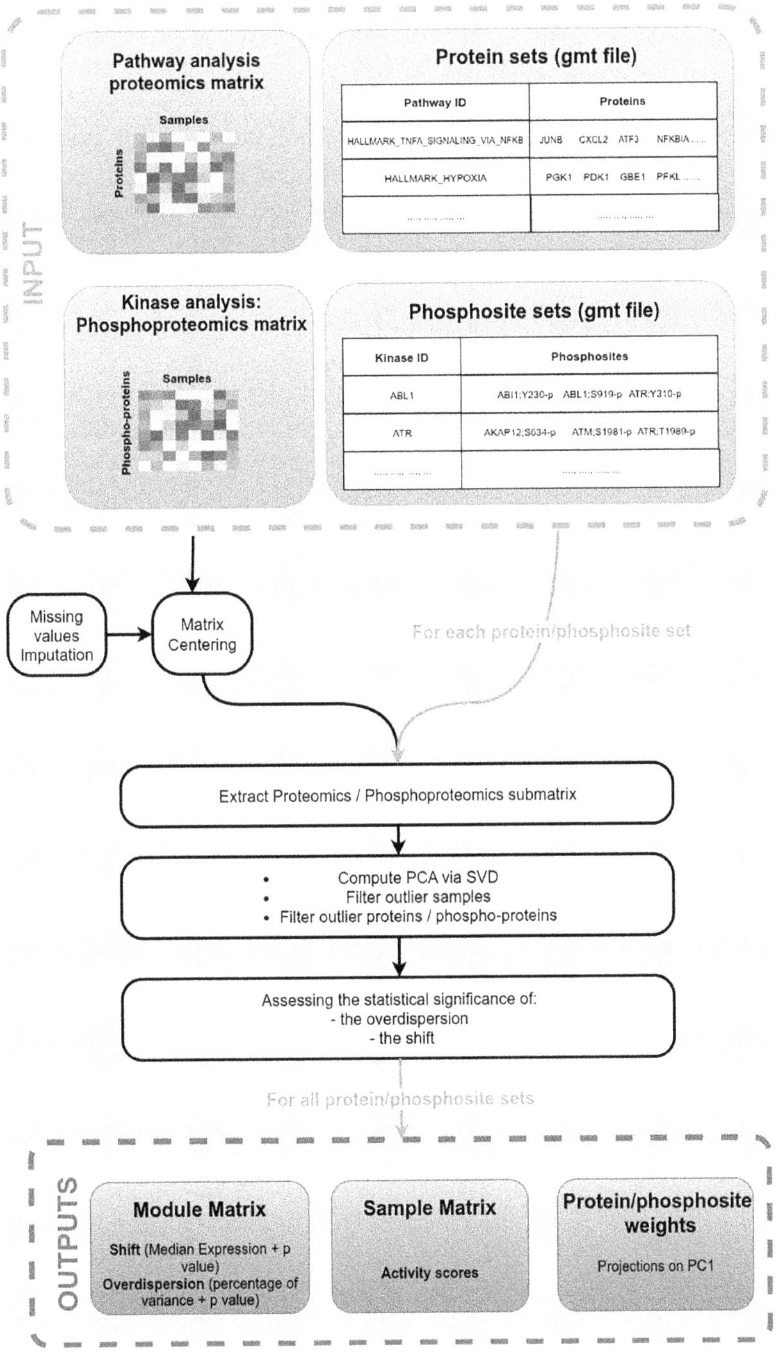

Fig. 1 rRoma workflow

metrics are then computed: (1) the percentage of variance explained by the PC1, referred to as L1 and (2) the median projection of gene expression onto PC1, referred to as Median Exp.

These metrics are used to assess pathway activity significance in two complementary ways, by evaluating shift and overdispersion:

- A **shifted gene set** is one in which the median expression (Median Exp) differs significantly from the overall background distribution. This suggests that the gene set shows elevated expression in at least one sample.
- An **overdispersed gene set** exhibits significantly greater variance explained by PC1 (L1) compared to random gene sets of the same size. This indicates higher variability in expression patterns across samples.

To enhance robustness, the algorithm includes an automatic procedure for detecting and removing outlier genes and samples prior to PC1 computation

The detailed method is illustrated in Fig. 1 and described in the publication by Najm et al [5], where the algorithm was originally applied to transcriptomic data. In this study, we have extended the model to infer pathway activity from protein expression, as well as to infer kinase activity based on associated sets of phosphosites.

Here, we showcase the interest of application of rROMA to the analysis of mass-spectrometry-derived proteomic and phosphoproteomic datasets. rROMA has been developed as a gene set analysis method, designed to identify and characterize coordinated activity within predefined groups of genes or proteins. Its strength lies in its ability to integrate biological knowledge encoded in gene sets, making it particularly suitable for functional interpretation of omics data.

In this chapter, we describe a protocol for successfully applying rROMA to two key areas: pathway analysis using proteomics data and kinase activity inference using phosphoproteomics data. We have detailed the steps required to perform each analysis, including data preparation, commands to run the analysis, and interpretation of the results, allowing researchers to gain deeper insights into molecular signaling and regulatory mechanisms.

In the first part, we apply rROMA to identify active biological pathways based on protein expression data in primary breast cancer tissues from the Clinical Proteomic Tumor Analysis Consortium (CPTAC) [7]. In the analysis of proteomics data, gene sets are composed of the genes encoding the proteins involved in a given pathway of interest. In the second part, we use rROMA to infer the activity or dysregulation of kinases using phosphoproteomic data from the same breast cancer tissues. In this case, we use the software to predict active kinases based on sets of phospho-peptides known to be substrates of specific kinases.

2 Materials

2.1 Datasets

Proteomics and phosphoproteomics used in this study have been downloaded from a comprehensive proteogenomics resource from prospectively collected breast tumors [8]. This is a proteomic and phosphoproteomics dataset of 122 treatment-naive primary breast tumor samples. It provides precise relative quantification of proteins and phosphorylation sites of breast cancer tissues, processed using Tandem mass tag (TMT)-based mass spectrometry. The final dataset includes 10,107 quantified proteins and 38,968 phosphorylation sites.

rRoma was applied to predict pathway activity and kinase activity in different subtypes of breast cancer (Basal, Her2-Enriched, Luminal-A, and Luminal-B, Normal-like). The PAM50 classification for these subtypes was provided in the original publication as follows: Basal-like (23.8%), Her2-Enriched (11.5%), Luminal A (46.7%), Luminal B (13.9%), and Normal-like (4.1%).

The pathway annotation file for the hallmarks of cancer [9] was downloaded from the Molecular Signatures Database [10]. It includes 50 annotated pathways. For the kinase–substrate annotation, we used the Kinase Substrate Relationship resource described in Sect. 3.4.

2.2 The rROMA Algorithm

Installing the rROMA Software

The rROMA software is freely available at https:://sysbio-curie.github.io/rROMA/ and depends on the installation of a few additional packages. The rRoma package relies on the biomaRt [11] and Scater [12] packages, which are available on BioConductor. It also requires ggplot2 [13] for descriptive plots. These packages can be installed with the following commands:

```
if (!require("BiocManager", quietly = TRUE))
....install.packages("BiocManager")
BiocManager::install("biomaRt")
BiocManager::install("scater")
if(!requireNamespace("ggplot2")){
....install.packages("ggplot2")
}
if(!requireNamespace("dplyr")) {
....install.packages("dplyr")
}
if(!requireNamespace("tidyverse")) {
....install.packages("tidyverse")
}
if(!requireNamespace("RColorBrewer")) {
....install.packages("RColorBrewer")
}
rRoma can then be installed using devtools [14]
```

```
if(!requireNamespace("devtools")) {
....install.packages("devtools")
}
if(!requireNamespace("rRoma")) {
....devtools::install_github("sysbio-curie/rRoma")
}
```

rRoma uses the mice package [15] to impute missing values. This package is required if the dataset contains missing values. However, we recommend using rROMA with datasets where missing values have already been imputed.

```
if(!requireNamespace("mice")) {
.... install.packages("mice")}
```

The additional code that supports this study is available at https://zenodo.org/records/16780702. The additional code must be saved locally and loaded into the R or RStudio working session.

Data and Code Availability

The computational analyses were performed in R 4.4.3, and all the code is available under a GNU General Public License V3 in a GitHub repository at the following URL: https://github.com/sysbio-curie/rROMA. The datasets used in this study are available at https://zenodo.org/records/16780702.

3 Methods

3.1 Using the rROMA Software

The package and additional code can be loaded with the following commands:

```
library(rRoma)
library(dplyr)
library(tidyverse)
library(ggplot2)
library(mice)
library(RColorBrewer)
source("src/roma_pheatmap.R")
source("src/plotGenesetsvsSampled.R")
source("src/PlotGeneWeight.R")
```

3.2 Predicting Dysregulated Pathways in Breast Cancer Subtypes from Proteomic Data

This section provides a step-by-step description of how to perform pathway analysis from proteomic data in a given dataset of interest. The basic steps for running an analysis in rROMA are as follows:

step 1 Prepare your data files:

Expression dataset file (tsv, csv, txt)

Pathway annotation gene set file (gmt)

Phenotype labels file (txt, optional)

step 2 Load your data files into your Rstudio session.

step 3 Set the analysis parameters and run the analysis.

step 4 View the analysis results

Preparing Your Data Files

rRoma requires a normalized proteomic matrix, with column names indicating samples and row names indicating gene IDs. The recommended gene IDs (official gene symbol) correspond to the gene name encoding the quantified protein. Protein to gene ID mapping can be done via the batch retrieval service ("Retrieve/ID mapping") in UniProtKB [16]. It is important to ensure that the input matrix for rROMA contains unique gene identifiers, as the algorithm does not support redundant or duplicated gene IDs. The software proceeds with the imputation of any missing values in the matrix, if present. MICE [15] is the imputation method by default. We, however, recommend performing your own imputation beforehand. The MICE (Multiple Imputation by Chained Equations) algorithm is a method for handling missing data by iteratively imputing missing values using regression models. For each variable with missing values, MICE builds a model based on the other variables and fills in the missing data. This process is repeated multiple times (chained equations), creating several complete datasets. The final estimates are obtained by averaging across these datasets, accounting for the uncertainty in the imputations

Loading Your Data Files

In your Rstudio session, you need to load the data matrix file:

```
matData.proteo <- read.table(
"./Datasets/breast_cptac_proteome.csv", header=TRUE
sep= "\t"
dec="."
row.names=1)
```

The matrix may contain missing values; in this case, imputation will be performed automatically within the rROMA analysis function. Alternatively, a preimputed matrix can be provided as input. Since the imputation of missing data is time-consuming, we preprocessed the matrix using the MICE package and saved the imputed version as an RData file.

The code used to impute missing values and saving the imputed matrix were the following:

```
FillNAMethod = list()
imp = do.call(what = mice, args = append(list(data
= matData.proteo),FillNAMethod))
matDataMice.proteo <- complete(imp)
save(matDataMice.proteo, file = "matDataMice_
proteo.RData")
```

We load the imputed proteomic matrix:

```
load("./Datasets/matDataMice_proteo.RData")
```

If available, we can load sample labels to enhance the visualization of analysis results. This file must be in txt format, with the rows containing the sample names and the columns containing the corresponding labels:

```
annot.proteo <- read.table(
"./Datasets/metadata
header= TRUE
sep= "\t")
```

sampleAnnot.proteo = annot.proteo$PAM50
names(sampleAnnot.proteo) = annot.proteo$SampleID
We created a list of customized colors for breast cancer subtypes to be displayed on heatmap.

```
annoCol <- list(PAM50 = c(Basal="red",
 Her2="pink",
 LumA="blue",
 LumB="lightblue",

"Normal-like"="forestgreen"))
```

To run the analysis, a module file in GMT format containing information on the pathways (also called modules) that need to be tested is required. The package has integrated the gene sets corresponding to the Hallmarks of cancer [9] in *Homo sapiens*, which can be called with the *SelectFromMSIGdb* function. The user can optionally upload a custom gene set GMT file via the function *ReadGMTFile*. Note that we chose the Hallmarks database for explanatory purpose, which includes 50 gene sets. To have a more complete view of dysregulated pathways, we recommend using rRoma with multiple databases such as Reactome and KEGG Pathway or others relevant to your own study.

```
gmtHallmark <- SelectFromMSIGdb("HALLMARK")
Optional: remove "HALLMARK_" from the names to
simplify labels
gmtHallmark <- lapply(gmtHallmark, function(x){
x$Name <- sub("HALLMARK_", "", x$Name)
x
})
```

Run the Analysis

You can now run rRoma with default parameters by simply specifying the proteome matrix and the modules you want to test. Here, we are fixing the seed to be used for random permutations. The algorithm accepts many hyperparameters that can be changed to better suit your needs. Detailed documentation about the hyperparameters is available on the GitHub section *"Performing Roma: changing hyperparameters"*.

```
set.seed(69)
rRomaProteo.output <- rRoma.R(matDataMice.proteo,
gmtHallmark)
```

View the Results

The output of the analysis is an object that contains multiple nested elements. The most important information can be found in the nested matrix called *ModuleMatrix*; here *rRomaProt.output$ModuleMatrix*. It contains p and q values corrected for multiple testing by the Benjamini and Hochberg method [17] for overdispersion (L1 parameter) and shift (Median Exp parameter) for all tested modules.

We are interested in two different types of modules:

- **Shifted modules**, whose genes behave differently from the rest of the genes in at least one sample. Corresponding p value is in **ppv Median Exp**. A q value is also calculated and given in q Median Exp. Here we will consider a module as shifted if $p < 0.05$. We do not look at q values here, as both the number of samples and the number of tested modules are small. Consider looking at q values for larger datasets.
- **Overdispersed modules**, for which the approximation to one PC is correct. Corresponding p value is in **ppv L1**. A q value is also calculated and given in q L1. Here we will consider a module as overdispersed if $p < 0.05$.

Shifted Modules

```
shifted.modules.proteo <- which(rRomaProteo.
output$ModuleMatrix[, "ppv Median Exp"] <= 0.05)
```

We want to see which samples are responsible for the shift. This can be done by looking at *Sample Scores*. As shown in Fig. 2, the function *Plot.Genesets.Samples* plots a heatmap of these scores:

```
Plot.Genesets.Samples(rRomaProteo.output,

Selected = shifted.modules.proteo, GroupInfo
= sampleAnnot.proteo, cluster_cols = TRUE,
GenesetMargin=8, annotCol=annoCol)
```

We can also visualize how much each module is shifted compared to the null distribution. In particular, we are interested in the PC1 median expression since we focus on shifted pathways:

```
Plot.Genesets.vs.Sampled(
rRomaProteo.output, Selected = shifted.modules.
proteo, Plot="PC1Median")
```

In Fig. 3, the most negatively shifted pathways correspond to "Epithelial mesenchymal transition," "Angiogenesis," "Allograft rejection," "Estrogen response early," and "Estrogen response late."

The most positively shifted pathways are "Coagulation" , "Interferon Alpha response", "E2F targets", "Myc target V2", "Oxydative phosphorylation,", "Interferon gamma response".

Overdispersed Modules

We can then focus on overdispersed modules. Note that overdispersion is calculated for all modules, including the shifted one. However, interpretation of overdispersion for shifted pathways might be

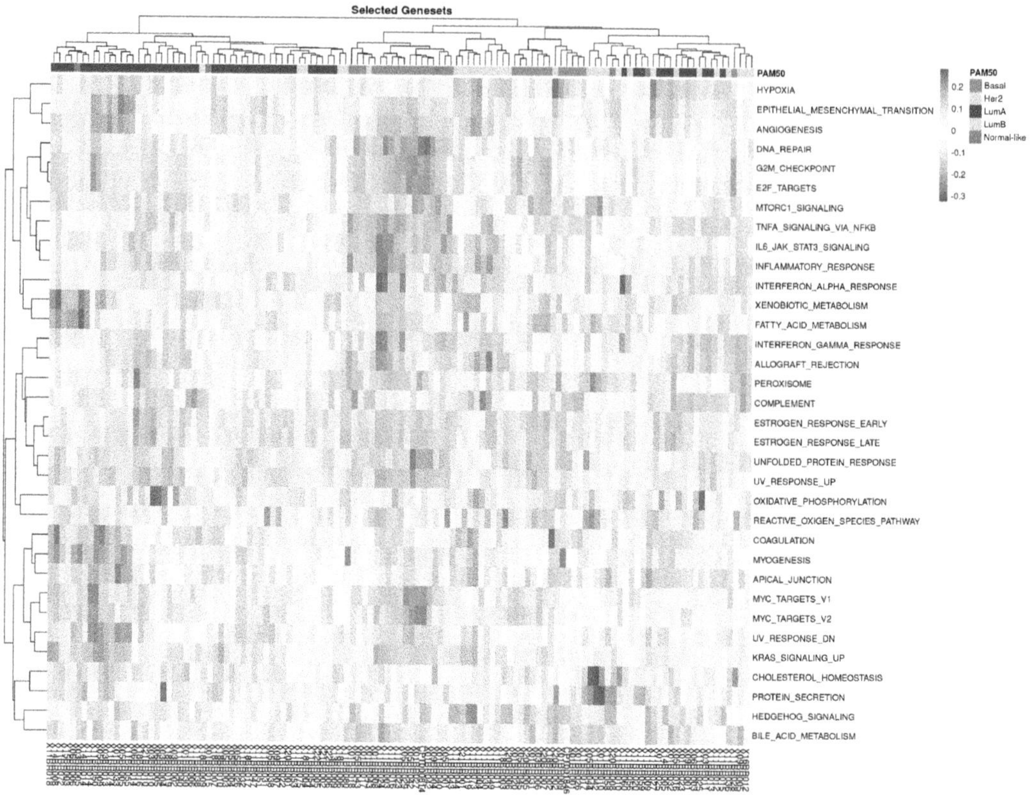

Fig. 2 Heatmap of activity scores for pathways identified as significantly shifted in the proteomic data

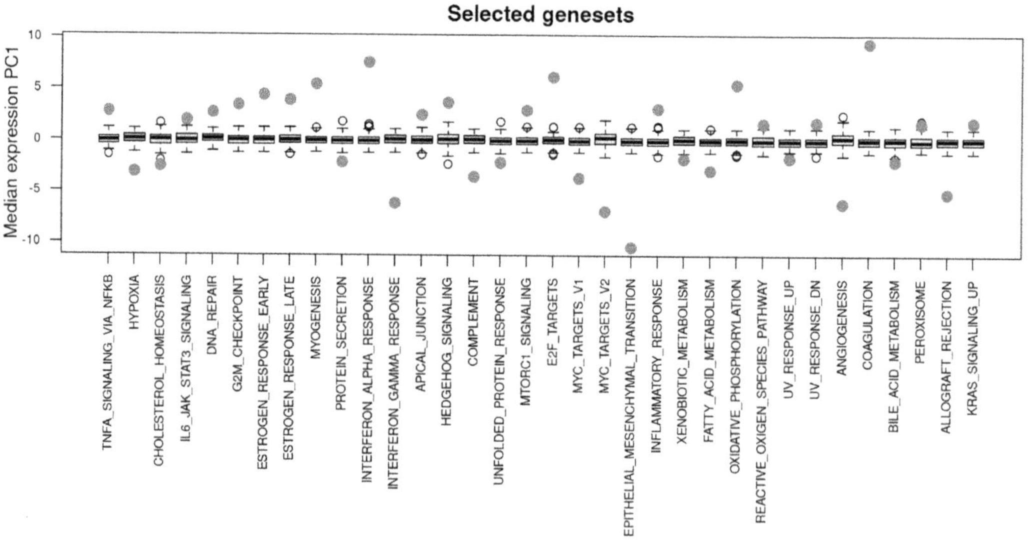

Fig. 3 PC1 median expression for significantly shifted pathways in the proteomic dataset. The boxplot corresponds to the distribution of median PC1 expression for random gene sets (null distribution), while the red dot corresponds to the value obtained for the tested pathway

misleading. Focusing on pathways that are overdispersed but not shifted will lead to more interpretable results.

```
overdispersed.modules.proteo=which(rRomaProteo.
output$ModuleMatrix[, "ppv L1"] <= 0.05 &
rRomaProteo.output$ModuleMatrix[, "ppv Median Exp"]
> 0.05)
```

For such pathways we are interested in both sample scores and gene scores. The former tells us in which sample the pathway activity is the highest, while the latter tells us which proteins are contributing the most to the activity score for this pathway. Overdispersed representation is particularly interesting in pathways that encompass both upregulated and downregulated proteins.

We first plot the same heatmap as before to visualize sample scores, as shown in Fig. 4:

```
Plot.Genesets.Samples(
rRomaProteo.output, Selected=overdispersed.modules.
proteo
GroupInfo=sampleAnnot.proteo
```

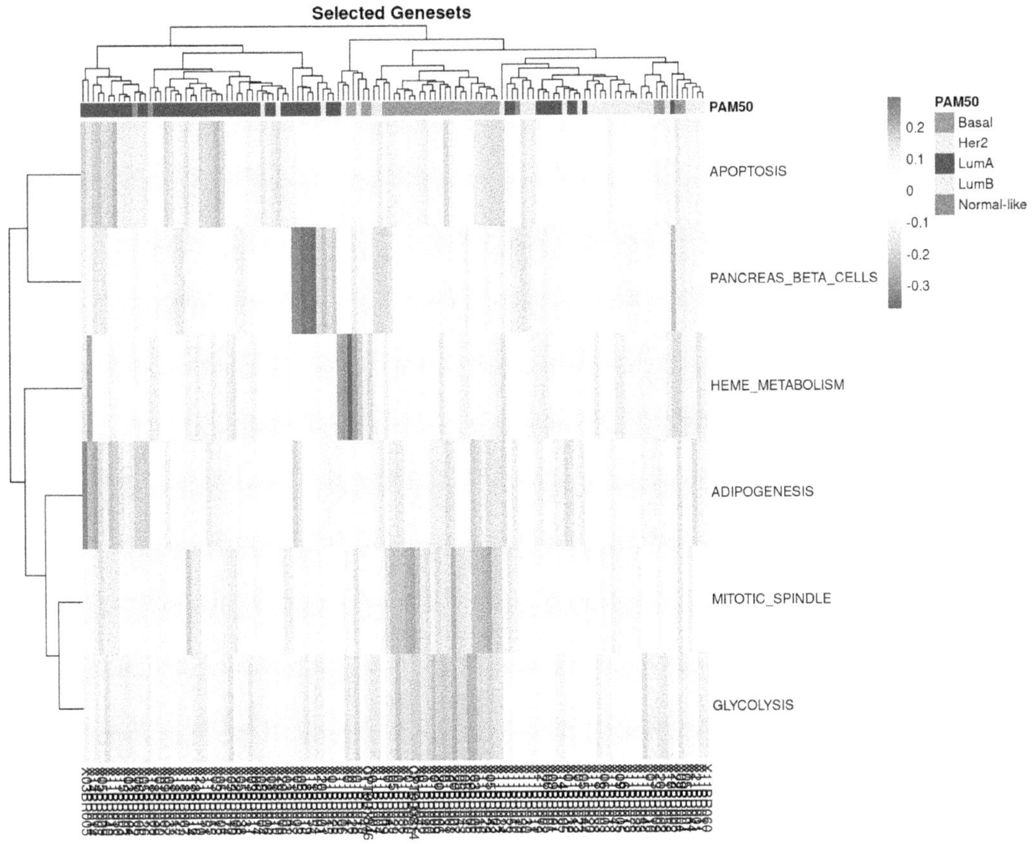

Fig. 4 Heatmap of activity scores for pathways identified as significantly overdispersed in the proteomic data

```
cluster_cols=TRUE
GenesetMargin=8
annotCol=annoCol)
```

Then we look at gene weights. To do this we use the *GetTopContrib* function. When using this function, the number of top contributing genes can be specified via the ratio of genes of the module (nGenes larger than 0 and smaller than 1) or via the number of top genes to return (nGenes integer and larger than 1).

Additionally, top contributing genes can be obtained via their correlation with the module score (Mode = "Cor") or via their weight (Mode = "Wei") (recommended).

To use correlation, the expression matrix needs to be specified (via the ExpressionMatrix parameter). The function will return a table (Table) with the computed information and the matrix used to produce the heatmap (VizMat).

The summary heatmap will be plotted only if Plot = TRUE.

Here, we will visualize the top 10% of genes that are most important within the module, as shown in Fig. 5.

```
GeneMatWei.proteo <- GetTopContrib(
 rRomaProteo.output,
 Selected = overdispersed.modules.proteo,
 nGenes = 0.1,
 OrderType = "Abs",
 Mode = "Wei",

Plot = TRUE)
```

Visualizing Modules in the PC Space

Among the most interesting plots to discriminate between shift and overdispersion is the PC1/PC2 projection of the genes of a given module against the background for null distribution. Such a plot is available with the *PlotPCProjections* function, as shown in Fig. 6.

```
modSelect=c(overdispersed.modules.proteo, shifted.
modules.proteo)
PlotPCProjections(
RomaData = rRomaProteo.output, Selected= modSelect
PlotPCProj = c('Points'))
```

Looking at the Orientation of the PCs

In the case of overdispersed modules that are not shifted, all components are computed with an undefined orientation sign. Several methods exist to try to orient the PCs (described in the GitHub section "*Choosing the PC orientation procedure*"). It is possible to visualize both gene weights and expressions for all genes in a given module in order to assess whether PC orientation is correct, as shown in Fig. 7.

```
PlotGeneWeight(
RomaData = rRomaProteo.output
```

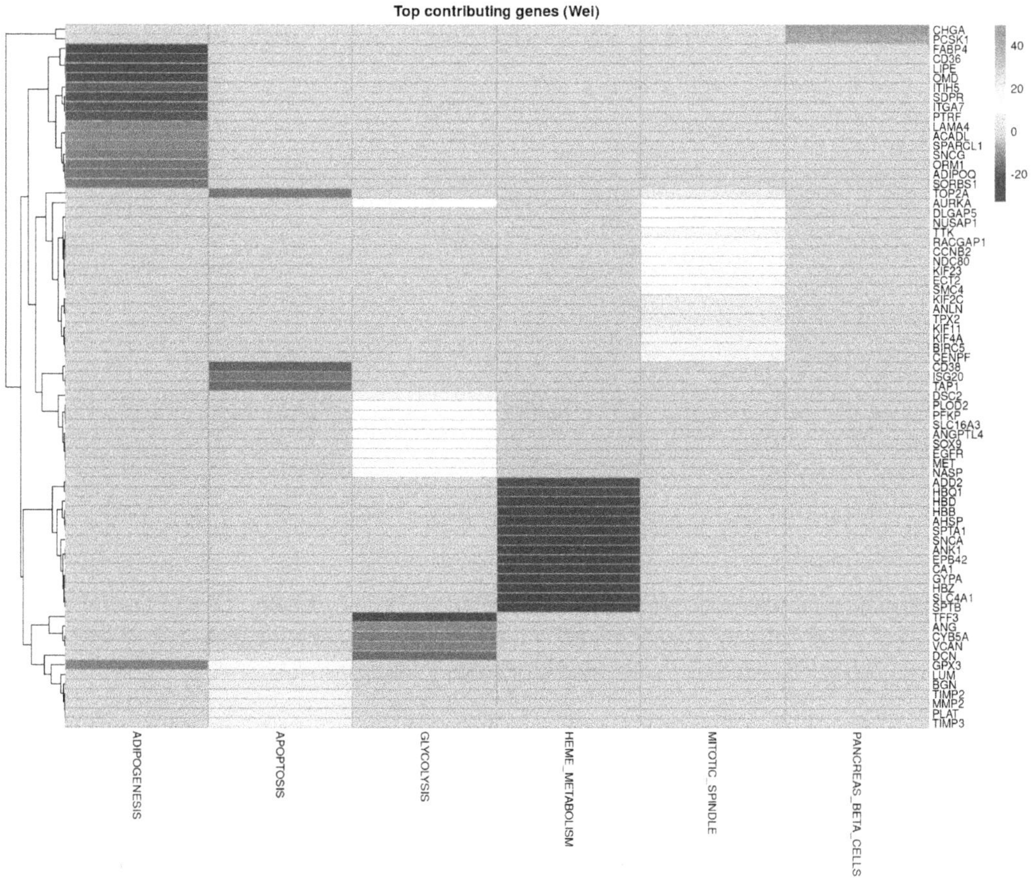

Fig. 5 Heatmap of the weights of the top contributing genes for pathways identified as significantly overdispersed in the proteomic dataset

```
PlotGenes = 30
ExpressionMatrix = matDataMice.proteo
LogExpression = FALSE
Selected = modSelect)
```

3.3 Predicting Active Kinases in Breast Cancer Subtypes from Phosphoproteomic Data

Below, we describe how rROMA is used to infer active kinases from a phosphoproteomic dataset. The steps are like those described for pathway analysis, but the required files and the interpretation of the results are specific to phosphoproteomic analysis.

Preparing and Loading Your Data Files

The required input is a normalized data matrix with column names indicating samples and row names indicating phosphosite IDs.

```
matData.phospho=read.table(
"./Datasets/breast_cptac_phospho.csv"
header=TRUE
sep="\t"
```

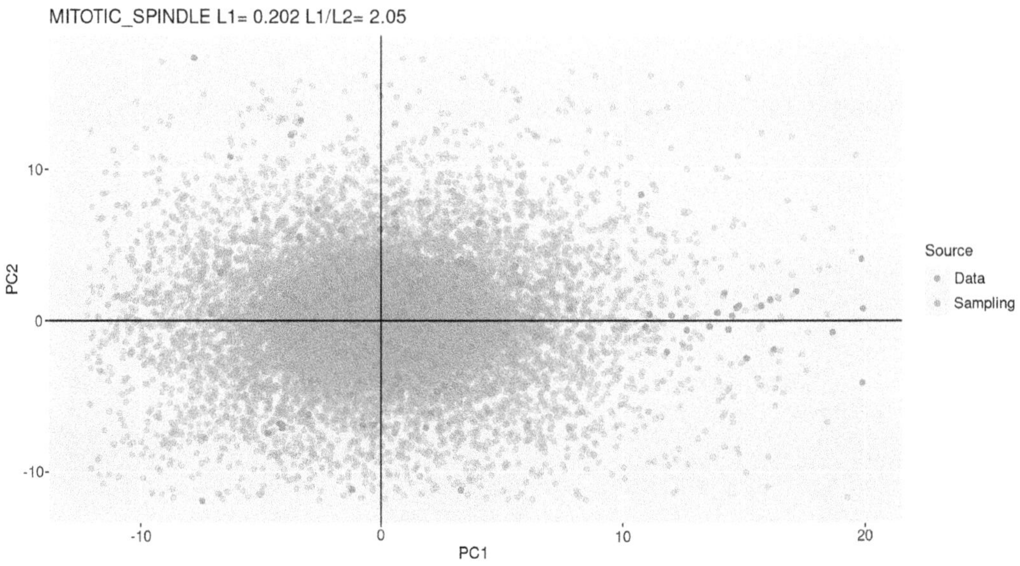

Fig. 6 Plot illustrating the contribution of proteins to the MITOTIC SPINDLE pathway activity score. The proteins of the MITOTIC SPINDLE pathway are represented in the PCA space. Red dots are proteins from the pathway and blue dots show randomly selected genes used to generate a null distribution

```
dec=","
row.names=1)
```

In case of a high rate of missing values, it is preferable to filter the dataset. We will perform some filtering of phosphosites so that only phosphosites with quantification for at least 66% of the replicates are retained. It means that we want a maximum of 34% missing values per phosphosite. The applied filtering leaves us with 23,103 phosphosites. In line with the literature [18], a missing value proportion of up to one third permits the application of various imputation methods with reliable performance. A high rate of missing values can substantially influence the conclusions drawn from the data and may introduce serious issues, including a reduction in statistical power, which affects downstream functional analyses, and inadequate representation of samples ("data holes").

```
dim(matData.phospho)
[1] 38500 122
threshold<-34
NA_threshold_Phos=(ncol(matData.
phospho)*threshold)/100
matDataFilter.phospho= matData.phospho %>
% filter(rowSums(is.na(matData.phospho)) <
NA_threshold_Phos)
dim(matData.phospho)
[1] 23103 122
```

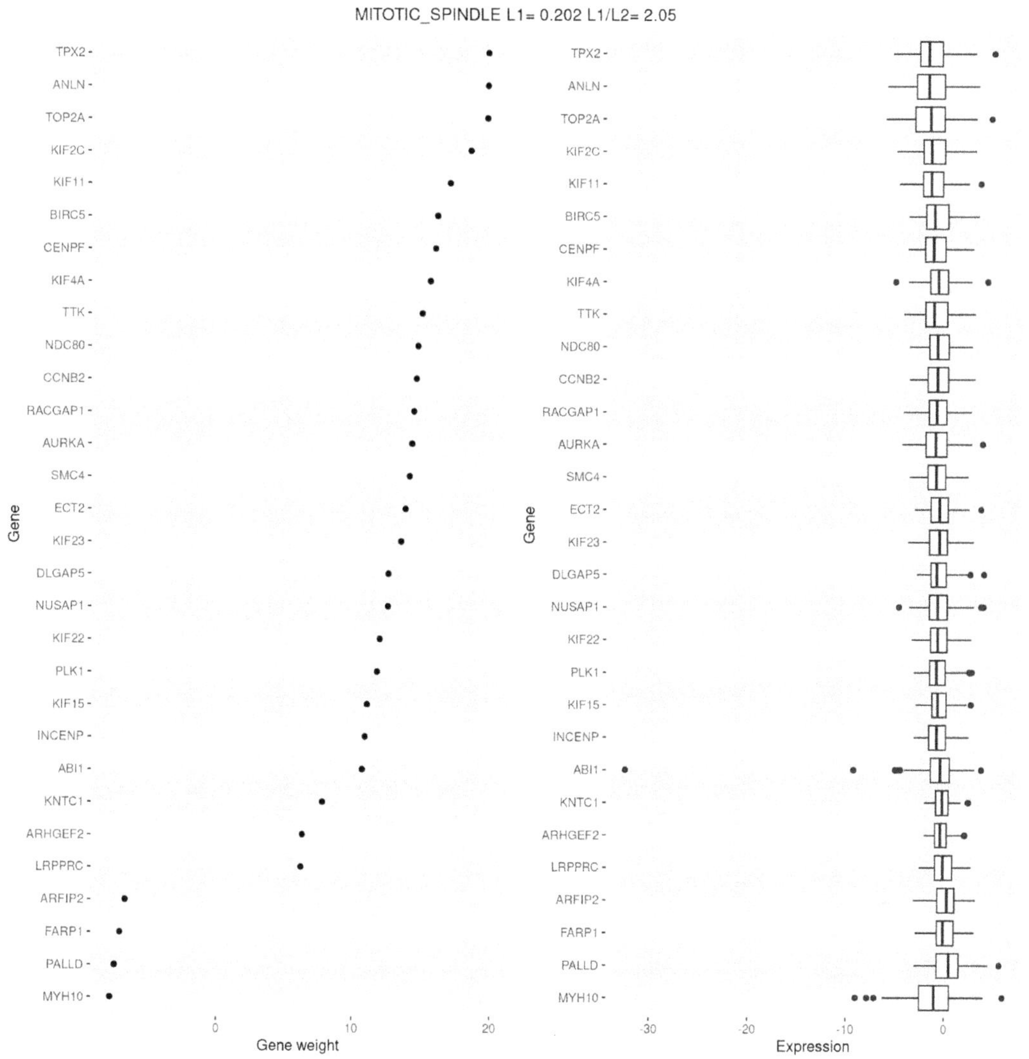

Fig. 7 Plot displaying gene weights (left panel) and corresponding expression levels (right panel) for all genes in the specified pathway

We can load the phosphoproteomics matrix already imputed with MICE:

```
load("./Datasets/matDataFilterMice.phospho.RData")
```

If available, we can load sample labels.

```
annot.phospho=read.table(
"./Datasets/metadata
header= TRUE
sep= "\t")
sampleAnnot.phospho<-annot.phospho$PAM50
names(sampleAnnot.phospho)<-annot.phospho$SampleID
```

To infer active kinases, rROMA requires a phosphosite annotation dataset in GMT format that maps substrates to their corresponding kinases. A manually curated database is PhosphoSitePlus [19], one of the most used libraries for kinase–substrate interactions, to link kinases to their downstream targets. However, only a limited number of phosphosites in these manually curated libraries can be linked to upstream kinases, which restricts the scope of kinase activity inference. For this reason, we have compiled a more extensive resource, called Kinase Substrate Relationship (KSR), that enables a more comprehensive analysis (see Sect. 3.4) including 701 kinases and 20,112 phosphosites.

Here, we import this database to use it in the analysis.

```
gmtKSR = ReadGMTFile("./Datasets/KSR_modPhosphoNet.
gmt")
```

Run the Analysis

We can now run rRoma by specifying the phosphoproteomic dataset and the kinase–substrate modules we want to test.

```
set.seed(69)
rRomaPhospho.output = rRoma.R(matDataFilterMice.
phospho,gmtKSR)
```

View the Results

The most important information can be found in the module matrix, here in *rRomaPhospho.output$ModuleMatrix*. It contains p and q values for overdispersion (L1) and shift (Median Exp) for all tested modules.

Shifted Modules

```
head(rRomaPhospho.output$ModuleMatrix)
shifted.modules.phospho=which(rRomaPhospho.
output$ModuleMatrix[, "ppv Median Exp"] <= 0.05)
```

We want to see which samples are responsible for the shift. This can be done by looking at Sample Scores. The function plots a heatmap of these score as shown in Fig. 8:

```
Plot.Genesets.Samples(
rRomaPhospho.output
Selected = shifted.modules.phospho
GroupInfo = sampleAnnot.phospho
cluster_cols = TRUE
GenesetMargin=8
annotCol=annoCol)
```

We can also visualize how much the module is shifted compared to random gene sets, as shown in Fig. 9. In particular, we are interested in PC1 median expression since we focus on shifted kinases:

```
Plot.Genesets.vs.Sampled(
rRomaPhospho.output
```

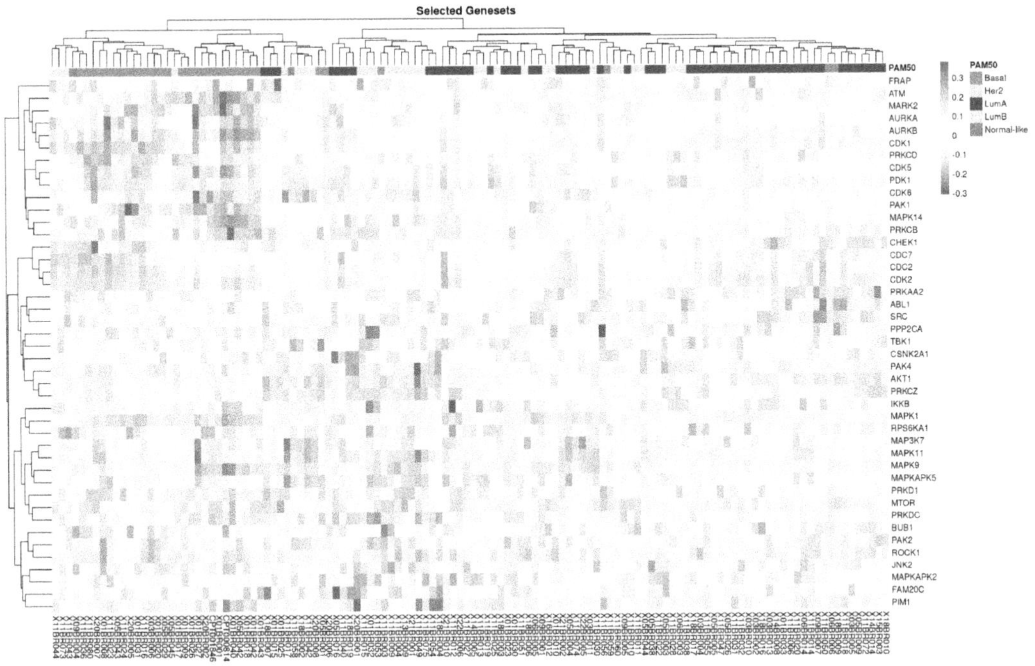

Fig. 8 Heatmap of activity scores for kinases identified as significantly shifted in the phosphoproteomic dataset

```
Selected = shifted.modules.phospho
Plot="PC1Median")
```

Figure 10 highlights the most positively shifted kinases: "PDK1," "ROCK1," "CDC2," "CDC7," "MAPKAPK5," "PAK2," and "CDK2." The most negatively shifted kinases are "ABL1," "FRAP," "SCR," "TBK1," "PRKCB," and "PRKCD."

Overdispersed Modules

overdispersed.modules.phospho=which(rRomaPhospho.output$-ModuleMatrix[,"ppv L1"] <= 0.05 & rRomaPhospho.output$-ModuleMatrix[,"ppv Median Exp"] > 0.05)

For such kinases we are interested in both sample scores and peptide scores. The former tells us in which sample the activity of the kinase is the highest, while the latter tells us which peptides are the most important for the activity score of this kinase.

We first plot the heatmap to visualize sample scores, illustrated in Fig. 10:

```
Plot.Genesets.Samples(
rRomaPhospho.output
Selected = overdispersed.modules.phospho
GroupInfo = sampleAnnot.phospho
cluster_cols = T
GenesetMargin = 8
```

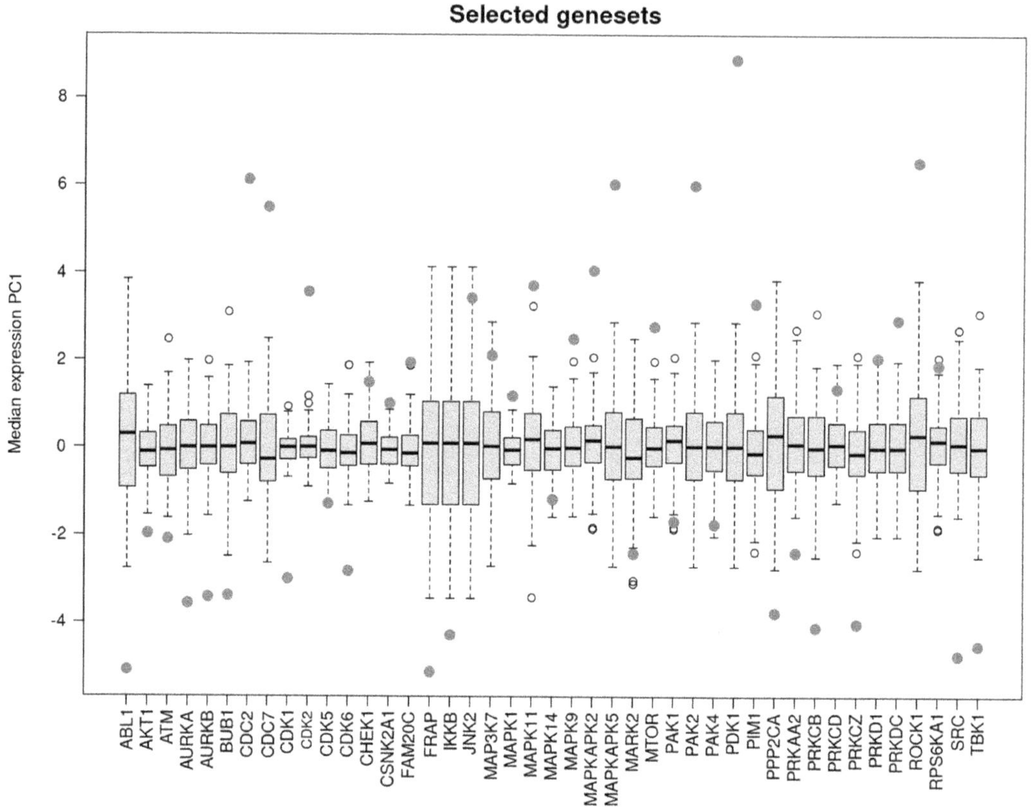

Fig. 9 PC1 median expression for significantly shifted kinases in the phosphoproteomic dataset. The boxplot corresponds to the distribution of median PC1 expression for random gene sets (null distribution), while the red dot corresponds to the value obtained for the tested kinase

```
annotCol=annoCol)
```

Then we look at phosphosite weights, as shown in Fig. 11. To do this we use the *GetTopContrib* function.

```
GeneMatWei.phospho=GetTopContrib(
rRomaPhospho.output
Selected = overdispersed.modules.phospho
nGenes = 0.1
OrderType = "Abs"
Mode = "Wei"
Plot = TRUE)
```

3.4 The Kinase Substrate Relationship (KSR) Resource

The impact of kinase–substrate libraries on inference methods has been evaluated recently in a benchmark study [20]. The choice of the knowledge resource impacts the inferred activities with a combination of manually curated libraries demonstrating superior performance in recapitulating kinase activities.

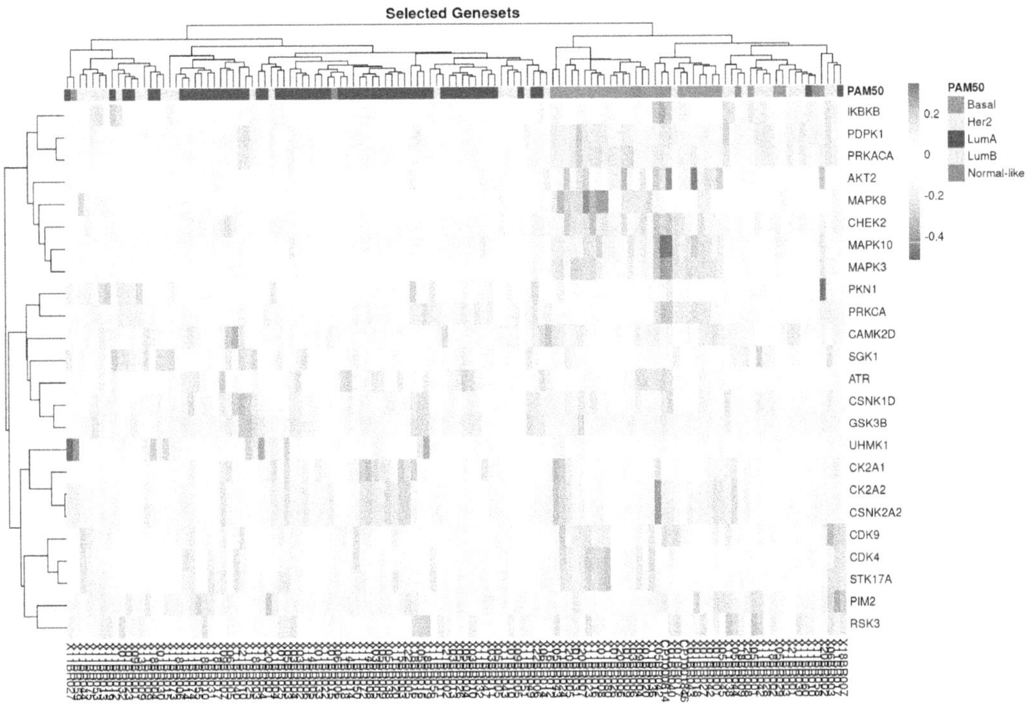

Fig. 10 Heatmap of activity scores for kinases identified as significantly overdispersed in the phosphoproteomic dataset

For this reason, we have exploited a published resource of kinase signaling knowledge bases [21], which outlines a network of phosphorylation site and kinase–substrate interaction databases.

We extracted information from the primary data source of multiple databases described in this study as the most interconnected and well-maintained databases and collected the information into a single resource that we have named Kinase Substrate Relationship (KSR). We have integrated and curated data from PhosphoSitePlus [19], SIGNOR [22], PhosphoELM [23], Swiss-Prot [24], and HPRD [25]. The following processing steps have been applied:

- Species identification: Homo sapiens (human) data have been selected.
- Protein IDs have been mapped to gene names according to UniProt: gene names associated with Homo sapiens are mapped to their corresponding entries in the UniProt database. The format of phosphosite IDs consists of the mapped gene name followed by the specific phosphorylation site (e.g., AKT1;S473-p), indicating serine 473 phosphorylation of the AKT1 protein.
- Filtering redundant information: redundant or duplicate information is filtered out to ensure the KSR resource contains only unique entries.

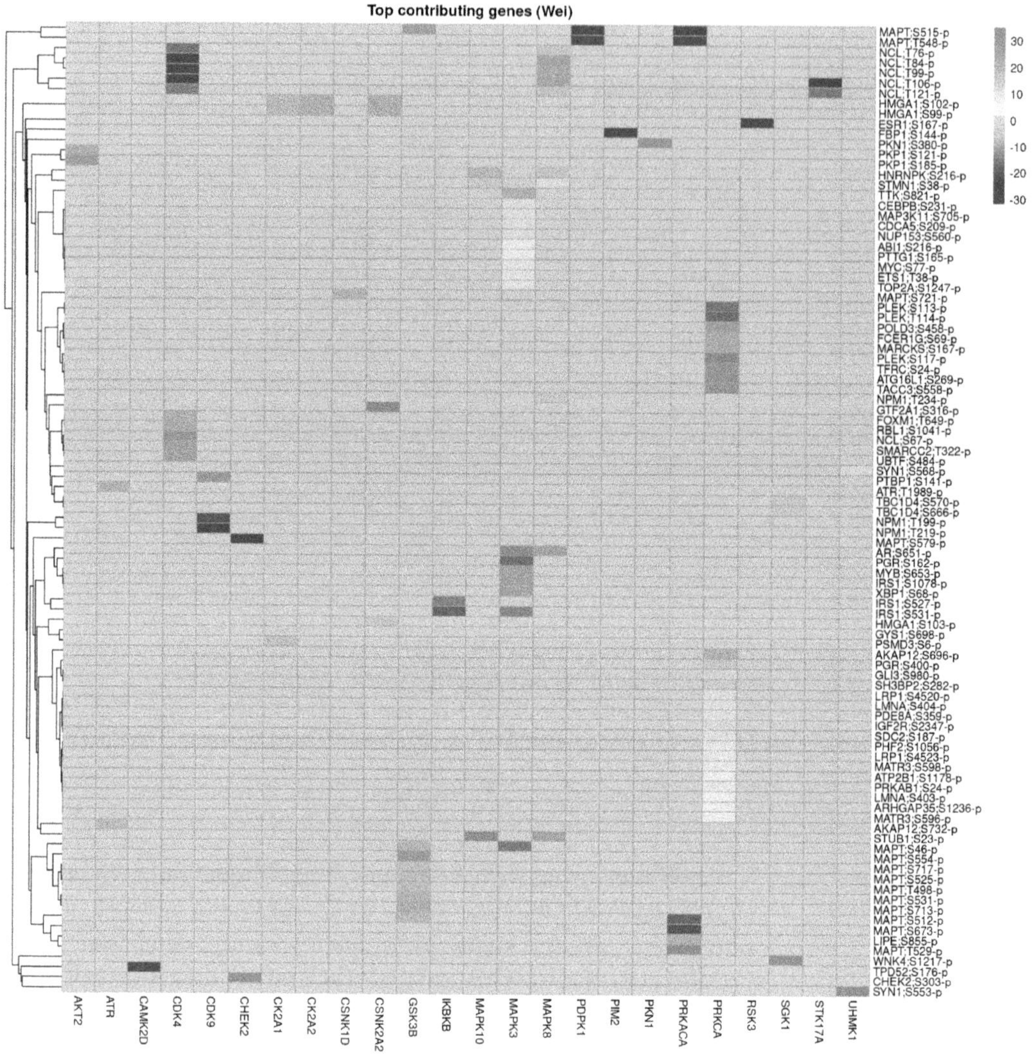

Fig. 11 Heatmap of the weights of the top contributing phosphosites for kinases identified as significantly overdispersed in the phosphoproteomic dataset

These processing steps are crucial for refining the dataset and preparing it for further analysis or downstream applications. This allowed us to obtain a resource that includes 701 kinases and 20,112 phosphosites.

4 Notes

1. Pathway analysis described in this chapter relies on the MsigDB v6.2 database. Users can update or expand the resource by uploading a custom .gmt file.

2. By default, analyses are performed on *Homo sapiens* data. However, other species can be supported if appropriate resources for pathway annotation and kinase–substrate annotation are provided. In this case, particular care must be taken to ensure the exact correspondence between the gene or protein IDs used in the input matrix and those in the pathway annotation file.
3. Coverage check: before running the rROMA analysis, it is important to verify that there is sufficient overlap between the input omics dataset and the pathway or kinase–substrate database entries and to ensure meaningful results. The parameters *MinGenes* and *MaxGenes* define the minimum and maximum number of features that must be present in the expression dataset for the module to be included in Roma's analysis. If coverage is insufficient, the software returns an error message.
4. Filtering of outlier samples: some samples may contain measurement errors that can bias rROMA results. By default, no sample filtering is applied, assuming valid input. However, outlier detection can be enabled via hyperparameters, triggering a two-step procedure: (i) checking for comparable numbers of detected genes and (ii) PCA-based identification of deviating samples. The number of PCs and the distance threshold are user-defined. Either step can be applied independently.
5. Filtering of outlier proteins: the calculation of PC1 may be distorted by the presence of an outlier protein in the dataset. Such an outlier can artificially influence the result. To enhance robustness, rROMA applies a 'leave-one-out' cross-validation strategy for PC1 computation as described in [26]. In the results, any identified outlier genes are tracked, allowing users to examine whether and which proteins were flagged as outliers.

Acknowledgments

This work was done with financial support from ITMO Cancer Aviesan within the framework of the 2021–2030 Cancer Control Strategy, on funds administered by Inserm (ITMO MIC project ModPhosphoNet).

Conflict of Interest The authors declare no conflicts of interest that are relevant to the content of this chapter.

References

1. Nefedov AV et al (2011) Bioinformatics tools for mass spectrometry-based high-throughput quantitative proteomics platforms. Curr Proteom 8:125–137
2. Wang K et al (2010) Analysing biological pathways in genome-wide association studies. Nat Rev Genet 11:843–854
3. Zhang Y et al (2022) Proteogenomic characterization of 2002 human cancers reveals pan-cancer molecular subtypes and associated pathways. Nat Commun 13(1):2669
4. Martignetti L et al (2016) ROMA: representation and quantification of module activity from target expression data. Front Genet 7:18
5. Najm M et al (2024) Representation and quantification of module activity from omics data with rROMA. NPJ Syst Biol Appl 10(1):8
6. Golub GH, Reinsch C (1970) Singular value decomposition and least squares solutions. In: Wilkinson JH, Reinsch C (eds) Handbook for automatic computation. Volume II: Linear algebra. Springer Berlin Heidelberg, Berlin, Heidelberg, pp 134–151
7. Edwards NJ et al (2015) The CPTAC data portal: a resource for cancer proteomics research. J Proteome Res 14(6):2707–2713
8. Krug K et al (2020) Proteogenomic landscape of breast cancer tumorigenesis and targeted therapy. Cell 183(5):1436–1456
9. Liberzon A et al (2015) The molecular signatures database hallmark gene set collection. Cell systems 1(6):417–425
10. Liberzon A et al (2011) Molecular signatures database (MSigDB) 3.0. Bioinformatics 27(12):1739–1740
11. Durinck S et al (2005) BioMart and bioconductor: a powerful link between biological databases and microarray data analysis. Bioinformatics 21(16):3439–3440
12. McCarthy DJ et al (2017) Scater: pre-processing, quality control, normalization and visualization of single-cell RNA-seq data in R. Bioinformatics 33(8):1179–1186
13. Ginestet C (2011) ggplot2: Elegant Graphics for Data Analysis. Journal of the Royal Statistical Society: Series A (Statistics in Society), 174(1):245–246
14. Wickham H et al (2022) devtools: tools to make developing R packages easier. https://devtools.r-lib.org/https://github.com/r-lib/devtools
15. van Buuren S, Groothuis-Oudshoorn K (2011) mice: multivariate imputation by chained equations in R. J Stat Softw 45(3):1–67
16. The UniProt Consortium (2025) UniProt: the universal protein knowledgebase in 2025. Nucleic Acids Res 53(no D1):D609–D617
17. Benjamini Y, Hochberg Y (1995) Controlling the false discovery rate: a practical and powerful approach to multiple testing. J Royal Stat Soc: B (Method) 57(1):289–300
18. Kong W et al (2022) Dealing with missing values in proteomics data. Proteomics 22(23-24):2200092
19. Hornbeck PV et al (2015) PhosphoSitePlus, 2014: mutations, PTMs and recalibrations. Nucleic Acids Res 43(D1):D512–D520
20. Müller-Dott S et al (2025) Comprehensive evaluation of phosphoproteomic-based kinase activity inference. Nat Commun 16(1):4771
21. Savage SR, Bing Z (2020) Using phosphoproteomics data to understand cellular signaling: a comprehensive guide to bioinformatics resources. Clin Proteomics 17(1):27
22. Perfetto L et al (2016) SIGNOR: a database of causal relationships between biological entities. Nucleic Acids Res 44.D1:D548–D554
23. Dinkel H et al (2010) Phospho.ELM: a database of phosphorylation sites-update 2011. Nucleic Acids Res 39(suppl_1):D261–D267
24. Gasteiger E, Jung E, Bairoch A (2001) SWISS-PROT: connecting biomolecular knowledge via a protein database. Curr Issues Mol Biol 3(3):47–55
25. Peri S et al (2004) Human protein reference database as a discovery resource for proteomics. Nucleic Acids Res 32(suppl_1):D497–D501
26. Hastie T, Friedman J, Tibshirani R (2001) The elements of statistical learning, Springer Series in Statistics, Springer New York, NY. https://doi.org/10.1007/978-0-387-21606-5

Check for updates

Chapter 10

Software Pipelines for Targeted and Untargeted Proteomics in Bioengineering Applications

Yuqian Gao and Aivett Bilbao

Abstract

This chapter presents concepts, guidelines, and data processing procedures for the analysis of targeted and untargeted proteomics data in bioengineering contexts. The pipelines have been successfully demonstrated in microbial systems such as *Rhodotorula toruloides*, *Pseudomonas putida*, and *Aspergillus niger*. The methods described in these workflows involve popular open-source software tools and downstream algorithms that enhance data annotation, quantification, and reproducibility. These procedures can be used to quantify metabolic pathway enzymes and validate engineered strains. While focused on bioengineering applications, most of these procedures apply to any proteomics study. The chapter uses published data and step-by-step guides to allow researchers unfamiliar with the approach to successfully execute the techniques.

Keywords Mass spectrometry, LC, SRM, DIA, DDA, Bioengineering, FAIR data, FAIR software

1 Introduction

Proteomics, the large-scale study of proteins in a biological system, provides critical insights into cellular mechanisms underlying biological systems [1, 2], including bioengineering and metabolic design. High-throughput mass spectrometry (MS)-based proteomics approaches combined with liquid chromatography (LC) separations have become indispensable in quantifying protein abundance and their modifications with high precision. In microbial bioengineering, proteomics enables the quantitative assessment of enzymes involved in metabolic pathways and facilitates the design–build–test–learn (DBTL) cycle that drives synthetic biology [3].

Targeted proteomics techniques such as selected reaction monitoring (SRM) provide sensitive and accurate quantification of defined proteins [4], ideal for hypothesis-driven studies such as quantifying specific enzymes in engineered microbes. In contrast, untargeted proteomics approaches using data-dependent acquisition (DDA) and data-independent acquisition (DIA) enable broad, discovery-oriented analyses to profile complex proteomes and detect unknown changes.

The pipelines described in this chapter combine best practices from SRM analyses for quantifying pathway proteins to untargeted

Frédérique Lisacek (ed.), *Bioinformatics for Mass Spectrometry*, Methods in Molecular Biology, vol. 3071,
https://doi.org/10.1007/978-1-0716-5519-1_10,

or discovery proteomics [5]. Figure 1 illustrates the differences between targeted and discovery proteomics and the kinds of spectra acquired in these workflow.

The integration of targeted SRM approaches with untargeted DDA and DIA workflow offers a powerful and complementary framework for quantitative and discovery-driven proteomics in biological systems. By following the procedures outlined in this chapter, researchers can confidently perform proteomics analyses that quantify key pathway enzymes, evaluate the performance of engineered strains, and uncover unanticipated molecular changes that inform the DBTL cycle. While focused on bioengineering applications, most of these procedures apply to any proteomics study.

2 Materials

The following software tools and datasets are publicly available (*see* **Note 1**) and required to complete the practical sections of this chapter:

1. A computer with Windows operating system.
2. FragPipe software downloaded and installed from https://fragpipe.nesvilab.org.
3. Skyline software downloaded and installed from https://skyline.ms/home/software/Skyline.
4. An FTP client such as WinSCP downloaded and installed from https://winscp.net.
5. Untargeted proteomics data. The proteomics data for this demonstration were acquired to study ash utilization by *Rhodosporidium toruloides* [6]. Using the FTP client, download the mzML files in the repository: ftp://massive-ftp.ucsd.edu/v05/MSV000091202.
6. Protein sequence database (FASTA file) downloaded from https://www.uniprot.org/proteomes/UP000016926.
7. Targeted proteomics data and Skyline project. The targeted proteomics data for this demonstration were acquired to study *Peudomonas putida* adaptation to different media or carbon sources [6, 7]. Download the study in the Panorama data repository: https://doi.org/10.6069/6j7y-t592.

3 Methods

3.1 Untargeted LC–MS/MS Pipeline

DDA and DIA untargeted or global pipelines acquire MS/MS spectra that are matched against protein databases or spectral libraries for peptide identification. In the context of bioengineering and genetic

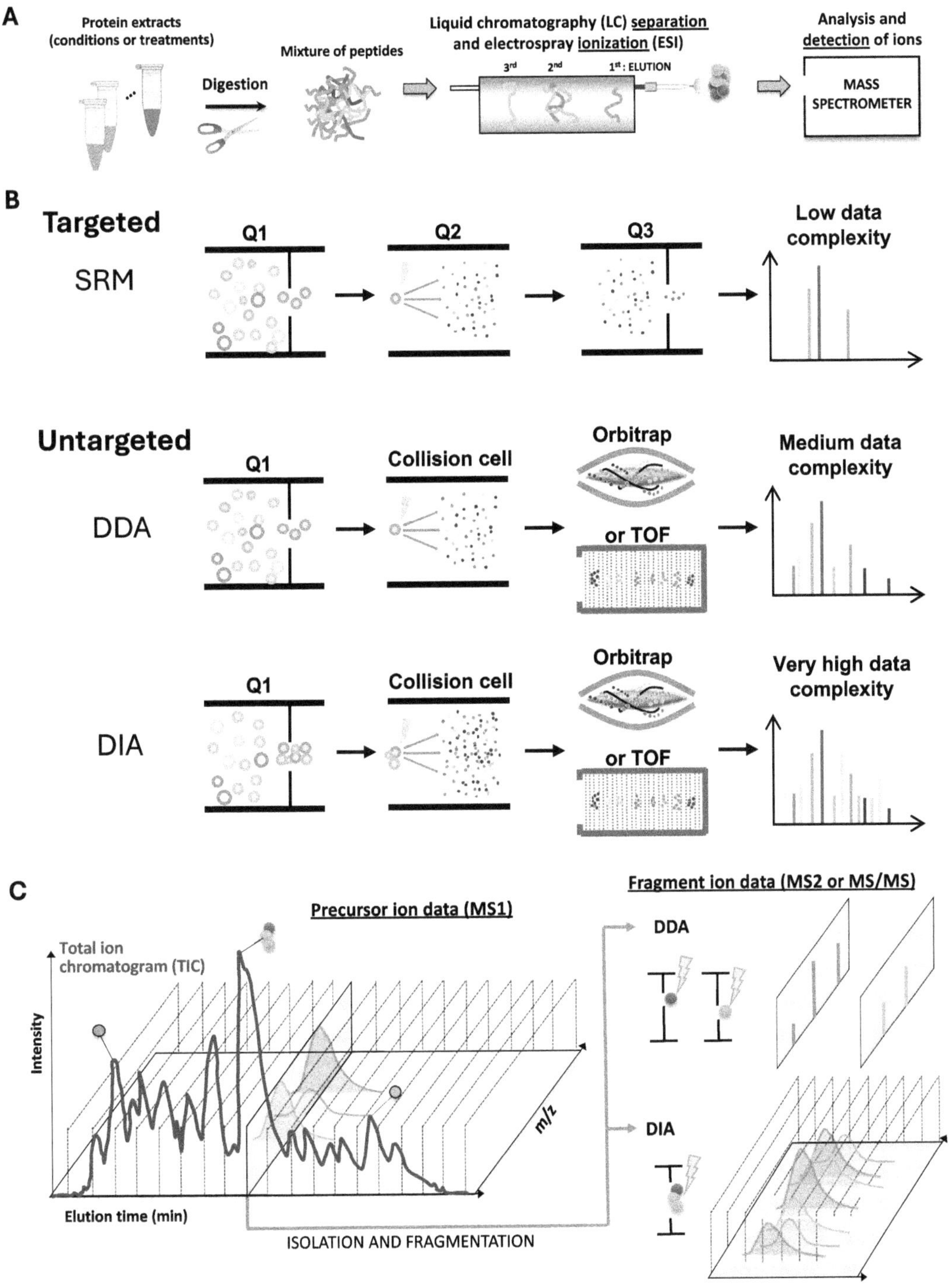

Fig. 1 Schematic diagram of targeted and untargeted proteomics approaches. (**A**) Common proteomics workflow typically begins with the digestion of proteins into peptides, often using trypsin, followed by subsequent analysis by LC–MS. (**B**) Instrumentation in targeted and untargeted approaches. SRM-based targeted proteomics utilizes triple quadrupole mass spectrometers and is focused on quantifying specific pathway enzymes in pathways of interest (e.g., glycolysis, TCA cycle), which are defined prior to the MS analysis.

modifications, untargeted proteomics offers valuable insights into cellular adaptations and engineered functionalities, including unexpected responses, requiring careful consideration of database configuration and search parameters:

Protein Database

In bioengineering applications, strains are frequently genetically modified to introduce new functionalities or improve metabolic pathways. Therefore, it is crucial to expand protein databases to include heterologous genes and engineered proteins. This would ensure a comprehensive proteomic analysis because proteins not included in the search database cannot be identified, even if they are present in the sample. By broadening the database, proteomics can detect genetically introduced proteins, even if only a fraction of them is expressed. In contrast, for genetically modified strains in which specific genes have been knocked out, it is acceptable to still include those knocked-out sequences in the protein database. This allows for comparisons across strains, accommodates incomplete knockout efficiency, and provides deeper biological insights.

Decoy and Contaminant Databases

False discovery rate (FDR) control is a critical step for ensuring the reliability of peptide and protein identification in untargeted proteomics. In DDA analysis, the FDR refers to the proportion of incorrect identifications (false positives) compared to the total number of identifications. It is used to assess confidence in spectra, peptide, and protein identifications. A commonly used approach to calculate the FDR is via a target-decoy search strategy. To do that, a decoy database (containing randomized, shuffled, or reversed sequences) is added to the target database. The rate at which decoys are matched serves as an estimate of the false positive rate. Typically, a 1% FDR is used in bioengineering applications, even though a 1%–5% FDR is generally acceptable in proteomics workflow.

In proteomics, contaminant databases are often included in the database searches. These contaminants are typically introduced during sample preparation, handling, or MS analysis, and they can interfere with the accuracy of the downstream analysis. Including contaminant databases helps reduce false positives caused by the presence of unrelated proteins. Common contaminants in proteomics

Fig. 1 (continued) Untargeted proteomics by DDA or DIA employs high-resolution MS, such as Orbitrap or TOF analyzers, with the goal to profile global proteome changes and discover novel responses. (**C**) Both DDA and DIA MS data typically contain precursor (MS1) and fragment (MS2) ion spectra (dashed lines) within a single LC–MS analysis. A first representation of the data is the MS1 total ion chromatogram (TIC) trace in blue, which is the sum of all ions detected per spectrum. Examples of MS2 spectra are shown. In DDA, the top-*n* most intense ions (red and blue LC–MS peaks) are selected for isolation and fragmentation to collect individual MS2; the low abundance one (green LC–MS peak) is undetected in this example. In DIA, multiple coeluting ions are fragmented simultaneously, collecting continuous fragment ion traces in convoluted or multiplexed MS2 spectra. Signals from the three fragmented peptide ions are detected, including the low abundance one (green LC–MS/MS peaks). Figure adapted with permission of Elsevier include proteins frequently found in proteomics workflow, such as keratins, trypsin, and bovine serum albumin. Even though these contaminants are included in the database searches, they are excluded from the final data output, making biological interpretation cleaner and more focused on the samples of interest.

Search Parameters

- Enzyme and Cleavage Specificity: Most proteomics experiments use trypsin as the digestion enzyme, assuming cleavage at the carboxyl side of lysine (K) and arginine (R) residues. When proline (P) is followed by K or R in the peptide sequence, cleavage by trypsin is inhibited, so exceptions for "KP" and "RP" need to be set in the search parameters. There can be variations in enzymatic digestion efficiency during sample preparation, which result in peptide bonds that remain uncleaved after enzyme digestion. To account for this, "missed cleavage" settings are configured to determine how many missed cleavage events are allowed during database searching and peptide identification. Most workflows allow for up to four missed cleavages during database searching (*see* **Note 2**).
- Modifications: The selection of modifications highly depends on the experimental design and biological context. Common modifications include static alkylation of cysteine residues, dynamic oxidation of methionine, dynamic acetylation of N-terminus, etc.
- Precursor and Fragment Mass Tolerance: Modern DDA workflows typically use high-resolution mass analyzers, such as Orbitrap or time-of-flight (TOF) instruments. Accurate mass tolerance settings are essential for minimizing mismatched identification and improving confidence in peptide-spectrum matches (PSMs). A precursor mass tolerance and fragment ion tolerance of ±20 parts per million (ppm) is often used for accurate assignment of spectra to peptide sequences.
- Match Between Runs: Tools leverage "match between runs" settings for improving identification consistency across multiple experiments. This feature aligns retention times and matches peptides across samples based on their precursor and fragmentation patterns, boosting detection sensitivity.

Running a DDA Search

There are numerous software tools available for processing DDA proteomics data, such as MSGF+ [8], MSFragger [9], and MaxQuant [10]. These tools are instrumental in identifying and quantifying proteins in complex biological samples. Although these tools employ different algorithms, the described shared best practices and settings in the previous sections ensure consistent, high-quality results.

Here the FragPipe software will be used, which integrates MSFragger and other tools into a complete workflow for a variety of proteomics experiment types. The software configuration can be customized through its graphical user interface. The general steps involve loading your data, selecting the workflow, specifying search parameters, and running the analysis.

1. Launch FragPipe. Open the FragPipe application on computer.
2. Go to the Workflow tab.
3. Load the DDA built-in Workflow. Select the Label-Free Quantification (LFQ-MBR) workflow from the drop-down menu, which is commonly used for DDA data analysis.
4. Load LC–MS Files. Locate the input files panel. Click Add files to select the raw or mzML spectral files downloaded in Materials (Sect. 2). Ensure the Data Type column for all added files is set to DDA.
5. Assign appropriate Experiment and Bioreplicate numbers to organize the results.
6. Go to the Database tab.
7. Load the FASTA file containing the protein sequences to search against, downloaded in Materials (Sect. 2).
8. Ensure options like Add decoys (for FDR estimation) and Add common contaminants are selected.
9. Configure Search Parameters (optional).
10. Tabs like MSFragger, MSBooster, Percolator, IonQuant, etc., allow you to fine-tune parameters such as modifications, mass tolerances, and quantification settings. For a standard DDA search, the default parameters in the loaded workflow are often a good starting point.
11. Go to the Run tab.
12. Specify the Output directory where all results will be saved.
13. Click the Run button to start the DDA search and analysis pipeline.
14. Once the process is complete, the results, including PSM, peptide, and protein reports, will be available in the specified output directory.

PostSearch Workflow

After performing database searches, several important postprocessing steps should be undertaken to ensure accurate interpretation and utilization of the results:

- Quality Control and Filtering: Critical postsearch steps in proteomics workflow to ensure accuracy and reliability in identified proteins and peptides. As mentioned above, typically a 1% FDR threshold is typically set to exclude incorrect PSMs or false identifications (*see* **Note 3**). Contaminants also need to be removed because they are unrelated to the experiment.

Additional filtering, like mass accuracy tolerance (ppm), will also help reduce false discovery.

- Normalization: An essential step for quantitative proteomics workflow, ensuring that protein intensity measurements are not biased by experimental variability, technical noise, or sample-to-sample variations. Proper normalization enhances comparability and statistical robustness of quantification results. Methods such as global median centering [11] are typically used to normalize peptide or precursor ion intensities across all samples. Additional batch correction or cross-run matching can also be used to improve quantification and reduce bias.
- Protein rollup: The process of aggregating peptide intensities to represent corresponding protein-level abundance quantitatively. Protein rollup can be performed in various ways, such as RRollup [12] and MaxLFQ [13].

Untargeted DIA LC–MS/MS Pipeline

DIA is designed for comprehensive proteomic profiling, enabling the simultaneous acquisition of fragment ion spectra from all detectable precursor ions within a given m/z range. Unlike DDA, DIA eliminates the bias of precursor selection, ensuring high reproducibility and deeper coverage of the proteome. It requires robust computational tools to decode complex datasets [5, 14].

Spectral libraries are critical for identifying peptides and proteins in DIA-based workflows. These libraries serve as reference datasets, allowing DIA fragment ion spectra to be matched to known precursor-fragment ion pairs. Spectral libraries are typically generated from DDA experiments with high-resolution spectra for known peptides. Some software tools enable library-free DIA analyses by utilizing theoretical fragmentation spectra and leveraging high-end computational methods.

Software tools such as DIA-NN [15], OpenSWATH [16], and Skyline [17] are nowadays widely utilized for analyzing DIA-based proteomics data because of their robust capabilities and tailored workflow. These tools integrate advanced algorithms for spectral matching, quantification, and FDR control, ensuring high-confidence peptide and protein identification. Each software employs its own unique workflow, offering varied features such as library generation, library-free searching, and real-time data processing for increased flexibility and precision. Similar to DDA, a well-curated protein database is essential for DIA analysis. The inclusion of engineered proteins, heterologous sequences, and knockout genes in the database is particularly critical for bioengineering applications. The selection of precursor and fragment mass tolerances, the number of missed cleavages, trypsin cleavage specificity, and the inclusion of peptide modifications in DIA workflow are similar to those used in DDA settings.

3.2 Targeted SRM Pipeline for Pathway Quantification

SRM-based or targeted proteomics is a powerful MS technique designed for the precise quantification and validation of specific proteins or peptides within complex biological samples, which are known a priori. SRM operates by selectively monitoring predefined precursor-to-fragment ion transitions, ensuring high sensitivity and reproducibility in detecting targeted peptides. This method is particularly suited for bioengineering and genetic modifications, where quantifying proteins is involved in pathways of interest. SRM-based targeted proteomics requires careful assay development, transition optimization, and rigorous validation to achieve accurate and high-specificity results, making it a gold standard for hypothesis-driven proteomic experiments.

The following outlines the pipeline workflow, including peptide selection, assay development, data processing, quality control, and data interpretation.

Peptide Selection and Assay Development

Peptide selection is the first step in SRM-based targeted proteomics. Strict criteria are used to choose peptides that are unique to the targeted protein to avoid interference from homologous or overlapping proteins. Peptides of strong MS signals evidenced by previous untargeted studies or predicted by software tools such as PREGO [18] are chosen. Ideally, the more suitable peptides are of 7–25 amino acids in length and not too hydrophobic or hydrophilic. Peptides that are prone to missed cleavages or chemical instability (e.g., methionine oxidation) should be avoided (*see* **Note 4**).

Heavy isotope-labeled internal standard peptides are commonly synthesized and spiked into the sample peptide digest to serve as a reference for SRM monitoring. To develop the SRM assays, heavy isotope-labeled internal standard peptides are spiked in a representative sample peptide digest, and all potential precursor-fragment ion transitions are monitored by LC-SRM. The top three precursor-fragment ion transitions with the highest signal strength and minimal interference are selected. The collision energy can be calculated by the Skyline software.

Data Analysis

SRM data are commonly analyzed using the Skyline [19] software. Raw data files are imported into Skyline, followed by manual inspection to ensure proper peak assignment and adjust peak boundary selection if necessary. Peak integration is automatically executed by the software. After integration, peptide quantification is carried out by calculating the ratio of light peptide intensities relative to their isotopically labeled heavy peptide standards. Protein rollup from peptide ratios is generally performed following similar approaches as untargeted proteomics.

Inspecting SRM Data

The following steps allow visualization of SRM raw data using Skyline.

1. Launch Skyline. Open the Skyline application on computer.

2. Open the Skyline project file downloaded in Materials (Sect. 2).
3. Skyline will load the Targets tree, which can be navigated on the left panel: proteins → peptides → transitions.
4. Open to overlay all runs: View → Chromatograms → Replicate Comparison.
5. Open to see transitions separated by ion type (precursors/fragments): View → Chromatograms → Split Graph.
6. Visualize Peak Areas (Quantification): View → Peak Areas → Replicate Comparison.
7. Visualize Mass Errors: View → Mass Errors → Histogram.
8. Export Reports. Use File > Export > Report to access custom reports.
9. Select Report Type: Choose pre-built reports like "Quantitation" to export tabulated peak abundances in comma-separated values (CSV) format or create a custom report to include specific data like retention times, intensities, or other metrics for downstream analysis.

3.3 AI and Omics-Agnostic Tools

As MS instrumentation, software, and artificial intelligence (AI) methods continue to advance, proteomics together with other omics techniques will enable increasingly comprehensive, high-resolution views of natural and engineered biological systems.

Omics-agnostic tools able to process various types of MS-based omics datasets, such as proteomics, metabolomics, lipidomics, and glycomics, and from different MS instrumentations could advance multiomics applications and contribute to improve standardization, harmonization, and reproducibility of results. For instance, Skyline was originally designed for proteomics and was later adapted to support metabolomics [20]. A recent omics-agnostic tool developed at Pacific Northwest National Laboratory (PNNL) moving in this direction is PeakQC [21], which provides user-friendly, accessible, and comprehensive quality control (QC) for raw MS data from various instrumentations, including LC–MS, DDA, DIA, ion mobility-MS (IM-MS), and infusion-MS data. PeakQC has been expanded into the IonToolPack suite (https://github.com/pnnl/IonToolPack), which houses various algorithm and software tools implemented in Python to provide raw data QC and visualization, and prototype tools for annotation and quantitation in progress. Raw MS data from various instrument vendors are automatically converted to MZA [22] format in a way that is transparent to the user, which is the base layer supporting the development of omics agnostic software workflow at PNNL.

IonToolPack provides a simplified and intuitive graphical user interface with each tool organized into its own tab with a single text box for parameter input, which avoids large and complicated sets of controls (dropdowns and checkboxes) that hamper reproducibility and automation. Other tools in progress incorporate modern AI approaches. While originally demonstrated for metabolomics [6], the new version of PeakDecoder enables sophisticated molecular annotation and quantitation of multidimensional MS datasets (LC–IM–DIA–MS) by adapting computer vision AI. It accesses raw MS data directly in instrument format using MZA and can annotate protein, metabolite, and lipid signatures in an automated fashion, which will enable holistic and iterative multiomics analyses [23]. PTMdiscoverer incorporates generative AI to detect and annotate potential posttranslational modifications in proteomics data [24].

3.4 Compelling Case Studies in Bioengineering

Untargeted Proteomics Decipher the Molecular Mechanisms

Rhodotorula toruloides is a promising yeast for industrial bioprocessing because of its ability to convert lignocellulosic hydrolysates into valuable biofuels and bioproducts. Bisabolene is a sustainable aviation fuel precursor produced via engineered pathways in *R. toruloides*. However, feedstock variability, particularly in corn stover quality and composition, significantly influences microbial growth and bioconversion efficiency. In this study [25], researchers examined the effects of corn stover heterogeneity, stemming from anatomical fractions (stalk, husk, cob, whole stover) and storage conditions (self-heated biomass), on bisabolene production. Untargeted DDA global proteomics revealed key differences in protein expression across feedstocks. For example, there were significant differential expressions of proteins related to lipid metabolism, energy production, and inorganic ion transport, with notable downregulation of iron-sulfur (Fe-S) cluster proteins involved in critical biological processes, suggesting that iron played a vital role in bisabolene production from lignocellulosic feedstocks (see Fig. 2).

Targeted Proteomics Quantify the Gene Expression

Pseudomonas putida is a desirable organism for industrial bioprocessing with its versatile catabolism and tolerance to inhibitory chemicals. Stable chromosomal integration and optimized expression of heterologous genes are crucial to maintaining its metabolic adaptability and to ensure efficient production. In this study [26], researchers investigated the optimization of chromosomal insertion loci in *P. putida* KT2440 for heterologous DNA expression of a muconate transporter gene (MucK). MucK is a transmembrane importer protein responsible for the uptake of muconate. Targeted proteomics were performed to compare MucK expression levels among strains with MucK inserted at different loci. It revealed that the insertion site significantly affected growth rates and protein expression, with some loci exhibiting an inverse relationship between growth rate and MucK expression (Fig. 3).

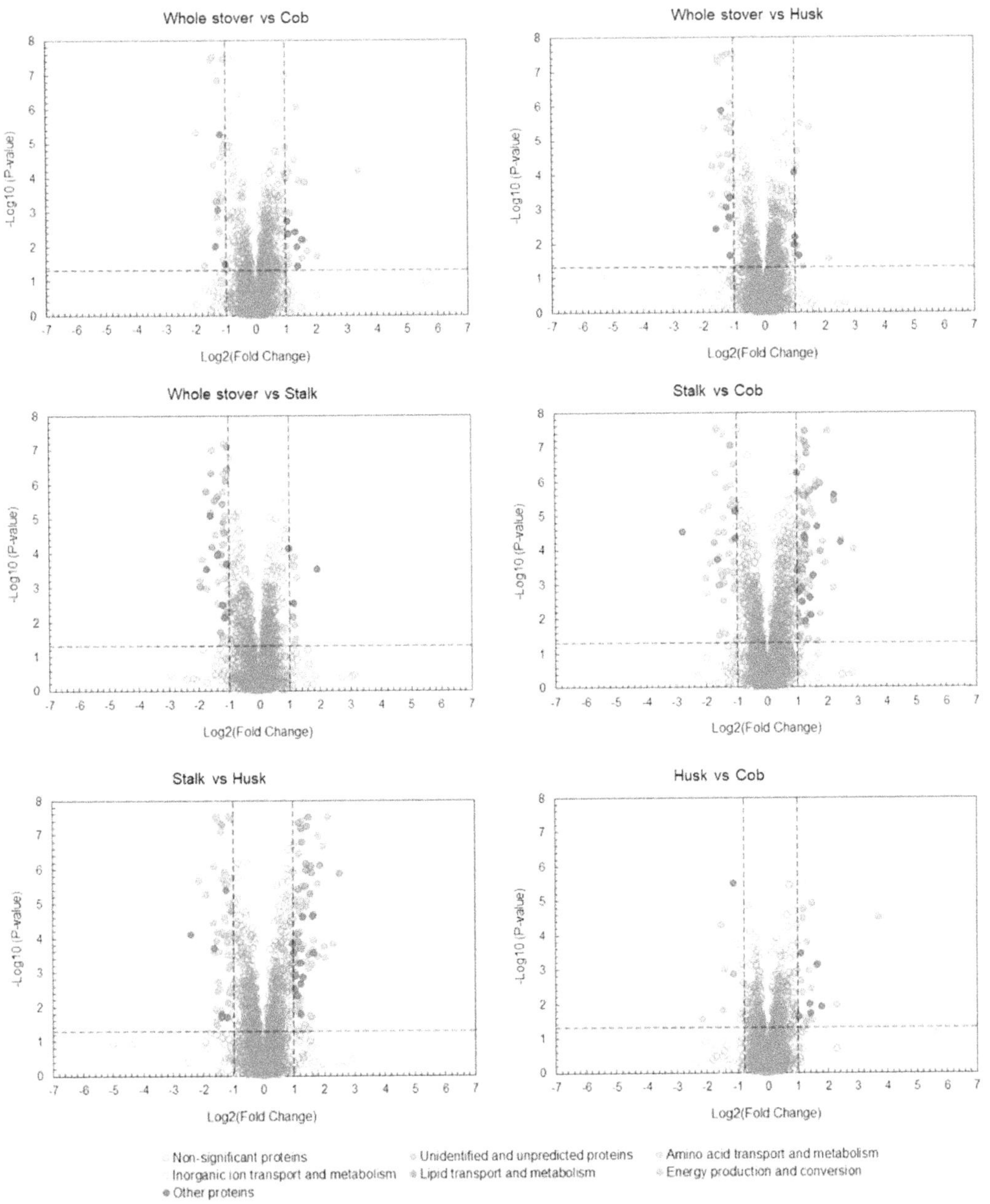

Fig. 2 Untargeted proteomics study of *R. toruloides* strains. Distribution of significantly upregulated and downregulated differentially expressed proteins, with an emphasis on proteins related to energy production and conversion, inorganic ion transport metabolism, and lipid transport metabolism, for their possible influence on bisabolene production. Plots of logarithmic fold change versus significance (i.e., volcano plot) show a pairwise comparison of protein expression of *R. toruloides* after 48 h of growth on hydrolysates from the four different anatomical fractions. In each volcano plot, the dotted horizontal line corresponds to a significance value of $p < .05$. The dotted vertical lines bound the minimal fold change for the most differentially expressed genes. To the right and left of the dotted vertical lines and above the horizontal line, functional categories for differentially expressed proteins are shown with solid circles in different colors (significantly upregulated and downregulated proteins). Reprinted with permission from the Journal of Industrial Microbiology & Biotechnology

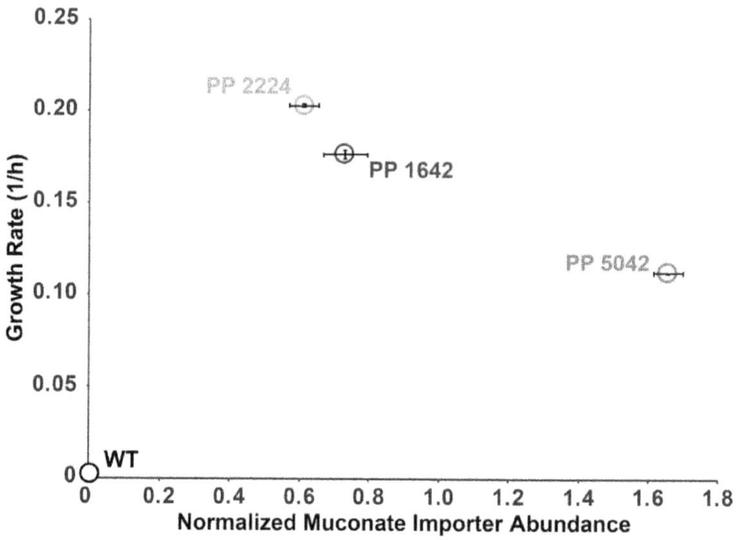

Fig. 3 Targeted proteomics study of *Pseudomonas putida* strains. *P. putida* KT2440 strains with a muconate transporter gene (MucK) inserted at three different loci were grown in M9 minimal medium supplemented with 30 mM cis, cis-muconic acid. Targeted proteomics quantified the amount of MucK in each strain (Y-axis error bars represent the standard deviation of three replicate samples) and is plotted with respect to the growth rate as measured by optical density at 600 nm (X-axis error bars represent standard error of four replicate samples).Reprinted with permission from the Metabolic Engineering Communications

4 Notes

1. Findable, Accessible, Interoperable, Reusable (FAIR) resources: The adoption of FAIR data and software practices [27] is paramount to ensure transparency, reproducibility, and long-term scientific value. Throughout the chapter, examples provided from our previous studies illustrate practical steps researchers can take to make their work FAIR-compliant:
   - Storing raw and processed data in instrument or open formats such as mzML and mzTab.
   - Depositing datasets in recognized ProteomeXchange repositories (e.g., PRIDE, MassIVE).
   - Providing complete metadata and parameter files.
   - Citing and contributing to open-source software that underpins modern proteomics analysis.
2. Enzyme specificity and missed cleavages: Issues related to enzymatic digestion are among the most common sources of

variability in proteomics workflows. Although trypsin is highly specific for cleaving at the C-terminal side of lysine (K) and arginine (R), several factors can compromise digestion efficiency and affect downstream peptide identification:

- Incomplete digestion: Variability in trypsin activity, suboptimal buffer conditions, or insufficient incubation time can leave peptide bonds uncleaved. This manifests as an excess of longer peptides, unusually large precursor masses, or identification results enriched in peptides containing internal K/R residues that should have been cleaved. Allowing up to two to four missed cleavages during database searching accommodates realistic digestion variability. Too few will under-identify peptides; too many can expand the search space and increase false positives.
- Blocked cleavage sites: When a proline (P) follows K or R (i.e., "KP" or "RP"), trypsin cleavage is sterically hindered. If search parameters do not account for these exceptions, spectra from legitimate peptides may fail to match the database, appearing as low identification rates or unexpected "no hit" spectra. Ensuring that search parameters explicitly encode proline-inhibited cleavage rules will avoid systematic under-identification of these peptides.
- Nonspecific or over-digestion: Conversely, harsh conditions or excess enzyme can lead to semi-tryptic or nonspecific cleavage. This is identified by a rise in peptides that do not begin or end with the expected tryptic residues. Using diagnostics in QC tools for visualization of peptide length distributions and missed-cleavage frequencies can help identify whether digestion is consistently efficient or if deviations suggest problems in the sample preparation protocol.

3. Inflated FDR: Even when applying a standard 1% FDR threshold, several issues can compromise the integrity of results if not carefully monitored. If the decoy database is not well designed or the scoring algorithm fails to separate the target from decoy distributions, the estimated FDR may appear acceptable while still allowing large number of incorrect matches. Warning signs include unusually high numbers of low-scoring PSMs passing the threshold or identical score distributions for target and decoy hits. Inspecting PSM score histograms, target-decoy separation, and q-value curves helps to identify issues. A clear separation between correct and incorrect matches is critical for a meaningful 1% FDR.
4. Selecting stable and reliable peptides for SRM: SRM workflows rely on precise, reproducible detection of predefined peptide transitions. Because SRM quantification depends on stable peptide behavior across runs, peptides prone to missed

cleavages or chemical instability can introduce significant variability and undermine assay performance:

- Missed-cleavage peptides reduce quantitative precision: Peptides containing cleavage sites adjacent to aspartic acid or glutamic acid (e.g., DK, EK, DR, or ERRP) or sequences that have consecutive basic residues (e.g., KK, RR, RK, or KR) result in mixed populations of fully cleaved and partially cleaved peptides, observable as unexpected precursor masses or multiple coeluting species that dilute signal intensity.
- Chemically unstable peptides can negatively impact signal consistency: Peptides with methionine (M), cysteine (C), or other oxidation-sensitive residues can spontaneously form oxidized variants. Oxidation of methionine produces additional precursor and fragment ions, splitting the signal across multiple species and causing unpredictable peak area ratios over time. Therefore, peptides containing methionine are typically avoided. On the other hand, cysteine-containing peptides can still be used as SRM targets because their oxidation can be easily reduced and further prevented through iodoacetamide alkylation during sample preparation. In fact, not only are iodoacetamide alkylation steps incorporated in preparation, but peptides are often synthesized with carbamidomethylation (CAM)-modified cysteine instead of unmodified cysteine.

Acknowledgments

The work described herein was performed at the Agile Biofoundry, funded by the Bioenergy Technologies Office in the Office of Critical Minerals and Energy Innovation of the US Department of Energy under Agreement 30038 at Pacific Northwest National Laboratory, operated by Battelle for the US Department of Energy under contract DE-AC05-76RL01830, and at the Environmental Molecular Sciences Laboratory (EMSL), a DOE Office of Science User Facility sponsored by the Biological and Environmental Research program under Contract No. DE-AC05-76RL01830. The views expressed herein do not necessarily represent the views of the US Department of Energy or the US Government.

Conflict of Interest The authors declare no conflicts of interest that are relevant to the content of this chapter.

References

1. Angel TE et al (2012) Mass spectrometry-based proteomics: existing capabilities and future directions. Chem Soc Rev 41(10):3912. https://doi.org/10.1039/c2cs15331a
2. Aebersold R, Mann M (2003) Mass spectrometry-based proteomics. Nature 422(6928):198–207. https://doi.org/10.1038/nature01511
3. Hillson N et al (2019) Building a global alliance of biofoundries. Nat Commun 10(1):2040. https://doi.org/10.1038/s41467-019-10079-2
4. Picotti P, Aebersold R (2012) Selected reaction monitoring–based proteomics: workflows, potential, pitfalls and future directions. Nat Methods 9(6):555–566. https://doi.org/10.1038/nmeth.2015
5. Bilbao A (2019) Proteomics mass spectrometry data analysis tools. In: Frederique Lisacek (ed) Encyclopedia of bioinformatics and computational biology. Elsevier, 84–95. https://doi.org/10.1016/B978-0-12-809633-8.20274-4
6. Bilbao A et al (2023) PeakDecoder enables machine learning-based metabolite annotation and accurate profiling in multidimensional mass spectrometry measurements. Nat Commun 14(1):2461. https://doi.org/10.1038/s41467-023-37031-9
7. Gao Y et al (2020) High-throughput large-scale targeted proteomics assays for quantifying pathway proteins in *Pseudomonas putida* KT2440. Front Bioeng Biotechnol 8:603488. https://doi.org/10.3389/fbioe.2020.603488
8. Kim S, Pevzner PA (2014) MS-GF+ makes progress towards a universal database search tool for proteomics. Nat Commun 5:5277. https://doi.org/10.1038/ncomms6277
9. Kong AT, Leprevost FV, Avtonomov DM, Mellacheruvu D, Nesvizhskii AI (2017) MSFragger: ultrafast and comprehensive peptide identification in mass spectrometry-based proteomics. Nat Methods 14(5):513–520. https://doi.org/10.1038/nmeth.4256
10. Tyanova S, Temu T, Cox J (2016) The MaxQuant computational platform for mass spectrometry-based shotgun proteomics. Nat Protoc 11(12):2301–2319. https://doi.org/10.1038/nprot.2016.136
11. Callister SJ et al (2006) Normalization approaches for removing systematic biases associated with mass spectrometry and label-free proteomics. J Proteome Res 5(2):277–286. https://doi.org/10.1021/pr050300l
12. Polpitiya AD et al (2008) DAnTE: a statistical tool for quantitative analysis of -omics data. Bioinformatics 24(13):1556–1558. https://doi.org/10.1093/bioinformatics/btn217
13. Cox J, Hein MY, Luber CA, Paron I, Nagaraj N, Mann M (2014) Accurate proteome-wide label-free quantification by delayed normalization and maximal peptide ratio extraction, termed MaxLFQ. Mol Cell Proteomics 13(9):2513–2526. https://doi.org/10.1074/mcp.M113.031591
14. Bilbao A et al (2015) Processing strategies and software solutions for data-independent acquisition in mass spectrometry. Proteomics 15(5–6):964–980. https://doi.org/10.1002/pmic.201400323
15. Demichev V, Messner CB, Vernardis SI, Lilley KS, Ralser M (2020) DIA-NN: neural networks and interference correction enable deep proteome coverage in high throughput. Nat Methods 17(1):41–44. https://doi.org/10.1038/s41592-019-0638-x
16. Röst HL et al (2014) OpenSWATH enables automated, targeted analysis of data-independent acquisition MS data. Nat Biotechnol 32(3):219–223. https://doi.org/10.1038/nbt.2841
17. Egertson JD, MacLean B, Johnson R, Xuan Y, MacCoss MJ (2015) Multiplexed peptide analysis using data-independent acquisition and Skyline. Nat Protoc 10(6):887–903. https://doi.org/10.1038/nprot.2015.055
18. Searle BC, Egertson JD, Bollinger JG, Stergachis AB, MacCoss MJ (2015) Using Data Independent Acquisition (DIA) to model high-responding peptides for targeted proteomics experiments. Mol Cell Proteomics 14(9):2331–2340. https://doi.org/10.1074/mcp.M115.051300
19. MacLean B et al (2010) Skyline: an open source document editor for creating and analyzing targeted proteomics experiments. Bioinformatics 26(7):966–968. https://doi.org/10.1093/bioinformatics/btq054
20. Adams KJ et al (2020) Skyline for small molecules: a unifying software package for quantitative metabolomics. J Proteome Res 19(4):1447–1458. https://doi.org/10.1021/acs.jproteome.9b00640
21. Harrison A et al (2024) PeakQC: a software tool for omics-agnostic automated quality control of mass spectrometry data. J Am Soc Mass Spectrom 35(11):2680–2689. https://doi.org/10.1021/jasms.4c00146
22. Bilbao A et al (2023) MZA: a data conversion tool to facilitate software development and artificial intelligence research in multidimensional mass spectrometry. J Proteome Res 22(2):508–513. https://doi.org/10.1021/acs.jproteome.2c00313
23. Bilbao A (2025) The future of a myriad of accelerated biodiscoveries lies in AI-powered mass spectrometry and multiomics integration.

J Mass Spectrom 60(8):e5157. https://doi.org/10.1002/jms.5157
24. Bilbao A et al (2025) Towards AI agents that accelerate and democratize PTM detection and interpretation of mass spectrometry-based multi-omics synthetic biology data. Proceedings of the 73rd ASMS Conference on Mass Spectrometry and Allied Topics June 2nd, Baltimore, MD.
25. Okonkwo O et al (2024) Corn stover variability drives differences in bisabolene production by engineered *Rhodotorula toruloides.* J Ind Microbiol Biotechnol 51:kuae034. https://doi.org/10.1093/jimb/kuae034
26. Chaves JE et al (2020) Evaluation of chromosomal insertion loci in the *Pseudomonas putida* KT2440 genome for predictable biosystems design. Metab Eng Commun 11:e00139. https://doi.org/10.1016/j.mec.2020.e00139
27. Perez-Riverol Y et al (2025) Open-source and FAIR research software for proteomics. J. Proteome Res 24(5):2222–2234. https://doi.org/10.1021/acs.jproteome.4c01079

Part III

Emerging Questions in Mass Spectrometry Applications

Check for updates

Chapter 11

Annotation and Dereplication of Peptidic Natural Products (PNPs)

Emma Ricart Altimiras

Abstract

Peptidic natural products (PNPs) represent a structurally diverse class of bioactive compounds with significant pharmaceutical potential. Their characterization using tandem mass spectrometry (MS/MS) is often challenging due to complex fragmentation patterns arising from nonlinear structures, modified residues, and diverse bond types. NRPro is a web-based computational tool designed to facilitate both dereplication and automatic annotation of MS/MS spectra of PNPs. By integrating theoretical fragmentation modeling with database matching and statistical validation, NRPro provides an interactive platform for identifying known compounds and interpreting spectral data. This chapter describes the principles, workflow, and practical use of NRPro, including data preparation, parameter selection, result interpretation, and troubleshooting. Particular attention is given to common pitfalls and best practices to ensure reliable annotation and identification of PNPs.

Keywords NRPro, Peptidic natural products, MS/MS, Dereplication, Spectral annotation, Mass spectrometry, Bioinformatics, Natural products

1 Introduction

Peptidic natural products (PNPs) are a major source of bioactive molecules, including antibiotics and anticancer agents, and play a critical role in addressing challenges such as antimicrobial resistance. Their structural diversity, characterized by cyclic scaffolds, branching, nonproteinogenic amino acids, and modifications such as glycosylation or lipidation, poses significant analytical challenges.

Tandem mass spectrometry (MS/MS) is widely used for PNP characterization due to its sensitivity and ability to provide structural information. However, interpreting MS/MS spectra of PNPs is complex and often requires specialized computational tools. Many software focus either on dereplication (identifying known compounds) [1,2] or annotation (assigning fragment ions) [3,4], but rarely integrate both functionalities effectively.

Frédérique Lisacek (ed.), *Bioinformatics for Mass Spectrometry*, Methods in Molecular Biology, vol. 3071,
https://doi.org/10.1007/978-1-0716-5519-1_11,

NRPro [5] was developed to address these limitations by combining automatic dereplication with detailed spectral annotation in a single platform. It relies on curated databases to enhance compound identification and annotation. These include NORINE [6], dedicated to nonribosomal peptides, ChEBI [7], which provides detailed information on chemical entities, and NPAtlas [8], a comprehensive resource of microbial natural products. NRPro uses theoretical fragmentation of candidate molecules, matches predicted fragments with experimental spectra, and evaluates statistical significance through a target-decoy approach. Additionally, the tool provides an interactive graphical interface that supports manual validation and editing of annotations.

This chapter presents the workflow and practical use of NRPro for MS/MS analysis of PNPs.

2 Materials

2.1 Software and Access

- NRPro is available as a web application through the NORINE database: https://bioinfo.lifl.fr/nrpro/
- Recommended browser: Google Chrome (*see* **Note 1**).
- Additional documentation and tutorials are accessible through the interface (*see* **Note 2**).

2.2 Input Data

MS/MS Spectra in

- MGF format (recommended).
- mzXML format (less tested).

MGF file requirements (see Fig. 1 example and Note 3)

- One spectrum per file.
- Header must include:
  - Precursor m/z (PEPMASS).
  - Charge (if not specified in parameters).
- Proper format tags: BEGIN IONS/END IONS.

Illustrative Example Dataset

To illustrate the NRPro workflow throughout this chapter, a representative MS/MS spectrum of Cyclosporin A is used as an example (Fig. 1). The dataset consists of a single spectrum in MGF format, including precursor information and fragment peaks. This example is used consistently in the following sections to demonstrate data submission, spectral annotation, and result interpretation.

2.3 Hardware

- Standard workstation with internet connection.

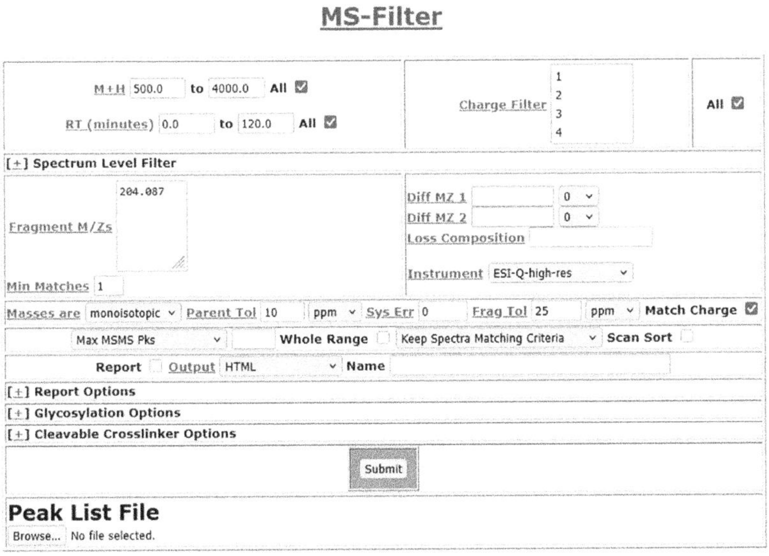

Fig. 1 Example input MGF file used as an illustrative dataset (Cyclosporin A)

2.4 Databases (internal to NRPro)

- Norine (https://norine.univ-lille.fr).
- ChEBI (https://www.ebi.ac.uk/chebi/).
- NPAtlas (https://www.npatlas.org/).

3 Methods

3.1 Overview of the Workflow

NRPro performs analysis in three main stages:

1. Generation of candidate fragmentation graphs.
2. Spectral matching and annotation.
3. Statistical scoring using a target-decoy strategy.

3.2 Data Submission

Uploading Spectra

- Upload up to 20 spectra files simultaneously.
- Prefer MGF format for reliability (*see* **Note 3**).

Parameter Selection

Key parameters include:

(a) Mass Tolerances

- Parent ion tolerance: ≤1 Da or ≤50 ppm.
- MS/MS tolerance: depends on instrument resolution.

(b) Ionization and Charge
- Specify ionization type (e.g., [M+H]+).
- Define precursor charge if not included in file.

(c) Neutral Losses and Adducts
- Common losses: H_2O, NH_3 (*see* **Note 4**).
- Additional losses/adducts affect annotation but not scoring.

(d) Fragment Ion Types
- Optional inclusion of c/z ions.

(e) Deisotoping
- Labels isotopic peaks without removing them (*see* **Note 5**).

(f) Decoy Option
- Enables p-value calculation.
- Increases computation time (*see* **Note 6**).

Errors may occur due to incorrect parameter settings or server instability (*see* **Note 7**).

3.3 Dereplication and Annotation

Fragmentation Modeling

NRPro simulates fragmentation by:

- Breaking chemical graphs at:
  - Amide bonds → b/y, x/a ions.
  - Ester bonds.
  - Glycosidic bonds.
- Generating fragmentation trees up to depth 4.

Spectral Matching

- Experimental peaks are matched to theoretical fragments within tolerance
- Peaks are:
- - Annotated : all experimental peaks that match one or more theoretical fragment ions within the specified mass tolerance.
- - Scored : only a subset of annotated fragment ions that meet the scoring criteria (as described in the following section) .

Scoring

- Based on:
  - Fragment intensity.
  - Annotation score (penalizing rare events).
- Requires:
  - Parent fragment support.
  - Expected fragmentation pattern.

Fig. 2 NRPro results table displaying candidate identifications (mass, formula, database links), number of annotated peaks, scores, and p-values (if decoy enabled). The table includes results from three submitted spectra: the illustrative example shown in Fig. 1 and two additional spectra, resulting in three candidate compounds

Statistical Validation

- Target-decoy approach generates p-values.
- Reflects confidence in peptide-spectrum matches.
- Multiple candidate identifications may be returned (*see* **Note 8**).

The approach is primarily limited to known compounds and may not fully capture highly novel structures (*see* **Note 9**).

3.4 Results Visualization

NRPro provides two main views:

Identification Table

After submission, results are displayed in a table listing candidate compounds along with their associated scores and metadata (Fig. 2).

Spectral Annotation View

The detailed annotation view for each spectrum can be accessed by clicking the “eye” icon located at the right side of each row (Fig. 2), which opens the annotation interface (Fig. 3) (*see* **Note 10** for troubleshooting).

3.5 Manual Curation and Export

- Users can:
  - Edit annotation.
  - Filter peaks.
  - Export results (CSV, XLS, JSON).
- Results are shareable via URL (*see* **Note 11**).

4 Notes

1. Browser Compatibility
   - NRPro performs best in Google Chrome. Issues have been reported with Firefox. If visualization fails, switch browsers.

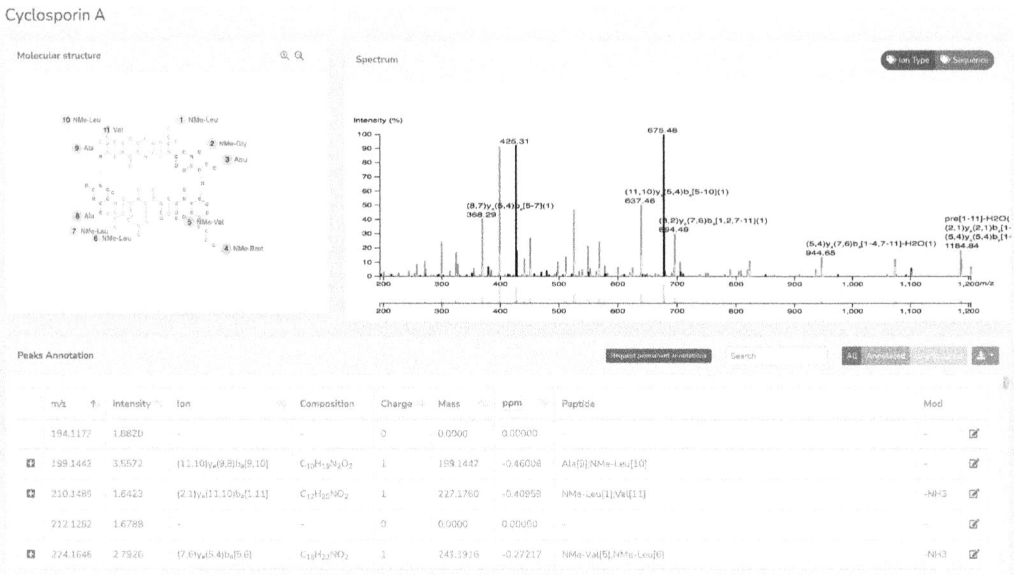

Fig. 3 NRPro detailed annotation interface showing the selected candidate (Cyclosporin A), its molecular structure, annotated MS/MS spectrum, and editable annotation table

2. Access to Documentation
   - Available via the "?" help button in the interface.
   - Provides:

   - Video tutorial illustrating the workflow.
   - Downloadable PDF manual (parameters, usage, and nomenclature).

3. Input File Issues
   Common errors include:
   - Multiple spectra per file.
   - Missing file extension.
   - Unsupported format.
   - Incorrect MGF structure.

   Solution: Ensure strict MGF compliance and one spectrum per file.
4. Impact of Neutral Losses
   - Including H_2O and NH_3 losses can significantly increase annotation coverage, especially for complex PNPs.
5. eisotoping Interpretation
   - With this option enabled, isotopic peaks are excluded from scoring but are highlighted in yellow in the spectral representation.

- Can affect perceived annotation coverage.

6. Sensitivity and Timescale
   - Runtime depends on:
     - Number of spectra.
     - Use of decoy option.
   - Decoy mode increases computation time but provides statistical confidence.
7. Unexpected Errors
   - May arise due to incompatible parameter settings. Try modifying the mass tolerances or enabling/disabling specific options, then resubmit the search
8. Interpretation of Results
   - Multiple candidates may be returned.
   - Structurally similar compounds can produce similar spectra: manual validation is recommended.
9. Limitations
   - Limited to known compounds in database.
   - Complex or novel structures may require manual interpretation.
   - Some fragmentation types (e.g., very high-order cuts) are not considered.
10. Access to Detailed Results View
    - The detailed annotation view is normally accessed via the "eye" icon in the results table (Fig. 2).
    - The icon may not be displayed due to rendering/browser issues.

    Solutions:

    - Modify the zoom of your browser (80% or 90% should be enough).
    - Alternative access: Use the left lateral panel to select the spectrum under "Input files".
11. Sharing and Editability
    - URLs are valid for ~1 month.
    - Edits are shared across users.
    - Annotations can be overwritten.

Conflict of Interest

The authors declare no conflicts of interest that are relevant to the content of this chapter.

References

1. Mohimani H, Liu WT, Kersten RD et al. (2017) Dereplication of peptidic natural products through database search of mass spectra. Nat Chem Biol 13:30–37
2. Mohimani H, Gurevich A, Shlemov A et al. (2018) Dereplication of microbial metabolites through database search of mass spectra. Nat Commun 9: 4035
3. Niedermeyer THJ, Strohalm M (2012) mMass as a software tool for the annotation of cyclic peptide tandem mass spectra. PLoS One 7:e44913
4. Novák J, Pluskal T, Kovalčíková A et al. (2015) CycloBranch: de novo sequencing of nonribosomal peptides from accurate product ion mass spectra. J Am Soc Mass Spectrom 26:1780–1786
5. Ricart E, Pupin M, Leclère V et al. (2020) Automatic annotation and dereplication of tandem mass spectra of peptidic natural products. Anal Chem 92:15862–15871
6. Flissi A, Ricart E, Campart C et al. (2020) Norine: update of the nonribosomal peptide resource. Nucleic Acids Res 48:D465–D469
7. Hastings J, de Matos P, Dekker A et al. (2013) The ChEBI reference database and ontology for biologically relevant chemistry: enhancements for 2013. Nucleic Acids Res 41:D456–D463
8. Poynton EF, Brecknell DJ, van Santen JA et al. (2025) The Natural Products Atlas 3.0: extending the database of microbially derived natural products. Nucleic Acids Res 53:D691–D699

Chapter 12

Glycopeptide Mass Spectrometry Data Analysis Using Protein Prospector

Robert J. Chalkley and Peter R. Baker

Abstract

The analysis of glycopeptide mass spectrometry fragmentation data has improved dramatically in the last 10 years, but there are still challenges with providing comprehensive while still reliable results. This chapter presents the data analysis pipeline in Protein Prospector as an example workflow. It talks about issues with estimating the reliability of results and provides tips for how to spot unreliable assignments. During the chapter, there is frequent reference to a video tutorial that presents analogous information in a different format.

Keywords Glycopeptide, Mass spectrometry, Data analysis, N-glycosylation, O-glycosylation, Database searching

1 Introduction

Glycosylation is the most common and varied modification of proteins, with over half of all proteins potentially being glycosylated. Most types of glycosylation occur extracellularly, either on secreted proteins or on the extracellular portion of membrane proteins [1]. These are most commonly N-linked through an asparagine residue in a consensus motif of N-X-S/T/C where X is anything other than proline, or O-linked through serine, threonine, or tyrosine residues. Glycosylation can also be found within the cell. These same types of N- and O-glycosylation occur on proteins within the lumen of the ER and Golgi, and a particular type of N-glycosylation involving mannose-6-phosphate residues is used to target misfolded proteins to lysosomes for degradation. Nuclear, mitochondrial, and cytosolic proteins may be found modified by a single N-acetylglucosamine residue (O-GlcNAcylation) [2]. Different to the other types of glycosylation, this is a dynamic posttranslational modification with enzymes responsible for its addition and removal and is often compared to phosphorylation in its roles.

Frédérique Lisacek (ed.), *Bioinformatics for Mass Spectrometry*, Methods in Molecular Biology, vol. 3071,
https://doi.org/10.1007/978-1-0716-5519-1_12,

Historically, glycosylated peptides have been difficult to analyze using mass spectrometry due to the different behaviors of the peptide and glycan portions upon fragmentation. Hence, many researchers have chosen to release the glycans from the protein, then perform glycomic analysis based on chromatographic separation, sometimes in combination with mass spectrometry [3]. This is great for surveying the types of glycosylation present, but if there is more than one glycosylation site on the protein/s then information about glycans on each site is lost.

Developments in mass spectrometry have now made intact glycopeptide analysis much more tractable. For analysis of N-glycopeptides, methods that combine fragmentation at different collision energies, sometimes referred to as stepped collision energy, allow optimal fragmentation of the glycan portion at a lower collision energy combined with improved peptide fragmentation at a higher collision energy [4]. For O-glycosylation, collision-based fragmentation is effective at identifying the peptide and glycan composition but eliminates the glycan from all fragments, thereby removing the ability to identify modification sites. Indeed, it can often be impossible to determine whether all glycan residues were attached to a single site or split across multiple modified residues. However, fragmentation methods that make use of radical chemistry, most commonly electron transfer dissociation (ETD), allow identification of both peptide and glycan attachment sites [5].

Glycopeptide analysis has improved dramatically in the last 15 years. This is through the production of more and better quality mass spectrometry fragmentation data, which has driven software development to extract as much information as possible. There are still questions over how reliable the results are from some of these software tools, which is not helped by the difficulty of creating a representative "ground truth" dataset that could directly assess performance. The Human Glycoproteomics Initiative conducted a community assessment of glycopeptide software that reported Protein Prospector being a top performer [6], but this study is now out-of-date. There is a new study underway, which hopefully will give a better assessment of where the field currently is.

In this chapter, the analysis pipeline within Protein Prospector (version 6.6.6) [7] (prospector.ucsf.edu) is presented as an example for analyzing mass spectrometry data of glycopeptides and is compared with other software approaches. There is discussion about the importance of different parameters and how changing these could affect results, and some commonsense tips for assessing results are provided. Similar content to this chapter is covered in a published video tutorial, so reference to this will be made on many occasions for an alternative presentation of the approach [8].

2 Materials

Glycopeptide mass spectrometry data are the input to the analysis software. This chapter does not talk about the results from a specific dataset, but the video tutorial that is heavily referenced [8] analyzed a fraction from a previously published dataset derived from a mouse brain [9], so this would be an appropriate dataset to use if one wanted to test the software approach.

Protein Prospector is the software package that is presented as the example for data analysis. This software is freely available through a public website at prospector.ucsf.edu. Several programs within this package will be used in a data analysis pipeline, including MS-Filter, Batch-Tag Web, Search Compare, MS-Viewer, and MS-Product. These programs can also be used independently of each other and are accessible from the Protein Prospector website homepage. Prior to submission of data to Protein Prospector, it is required to convert the raw data into centroided peak list files. A suitable software to do this is msconvert, part of the Proteowizard package [10].

Watching the related video tutorial [8] should help reinforce information presented in this chapter.

3 Methods

3.1 Identifying Potential Glycopeptide Spectra Using MS-Filter

Starting from a raw mass spectrometry file, the first step is to convert this into a peak list file. We suggest using msconvert, which is part of the free proteowizard package [10], and converting it into .mgf format. See **Note 1** for more details about the content of different peak list formats.

An initial step is to identify spectra that are potentially of glycopeptides by looking for specific masses of glycan oxonium ions, mostly notably a m/z 204.087 ion. See **Note 2** for more details. MS-Filter is software that can be used to filter peak list files based on mass, charge, presence of specific peaks, or presence of pairs of peaks with defined mass spacing [11]. Figure 1 shows a screenshot of MS-Filter. To filter peak list files for potential glycopeptide spectra:

1. Input "204.087" in the "Fragment M/Zs" field (one could optionally also input "138.055," although if your data is of high mass accuracy, then the m/z 204 peak alone is specific).
2. In the "Min Matches" field underneath input "1." This is defining how many of the masses you have input must be present in the spectrum.
3. Check the "Frag Tol" is set to a suitable tolerance. This should be the same as you would use for fragment mass tolerance in a database search of this data, so for Orbitrap data this is typically

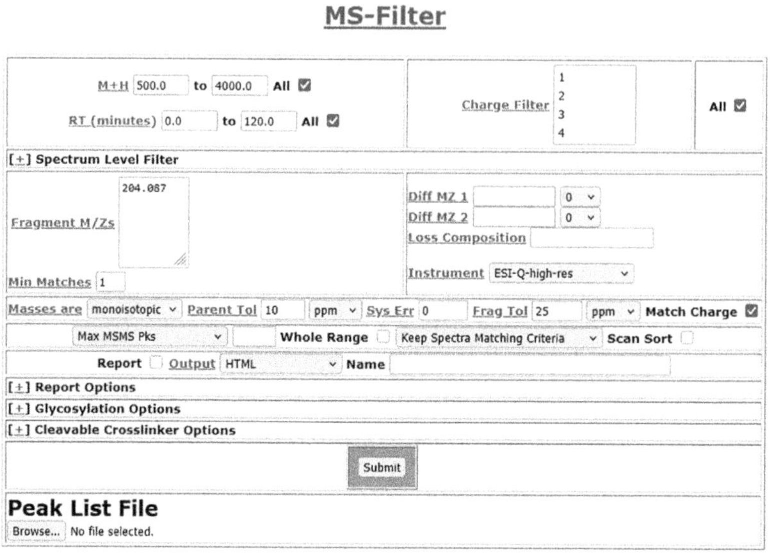

Fig. 1 MS-Filter can be used to identify potential glycopeptide spectra by filtering peak list files for the presence of the HexNAc oxonium ion at m/z 204.087

around 20 ppm; for data measured in an ion trap this could be 0.3 Da.

4. Optionally, one can enter a value in "Max MSMS Pks," which defines among how many of the most intense peaks in the spectrum the target m/z must be found. Specifying this as 20 is generally a good setting. If no value is entered, then it considers the 50 most intense peaks. See **Note 3** for more discussion of this value.
5. Upload the peak list file to be filtered. This could be in one of a variety of peak list formats, including .mzML; .mzXML; .mgf; .pkl; and .apl. If one wants to analyze multiple files together, they can be uploaded as a single compressed file, such as a .zip file.

The output file from MS-Filter (which is in .mgf format) can be used as input for glycopeptide searching by Protein Prospector, but also by other software.

3.2 Identifying Glycopeptides by Database Searching

Glycopeptide spectra contain three types of fragments: (1) those that only contain parts of the peptide (for collision-based fragmentation data these are mostly b and y ions); (2) those that only contain parts of the glycan, which are often collectively referred to as B ions, although they may include glycan parts that are internal to the glycan, so do not all technically meet the nomenclature definition [12]; (3) fragments that include peptide and glycan parts, known as Y ions if the whole peptide is present, or modified peptide fragment ions if the peptide backbone has been cleaved.

Most glycopeptide analysis software evaluate the identification of the peptide and glycan portions of the glycopeptide separately. In the case of Protein Prospector and software such as MS-Fragger [13], which are sometimes described as "peptide-first" software, the first step is to try to identify the peptide component. When doing this, the B ions are irrelevant; indeed, Protein Prospector can actively remove them for this step (see below). Y ions that contain the whole of the peptide are mostly not relevant, although they can be used to infer the mass of the unmodified peptide, which is valuable information. Indeed, the determination of the unmodified peptide mass is key for the alternative glycan-first software approach [14–16], which assigns the peptide mass based on the Y ion pattern and then only considers peptides of that mass that contain an N-glycosylation motif when performing the peptide identification step.

In collisional dissociation data of O-glycopeptides only unmodified b and y ions should be considered. For N-glycopeptides, the first GlcNAc attached to the asparagine is generally retained on peptide fragments, so those b and y ions that include the glycosylation site are assumed to be observed shifted in mass by a HexNAc residue. In ETD, for all types of glycosylation, the expectation is that the whole of the glycan will be retained on c and z ions that contain the glycosylation site.

Although the glycan is not being identified in this step, the list of glycans considered is still important for successful peptide identification. The mass difference between the unmodified peptide and the observed glycopeptide precursor must correspond to the mass of a glycan that is being considered. Hence, if the glycan present on the glycopeptide was not considered during the peptide identification step, then the spectrum will not be correctly identified.

Figure 2 shows a screenshot of Batch-Tag Web, which is the software in Protein Prospector that performs database searches of tandem mass spectrometry data. This software is used for the identification of peptides, both unmodified and with modifications such as glycosylation [17]. Hence, most of the parameters are the same as other peptide identification software, such as specifying a protein database to query, the enzyme that was used for cleavage, the mass tolerance to allow when matching parent/precursor and fragment ions. However, there are a couple of parameters specifically tailored for glycopeptide analysis. In the video tutorial [8], the selection of parameters in Batch-Tag Web is covered between times 1:15 and 5:30. The steps specific to glycopeptide analysis are:

1. Tick the box next to "Glycopeptide Peak Filtering." This automatically removes the masses of any known glycan "B" type ions while performing the peptide identification. See **Note 4** for why removing these peaks is beneficial.
2. In the Variable Mods section, select a set of glycan modifications to consider. Modifications to consider are grouped into a

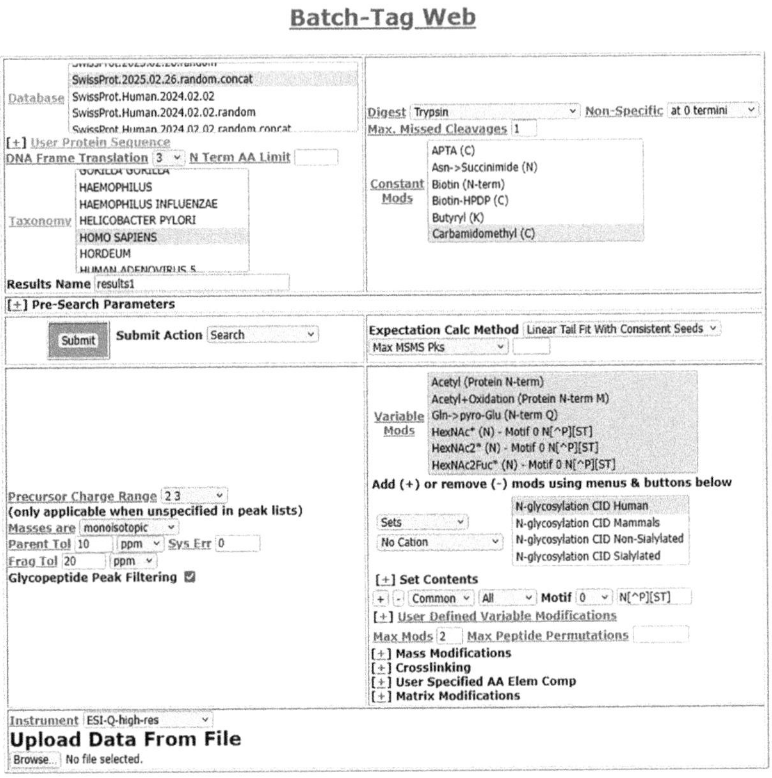

Fig. 2 Glycopeptide analysis using Batch-Tag Web. Parameter changes specific to glycosylation include selecting "Glycopeptide Peak Filtering" and adding sets of glycans for consideration, including specifying a modification motif for N-glycosylation

few different menu lists. The default is "Frequent," but there is another "Sets." This allows selecting many modifications in one go, and most of these sets are glycopeptide groups. The glycan options are segregated based on the type of fragmentation (which defines how the glycan will behave in the fragmentation spectra, i.e., whether they will be retained (ExD) or whether the N-glycans will only retain a single HexNAc and O-glycans will be completely eliminated (CID). They are also subdivided based on whether the sample was human or other mammalian (only the latter will consider glycans containing NeuGc), or one could only consider sialylated glycans (which may have been specifically enriched for). The contents of all the sets can be seen by clicking on the + next to "Set Contents." As a default the software will consider modifications on any Ser or Thr for O-glycosylation and any Asn for N-glycosylation. However, as there is a consensus modification motif for N-glycosylation one can specify that the Asn must be within the predicted motif. To do this, one specifies this in the motif area, then changes "Off" to "0." After selecting a set of glycans, e.g., "N-glycosylation CID Mammals," one must then click the + below "Set Contents" to add these glycans to the list of variable modifications considered.

3. Protein Prospector uses target:decoy database searching to estimate the reliability of peptide identifications [18], so select a .random.concat protein database to search against to allow this false discovery rate (FDR) calculation.

3.3 Identifying the Glycan Component of a Glycopeptide

Although there are potentially many hundreds of different glycosylation modifications of peptides, they are built from a smaller list of building blocks than peptides, especially if one considers that some sugar residues are structural isomers of each other so have the same mass. Mammalian glycans mostly consist of four types of sugar residues: deoxyhexoses (fucose); hexoses (mannose and galactose); N-acetylhexosamines (GlcNAc and GalNAc); and sialic acids (N-acetylneuraminic acid for humans, and also N-glycolylneuraminic acid for other mammals). These can be further decorated with acetyl, phosphate, or sulfate groups. Hence, if one knows the mass of the glycan, there is often only one glycan composition that this could correspond to, and most glycopeptide analysis software only assigns glycan compositions. As N-glycosylation is somewhat template-driven, with a common pentasaccharide core, then built up using known rules, software can make an educated guess at the topology of the glycan purely from the composition. See **Note 5** for comment on the ability to assign glycan topology.

There is often only one glycan composition considered that corresponds to the mass difference between the peptide mass and the observed precursor mass. Unfortunately, this does not always mean it is correct. Among the most common reasons for this is that the correct glycan assignment was not considered because there was a salt or metal adduct, or because the wrong monoisotopic peak was assigned. Glycopeptides have a higher propensity than unmodified peptides to pick up adducts such as ammonium or iron, and if these are present and not considered, then incorrect glycan assignments can be common [7]. Some glycopeptide software can specify looking for an ammonium adduct and one or two software, including Protein Prospector, can look for unexpected glycan modifications, as described below.

Once software has a list of candidate peptide/glycopeptide identifications, it needs to draw an acceptance threshold to control the reliability of reported assignments. Search Compare is the software used for displaying results from Batch-Tag and to filter to a defined peptide FDR. The complication for initial glycopeptide results in Protein Prospector is that it has not formally evaluated the reliability of the glycan identification at this point. It will report a glycan that is consistent with the mass difference between the peptide and precursor mass, but if there are alternative glycan assignments, these will not have been compared.

Protein Prospector can use MS-Filter at this stage to score glycan assignments and to try to identify additional glycoforms of those identified by Batch-Tag [11]. This is done by supplying a list of glycosylated peptides that have been identified (including any other posttranslational modifications such as methionine oxidation), then it will search the peak list files to look for potential Y0 and/or Y1 ions from these peptides. If one of these peaks is found, it will score the match to the peptide component, it will compare the precursor mass to the assigned peptide, and if the mass difference corresponds to glycan/s in the database it will score the glycan assignment to report the best matching glycan assignment. If the mass difference does not correspond to a known glycan, one has the option to ask it to report simply the mass or have it not report these assignments. The following section is covered in the video tutorial [8] from time 6:30 to 12:30.

1. In Search Compare, set the Max FDR thresholds you want to apply at the Protein and Peptide levels. A 1% FDR at both levels is most commonly used.
2. Expand the Glycosylation options by clicking on the [+], then select "MS-Filter Y0 List."
3. Select Report Type as "Peptide."
4. Hit "Submit."
   This will produce a peptide report that will list both nonglycosylated and glycosylated peptides.
5. Expand the MS-Filter Y0 List area near the top of the results report by clicking on the [+].
   This displays MS-Filter submission parameters with the "Glycosylation Options" section expanded and the Y0 List field is populated. An example of one of these entries is:
   AEFAVANDTGFVDIPQQEK P06797 215 8 157.4930 4.0460
   This information is the peptide, its protein accession number, the start residue of the peptide in the protein (i.e., the alanine in this example is residue 215 in Protein P06797), the number of peptides matched to the protein; an average retention time of glycosylated peptides identified to this peptide; and a retention time spread of glycosylated peptides to this peptide.
6. Select glycan set(s) to consider (select multiple by holding down ctrl on a PC)
   One could just select the same glycan database that was used in Batch-Tag, but one could also consider adducts. See **Note 6** for discussion of whether one should consider adducts. (Recommendation) Set "Precursor Offset" to "0 to 1."

The peak list file is intended to contain the mass of the monoisotopic peak (sometimes referred to as the first isotope peak). As the mass of a peptide increases, the monoisotopic peak becomes lower in intensity compared to the second and third isotopes. As glycosylated peptides are generally significantly heavier than unmodified peptides, mistaking the second isotope peak for the monoisotopic peak is more likely. The Precursor Offset setting of 0 to 1 means for each spectrum, the precursor mass in the peak list file will be considered as being the monoisotopic peak or the second isotope peak.

7. Select which Y ions to consider during filtering. We recommend for N-glycopeptides select only Y1; for O-glycopeptides select only Y0; select both if you are expecting both types of glycosylation.
8. (Optional) Alter the Score Threshold for Glyco to get more results or less. See **Note 7** for the effect of changing this threshold.
   If this field is left blank, then all spectra that contain a matching Y0/Y1 ion will be reported, even if no glycan in the database matches the mass difference between the unmodified peptide and observed precursor. From this output, if one compares these unexplained masses to those that have been assigned to a glycan, the difference may correspond to an adduct that could be considered (see time 16 min 50 s to 20 min 25 s in video tutorial [8]), or it may be the difference is the mass of a sugar residue, indicating a glycan composition missing from the database.
9. Select "N Motif"
10. Click "Submit."
    This will filter the spectra and report how many contained a potential Y ion.
11. Click "MS-Viewer" to display the results.

3.4 Viewing and Assessing Glycopeptide Results in MS-Viewer

MS-Viewer displays the results as a large table [19]. There is one row for each result, and for glycopeptide results, there are over three hundred columns in the table, although most of them are hidden as a default. The default parameters sort the results by retention time and filter to unique glycopeptide identifications; i.e., if the same peptide with the same glycan modification is identified from multiple spectra, then it will only be listed once. Many of the columns are self-explanatory. Most of the later columns are all related to glycan assignments. Glyco Score is used to assign the glycan (you set a minimum threshold in MS-Filter). Glyco Sc Diff is how much better this glycan assignment scored compared to any alternative glycan assignment within the mass tolerance window: in most cases there is no alternative glycan assignment consistent with the mass, so this column is blank. It has columns that indicate the presence of Y0/Y1 ions at different charge states (the presence of at least one of these was set by MS-Filter). The default displayed columns then list the number of B ions consistent and inconsistent with the glycan assignment, then the same information for Y ions. See **Note 8** for a comment on observing unexpected B ions. The column "138/144" reports the intensity ratio for these two fragment ions, which can be used to decide whether the spectrum is of a N-glycopeptide or O-glycopeptide [20], and is most useful in O-GlcNAcylation analysis to differentiate between single GlcNAc and Tn antigen (O-GalNAc) glycosylations. See **Note 9** for comment on the content of other columns in this report.

To change the columns displayed or to apply different sorting or filtering, click on the [+] next to Display and Parameter Settings:

- The "Remove Column" field determines which columns are hidden from the table, so one can manually tailor the information provided.
- "Report Filtering" allows filtering results displayed based on the content of any defined column. One defines which column you want to filter using, then the content. As a default it filters for the content to "equal" the input; i.e., you would input the full value to get a match. One can also filter for "contains" and match part of the content, or for numerical columns you can filter based on a minimum or maximum value.
- Report Sorting defines the order in which results are reported. The default sorting is based on retention time, but one could alternatively, for example, sort based on peptide sequence or "Mod Mass," which is useful if you are interested in seeing whether there is a particular mass that is frequently observed that is not explained by a defined glycan modification, suggesting another modification should be considered.
- "Remove Replicates" allows removal of repeat identifications. As a default it is removing repeat identifications of the same peptide and glycan combination to produce a list of unique glycopeptide IDs. If one wanted the best glycopeptide ID for each spectrum one could, for example, sort the results by MSMS Info (scan number) then Glyco Score to get the best match per spectrum, then remove replicates of the scan number to leave on the best match for each spectrum.

For each identification, there are two hyperlinks available:

1. If you click on the "MSMS Info" column entry (which reports the spectrum identifier), this will bring up MS-Tag that allows you to re-search this particular spectrum. This allows one to see the next best peptide matches to this spectrum, or one could

change the search parameters to see if this leads to a different result.

2. If you click on the "DB Peptide," it will produce an annotated spectrum for that identification.

3.5 Viewing Annotated Spectra

If one is interested in a particular result, the ability to view an annotated spectrum to manually decide on the quality of the identification is important. While there is a standard nomenclature for labeling the peptide backbone fragment ions, the labeling of fragments containing all or part of the glycan is less consistent between software. Protein Prospector annotates glycan-containing fragments using the community standard symbols [21] for B ions and Y_0 (i.e., the unmodified peptide) with these symbols for Y ions. The user has options to display the whole annotated spectrum, or one can separate out the information for peptides and glycans separately, as shown in Fig. 3. The display is interactive, so one can zoom in on regions of the spectrum, which is often necessary when there are a lot of peaks annotated in a close mass range, so labels overlap, as in Fig. 3A. It is also possible to compare annotation for different interpretations of the spectrum.

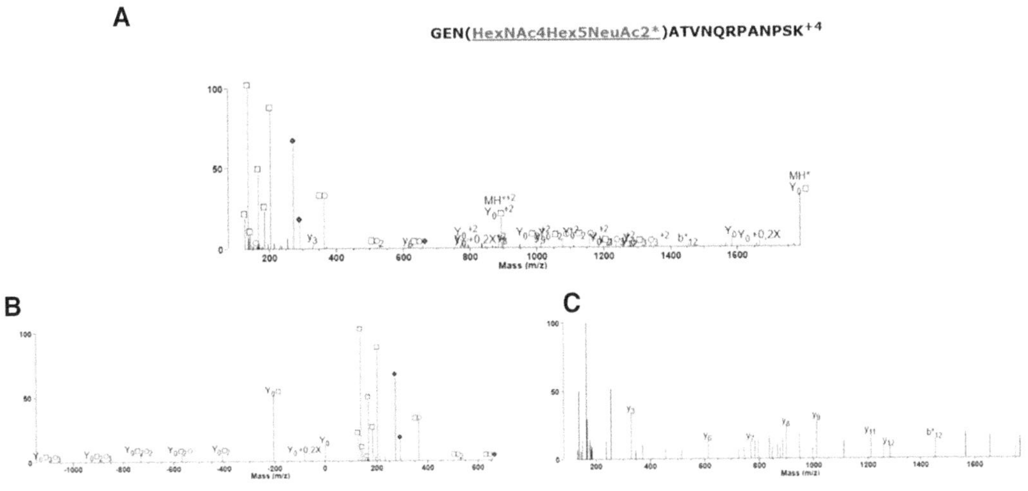

Fig. 3 Annotated spectra of a glycopeptide from bovine Fetuin-B. (**A**) The full spectrum shows most peaks are explained and are dominated by glycan-related fragments. (**B**) This plot shows only the B/Y fragments from panel (A). In this plot "0" is in the middle, then to the right of this the B ions are plotted and to the left Y ions are plotted with 0 representing the mass of the unmodified peptide (i.e., Y0). Note: the Y ions do not span the full length of the glycan. This is common and can be because singly-charged versions of these fragments are formed that are outside the m/z range the spectrum is acquired over. (**C**) Plot of spectrum in panel (A) with B/Y ions removed shows there are seven peptide y ions and a b ion (containing the modification site, so retaining a HexNAc modification) matched in this spectrum

3.6 Assessing the Reliability of Glycopeptide Results

Most glycopeptide software report FDR estimates for the peptide part of the glycopeptide identification, and some also for the glycan. The peptide FDR is based on mature methods for unmodified peptide identification, and most software can produce reasonably accurate estimates for the peptide reliability. However, it is highly questionable whether any method currently used to estimate the reliability of the glycan portion is accurate. See **Note 10** for comment on how Byonic software scoring differs from other software. Those software that attempt to report a FDR for the glycan portion mostly create decoy glycan spectra by shifting the masses of peaks to estimate how many peaks might be matched at random. Unfortunately, this does not accurately simulate incorrect glycan identifications, as essentially any N-glycopeptide identification, whether correct or incorrect, produces the same Y ion fragments from the conserved pentasaccharide core, so many fragment ions will be in common between correct and incorrect glycan assignments. The primary parameter used for glycan assignment is its mass. If there is only one glycan considered of a particular mass (within a certain mass tolerance), then this is going to be reported by practically all software. Hence, it is very important that the glycan list considered includes the correct identification (including allowing for adducts). *Providing* this is the case, most software will be able to correctly differentiate between potential glycan assignments when there is evidence for one compared to the other. However, no glycopeptide software currently reports multiple potential glycan assignments when it is ambiguous which is correct.

3.7 Common Sense Rules for Interpreting Results

- Glycans on a given site should be related to each other.
  Different glycans on a given site are created by adding/removing single sugar residues at a time. Hence, one expects that glycans found on a given site are related to each other. This concept is illustrated by GlyConnect Compozitor [22], which plots the relationship between reported glycans for given sites/proteins. If for a given peptide/site several glycans are identified, and most are of one type (e.g., oligomannose), but one is a sialylated complex glycan, then this latter identification is unlikely to be correct.
- Glycopeptide relative retention times can indicate incorrect glycan assignments.
  Most sugar residues are neutrally charged, so have minimal effect on glycopeptide retention in reverse-phase chromatography. However, sialic acid residues (and phosphate or sulfate groups) are highly acidic, so the addition of one of these significantly shifts the glycopeptide retention time. This means that all glycoforms of a given peptide with the same number of sialic acid residues (including none) should roughly coelute, whereas each addition of a sialic acid will shift to later

retention times. Hence, retention time is a useful parameter for verifying the reliability of glycan assignments, and, for example, can easily indicate whether two fucoses (+292 Da) or a sialic acid (+291 Da) in the glycan is correct. See **Note 11** for how MS-Filter can make use of retention time information.

- You are unlikely to detect an adducted version of a glycopeptide if you did not also identify the fully protonated version.

  Ammonium and metal adducts are nearly always of lower intensity than fully protonated glycopeptides. Hence, identifying an adduct of a glycopeptide when the protonated version is not also identified in another spectrum is a red flag. This is illustrated between times 14 min 45 s and 15 min 30 s in the video.

4 Notes

1. Raw data are often profile data where there are multiple datapoints across a peak. However, a search engine wants a list of apex peak masses and intensities, and only for monoisotopic peaks. Some software packages will do this conversion "behind the scenes," whereas others make use of tools specific for this task. The most universal software to perform this is msconvert, which is part of the free Proteowizard package [10]. This can read raw data files from most mass spectrometry instrument manufacturers and output files in a large variety of formats. The recommended format to use for MS/MS database searching is .mgf files. These provide all the information a search engine needs and only store MS2 data, with minimal metadata about how the data was acquired. This makes these files much smaller than the raw data, so faster to work with. The community standard format for peak lists, .mzML, is designed more as a substitute for the raw data, so contains additional information not needed by the search engine, and in many cases is larger than the original raw data file.
2. Glycosidic linkages are more labile than peptide backbones, so fragments corresponding to only sugar residues are prominent in practically all glycopeptide fragmentation spectra. The most intense of these is often a peak at m/z 204.087 that corresponds to a single N-acetylhexosamine, which in the case of N-linked glycans will be N-acetylglucosamine (GlcNAc), whereas for O-linked glycans (other than O-GlcNAc modification) it is either N-acetylgalactosamine (GalNAc) or a mixture of these two residues. At higher collision energies, these are further fragmented into smaller pieces, including two

peaks at m/z 138.055 and m/z 144.065 that can be used to differentiate between GlcNAc and GalNAc (and hence N- vs O-GalNAc glycosylation) [20]. The presence of the m/z 204 ion should be a prerequisite for considering a spectrum as potentially being N-glycosylated or bearing O-GalNAc-type glycosylation. It is not necessary to filter out spectra that are not of glycopeptides prior to glycopeptide database searching. However, doing so can reduce false positive glycopeptide identifications, and depending on how enriched the sample is for glycopeptides, it could significantly shorten database searching time.

3. The intensity of the sugar oxonium peaks is dependent on the acquisition collision energy used. In most datasets the m/z 204 peak will be among the twenty most intense peaks, whereas the peak may be present as "background" in spectra of non-glycosylated peptides if glycopeptide enrichment was performed. This is because there are often coeluting glycopeptides coisolated in the mass spectrometer with the unmodified peptides when the majority of the sample is glycopeptides. When deciding on how many peaks to consider, the decision is whether one wants to risk missing a few glycopeptides or consider some spectra from unmodified peptides. Allowing in more spectra may lead to a marginal increase in the software's error rate and will increase the search analysis time but could provide more correct identifications.

4. While the m/z 204 peak and its fragments are the most common B ions, there are over 75 different glycan oxonium ions that can be seen in mammalian glycopeptide spectra. These could be misinterpreted as peptide fragments (although they normally differ slightly from potential peptide fragment masses because the much larger number of oxygen atoms compared to amino acids makes them relatively mass deficient), but also if the search engine software uses an intensity threshold for which peaks to consider in the database searching (for example, Batch-Tag most commonly uses fifty of the most intense peaks), then by removing the glycan ions, the software will consider additional peaks as potentially being peptide fragments.

5. There are a few software that try to assign topology from the fragmentation data, such as StrucGP [16]. However, the data may often not contain enough information to determine this, and these software cannot express ambiguity in the topological assignment.

6. One could argue that identifying adduct spectra is not very important, as the fully protonated version should also have been identified, so it is not providing additional glycosylation information. However, the risk of not considering adducts is

that these can be falsely identified as additional glycoforms [7]. The most common occurrence of this is mistaking an ammonium adduct of a glycopeptide containing a fucose residue for a version containing a hexose instead (ammonium adduct is +17 Da; hexose replacing a fucose is +16 Da). A similar mistake can be made if one matches the wrong type of adduct: the difference in mass between a calcium or potassium adduct compared to a sodium adduct is also +16 Da. Adding an adduct you are confident is not present is one way to get an estimate of glycan FDR.

7. The glycan scoring is described in a previous publication [7]. The default score threshold is 3. This is a compromise value where results from small O-glycans that cannot produce many glycan fragments can still be reported (although it will probably filter out Tn antigen-modified peptides), but it does filter out obviously unreliable IDs. If one is focusing on N-glycosylation a score threshold of 5 is a good conservative threshold [23], but will mean single HexNAc-modified peptides and few O-glycosylations will be reported.
8. It is not unusual to observe B ions inconsistent with a correct glycan assignment in glycopeptide-enriched datasets. This is because there may be other glycopeptides co-isolated with the peptide of interest, and some oxonium ions, particularly those from sialic acids, are often found in spectra of nonsialylated glycopeptides [7].
9. The (large number of) hidden columns in the MS-Viewer table summarize B and Y ions observed, including their relative intensities, then list all considered B and Y ions, and report an intensity if this peak is seen in this spectrum. This information could be used to try to assign glycan topology or in the future, be used to develop a more sophisticated glycan scoring system.
10. The software Byonic [24] is currently unique in trying to assess the reliability of peptide and glycan assignments with one score. This compromises the normally well-controlled FDR estimation for the peptide portion. As a result, whereas most incorrect identifications by other software report the correct peptide but with the wrong glycan (as the peptide FDR is normally lower than the glycan FDR), Byonic is more likely to get both incorrect.
11. The MS-Filter input from Search Compare reports the average retention time for glycoforms of a given peptide, and it is possible in MS-Filter to filter glycopeptide results to be within a certain window of this retention time. However, this does sometimes filter out correct identifications if, for example, in the initial Batch-Tag database searching for a given

peptide, only singly sialylated glycoforms were identified, but MS-Filter is able to detect the presence of weaker doubly sialylated versions at a later retention time. Software such as GlycopeptideGraphMS makes use of retention time to detect additional glycoforms of glycopeptides in MS1 data [25]; the challenge is predicting how much of a retention time shift to expect. People have worked on predicting this [26], but it is chromatography column-dependent, so probably needs to be learned from the dataset under analysis.

Conflict of Interest

The authors declare no conflicts of interest that are relevant to the content of this chapter.

References

1. Colley KJ, Varki A, Haltiwanger RS, Kinoshita T, (2022) Cellular organization of glycosylation. In: Varki A, Cummings RD, Esko JD, (Eds.) Essentials of glycobiology. 4th edn. Cold Spring Harbor Laboratory Press, Cold Spring Harbor, NY, 1–6
2. Zachara NE, Akimoto Y, Boyce M, Hart GW, (2022) The O-GlcNAc modification. In: Varki A, Cummings RD, Esko JD,(Eds.) Essentials of glycobiology. 4th edn. Cold Spring Harbor Laboratory Press, Cold Spring Harbor, NY, 1–10
3. Lu J, Guo S, Liu Q et al. (2025) Recent advances in analytical methods and bioinformatic tools for quantitative glycomics. Anal Bioanal Chem 417:1947–1959. https://doi.org/10.1007/s00216-025-05778-3
4. Hinneburg H, Stavenhagen K, Schweiger-Hufnagel U et al. (2016) The art of destruction: optimizing collision energies in quadrupole-time of flight (Q-TOF) instruments for glycopeptide-based glycoproteomics. J Am Soc Mass Spectrom 27:507–519. https://doi.org/10.1007/s13361-015-1308-6
5. Riley NM, Malaker SA, Driessen MD, Bertozzi CR (2020) Optimal dissociation methods differ for N- and O-glycopeptides. J Proteome Res 19:3286–3301. https://doi.org/10.1021/acs.jproteome.0c00218
6. Kawahara R, Chernykh A, Alagesan K et al. (2021) Community evaluation of glycoproteomics informatics solutions reveals high-performance search strategies for serum glycopeptide analysis. Nat Methods 18:1304–1316. https://doi.org/10.1038/s41592-021-01309-x
7. Chalkley RJ, Baker PR (2025) Improving the depth and reliability of glycopeptide identification using protein prospector. Mol Cell Proteomics 24:100903. https://doi.org/10.1016/j.mcpro.2025.100903
8. Chalkley RJ, Baker PR (2024) Protein prospector: glycopeptide analysis using MS-filter with glycan scoring. In: Protein prospector: glycopeptide analysis using MS-filter with glycan scoring. https://vimeo.com/992202630. Accessed 30 Jul 2024
9. Liu M-Q, Zeng W-F, Fang P et al. (2017) pGlyco 2.0 enables precision N-glycoproteomics with comprehensive quality control and one-step mass spectrometry for intact glycopeptide identification. Nat Commun 8: 438. https://doi.org/10.1038/s41467-017-00535-2
10. Adusumilli R, Mallick P (2017) Data conversion with ProteoWizard msConvert. Methods Mol Biol 1550:339–368. https://doi.org/10.1007/978-1-4939-6747-6_23
11. Chalkley RJ, Medzihradszky KF, Darula Z et al. (2020) The effectiveness of filtering glycopeptide peak list files for Y ions. Mol Omics 16:147–155. https://doi.org/10.1039/c9mo00178f
12. Domon B, Costello CE (1988) A systematic nomenclature for carbohydrate fragmentations in FAB-MS/MS spectra of glycoconjugates. Glycoconj J 5:397–409. https://doi.org/10.1007/BF01049915
13. Polasky DA, Yu F, Teo GC, Nesvizhskii AI (2020) Fast and comprehensive N- and O-glycoproteomics analysis with MSFragger-Glyco. Nat Methods 17:1125–1132. https://doi.org/10.1038/s41592-020-0967-9

14. Zeng W-F, Cao W-Q, Liu M-Q et al. (2021) Precise, fast and comprehensive analysis of intact glycopeptides and modified glycans with pGlyco3. Nat Methods 18:1515–1523. https://doi.org/10.1038/s41592-021-01306-0
15. Fang Z, Qin H, Mao J et al. (2022) Glyco-Decipher enables glycan database-independent peptide matching and in-depth characterization of site-specific N-glycosylation. Nat Commun 13: 1900. https://doi.org/10.1038/s41467-022-29530-y
16. Shen J, Jia L, Dang L et al. (2021) StrucGP: de novo structural sequencing of site-specific N-glycan on glycoproteins using a modularization strategy. Nat Methods 18:921–929. https://doi.org/10.1038/s41592-021-01209-0
17. Chalkley RJ, Baker PR, Medzihradszky KF et al. (2008) In-depth analysis of tandem mass spectrometry data from disparate instrument types. Mol Cell Proteomics 7:2386–2398. https://doi.org/10.1074/mcp.M800021-MCP200
18. Elias JE, Gygi SP (2007) Target-decoy search strategy for increased confidence in large-scale protein identifications by mass spectrometry. Nat Methods 4:207–214. https://doi.org/10.1038/nmeth1019
19. Baker PR, Chalkley RJ (2014) MS-viewer: a web-based spectral viewer for proteomics results. Mol Cell Proteomics 13:1392–1396. https://doi.org/10.1074/mcp.O113.037200
20. Halim A, Westerlind U, Pett C et al. (2014) Assignment of saccharide identities through analysis of oxonium ion fragmentation profiles in LC-MS/MS of glycopeptides. J Proteome Res 13:6024–6032. https://doi.org/10.1021/pr500898r
21. Varki A, Sharon N, (2009) Historical background and overview. In: Varki A, Cummings RD, Esko JD, (Eds.) Essentials of glycobiology. 2nd edn. Cold Spring Harbor Laboratory Press, Cold Spring Harbor, NY, 1–11
22. Robin T, Mariethoz J, Lisacek F (2020) Examining and fine-tuning the selection of glycan compositions with GlyConnect Compozitor. Mol Cell Proteomics 19:1602–1618. https://doi.org/10.1074/mcp.RA120.002041
23. Hogan RA, Pepi LE, Riley NM, Chalkley RJ (2025) Comparative analysis of glycoproteomic software using a tailored glycan database. Anal Bioanal Chem 417:1985–2001. https://doi.org/10.1007/s00216-025-05780-9
24. Bern M, Kil YJ, Becker C (2012) Byonic: advanced peptide and protein identification software. Curr Protoc Bioinformatics Chapter 13:Unit13 40:20. https://doi.org/10.1002/0471250953.bi1320s40
25. Choo MS, Wan C, Rudd PM, Nguyen-Khuong T (2019) GlycopeptideGraphMS: improved glycopeptide detection and identification by exploiting graph theoretical patterns in mass and retention time. Anal Chem 91:7236–7244. https://doi.org/10.1021/acs.analchem.9b00594
26. Klein J, Zaia J (2020) Relative retention time estimation improves N-glycopeptide identifications by LC-MS/MS. J Proteome Res 19:2113–2121. https://doi.org/10.1021/acs.jproteome.0c00051

Part IV

Mass Spectrometry in the Age of Artificial Intelligence Technology

Chapter 13

Artificial Intelligence and Machine Learning in Mass Spectrometry-Based Proteomics and Metabolomics

Kristoffer Basse, Beata Szeitz, Frank Suits and Peter Horvatovich

Abstract

Analytical chemistry, and particularly mass spectrometry (MS), plays a central role in the molecular characterization of biological and environmental samples. These data are essential for understanding the molecular mechanisms underlying healthy aging, disease onset and progression, and the complex chemical composition of environmental systems. Over the past decades, MS has undergone unprecedented technological advancements. Current state-of-the-art instruments enable the thorough characterization of deep proteomes at the single-cell level and achieve comprehensive coverage of metabolites and proteins. Innovations such as new mass analyzers (e.g., the Astral mass analyzer) and the integration of multidimensional separation techniques (e.g., trapped/cyclic ion mobility) have significantly expanded the analytical capabilities of the instrumentation. However, the volume and complexity of the data produced by these advanced platforms pose major challenges for interpretation and integration. To address these challenges, machine learning (ML) and artificial intelligence (AI) methods are emerging as indispensable tools in analytical chemistry. These approaches, particularly deep learning, generative AI, and large language models, offer powerful solutions for analyzing, integrating, and interpreting the high-dimensional, heterogeneous datasets generated in proteomics and metabolomics research. This chapter first introduces the fundamental principles and operational mechanisms of commonly used AI/ML methods in analytical chemistry. These include regression models, decision trees, *k*-nearest neighbors, support vector machines, neural networks, deep learning architectures, and generative AI. The first part of the chapter also discusses key requirements for model training, including the need for sufficient sample and molecular coverage, and evaluates how data noise affects model performance and reliability. In the second part, a series of representative applications are presented, demonstrating the use of AI/ML for peptide and metabolite observability prediction, retention time and ion mobility prediction, fragment ion spectrum generation, de novo compound identification, and improvements in compound quantification and scoring of identifications. The chapter concludes with practical recommendations for applying AI/ML approaches in proteomics and metabolomics studies, and for their integration into bioinformatics pipelines to enhance data interpretation, reproducibility, and scientific discovery.

Keywords Artificial intelligence, Machine learning, Mass spectrometry, Proteomics, Metabolomics, Deep learning

Statement on Use of AI

All text in this chapter has been written by the authors, and AI was only used to correct grammatical and stylistic errors, and changes were applied after thorough proofreading. No AI assistance was used for figures.

Frédérique Lisacek (ed.), *Bioinformatics for Mass Spectrometry*, Methods in Molecular Biology, vol. 3071, https://doi.org/10.1007/978-1-0716-5519-1_13,

1 Introduction

This chapter addresses the use of machine learning (ML) and artificial intelligence (AI) in analytical chemistry, and more particularly for mass spectrometry-based proteomics and metabolomics applications. Artificial intelligence has many definitions. For example, on December 18, 2018, in the declaration of a high-level expert group on artificial intelligence [1], the European Commission defined AI as "systems that display intelligent behavior by analyzing their environment and taking actions—with some degree of autonomy—to achieve specific goals. AI-based systems can be purely software-based, acting in the virtual world (e.g., voice assistants, image analysis software, search engines, speech, and face recognition systems) or AI can be embedded in hardware devices (e.g., advanced robots, autonomous cars, drones or Internet of Things applications)." The IEEE Standards Association has a position statement describing AI as "the combination of algorithms, data, and computing power that enables machines to exhibit human-like intelligence and perform tasks such as visual perception, speech recognition, decision-making, and language translation" or "Artificial Intelligence is that activity devoted to making machines intelligent, and intelligence is that quality that enables an entity to function appropriately and with foresight in its environment" [2]. The founding father of AI, Alan Turing, defined AI based on the Turing test, which assesses if a machine showing intelligent behavior mimics human responses sufficiently that an observer could not distinguish it from a human during a conversation [3]. These definitions are quite general and do not precisely describe the exact underlying computational approaches, but rather focus on just the behavior of a system.

ML is a subfield of AI, which aims to develop new statistical approaches and algorithms that allow computers to learn from data, and improve their actions (typically classification, prediction, and pattern recognition) through iterative processes rather than by direct updates to the underlying program code. The goal of ML is to generalize the actions of a learning agent during training on well-defined training data and to deploy to unseen/unknown data with high accuracy. To achieve good performance, ML requires adequate amounts of training data (Fig. c).

There are three distinct types of learning: supervised, unsupervised, and reinforcement learning. In the case of supervised learning, the ML algorithm is fed human-labeled data, and it learns how to make predictions based on those labels. On the other hand, unsupervised learning is a process in which unlabeled data is introduced to the ML system, and the underlying structures/relationships in the data must be uncovered by the algorithm. Lastly, in the case of reinforcement learning, the ML system learns through rewards and

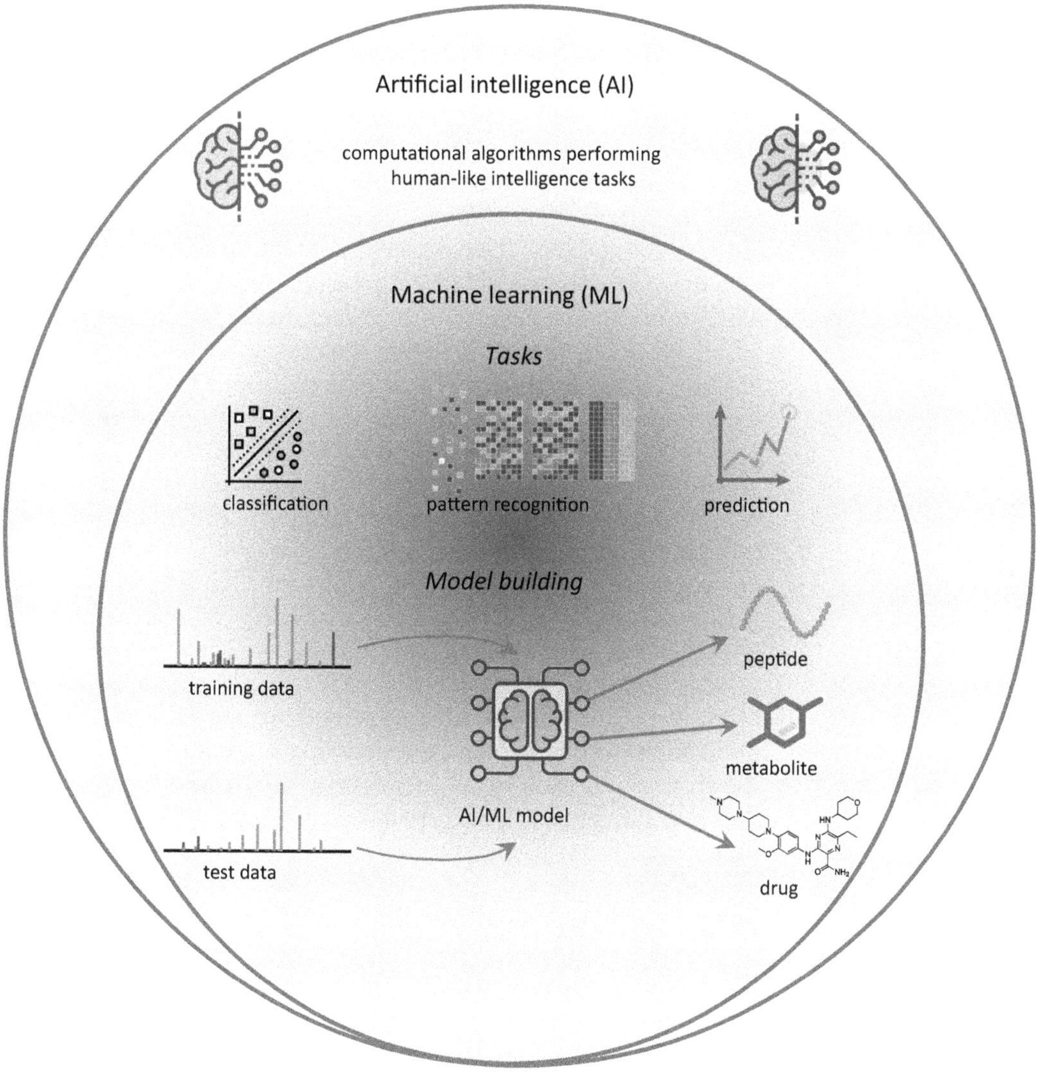

Fig. 1 ML/AI methods overlap. This scheme illustrates that Machine Learning is a subfield of Artificial Intelligence

punishments, which helps it to devise an action model based on input for a specific problem.

ML algorithms have proven to be highly valuable in analytical chemistry and are most commonly used chemometric approaches for prediction and classification in analytical chemistry and biochemistry [4]. One hallmark of ML is the need for large training datasets for good performance on classification and prediction tasks [5], which is a criterion that is sometimes difficult to meet even in this data-intensive field. Recently, datasets grew large enough in many fields of analytical chemistry to move in the direction of deep learning (DL), which requires even larger training datasets than ML. In some subfields, however, there is still an ongoing debate over whether sufficient data is available to train and test the ML/AI models, so a thorough data availability assessment is required.

DL is a subfield of ML, where neural networks, which attempt to mimic the behavior of the human brain, are used to model and solve complex problems. The major difference in the usage of ML and DL is the way in which the raw data is handled. Most often, ML requires a well-defined list of features carefully curated from the raw data through a process called feature engineering, whereas DL eliminates this step and requires only the raw data. While this is a clear advantage of DL and results in better performance [6, 7], the models built by DL are essentially "black boxes," from which it is difficult, if not impossible, to decrypt which features were considered for the prediction and why a certain prediction was made.

Analytical chemistry is an enabling technology that aims to measure the quantities, identity, structure, and physicochemical properties of compounds in samples, which can have biological, chemical, or environmental origins. As such, analytical chemistry is driven by well-defined but complex physicochemical properties and laws, which have been extensively modeled with exact approaches with medium to good accuracy, but the model complexity and large parameter space often cause insurmountable obstacles. Current ML/AI statistical modeling is well-suited for problems in analytical chemistry due to the close connection between physicochemical properties and the data generated by analytical chemistry instrumentation [4], such as the mass analyzers used in mass spectrometry-based proteomics and metabolomics approaches. Recent advances in the field have resulted in higher sample analysis throughput, higher data density in terms of spectra acquisition speed, increased peak capacity (resolution), wider dynamic range, and improved fragmentation methods through the development of improved and new types of mass analyzers, and the addition of an extra separation dimension such as ion mobility. Figure 2 provides an overview of how ML and AI are integrated into analytical chemistry, from sample preparation and raw data acquisition to the extraction of meaningful information for scientific and societal applications.

In the following sections, we first provide an overview of the main categories of ML and AI methods relevant to analytical chemistry, with a particular focus on mass spectrometry-based proteomics and metabolomics. We then highlight current applications that demonstrate how these approaches are used in practice, and conclude with practical recommendations for analytical chemists who wish to incorporate ML and AI into their own research workflows.

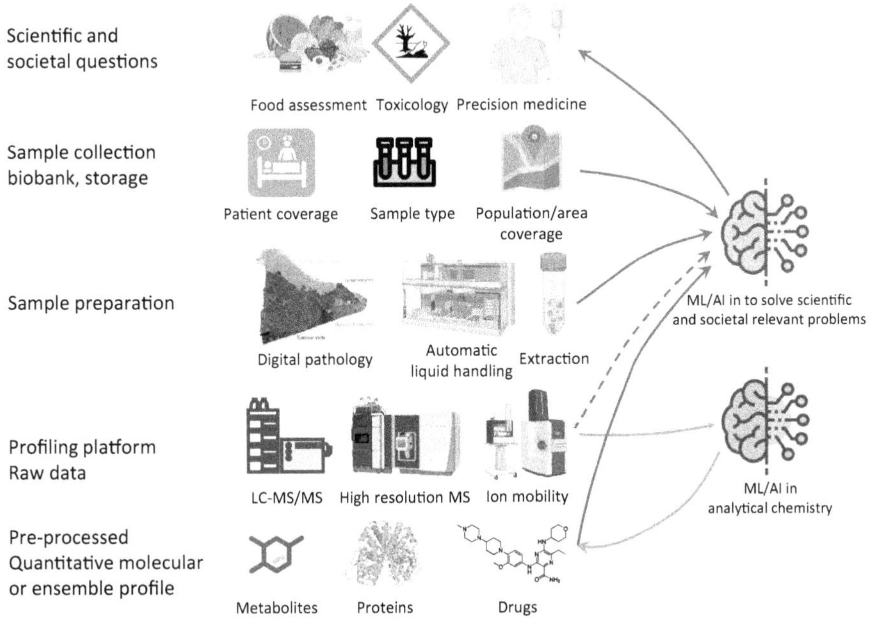

Fig. 2 ML/AI methods usage. This scheme illustrates how ML/AI methods are used in analytical chemistry to solve scientific and societally relevant questions. Sample coverage, storage, and preparation are important aspects that influence the quality and information extractable from the samples by analytical chemistry methods. The generated raw data will contain signals from individual or an ensemble of biocompounds. ML and AI are mainly used to extract meaningful information from acquired raw data, and the extracted information is used in combination with sample metadata to address scientific and societally relevant questions such as biomarker detection for human diseases, predicting treatment efficacy, or precision diagnostics, assessment of food authenticity, fraud detection, and treatment or assess toxicity of the environment [8]

2 Data Space for ML and AI Applications

2.1 Training, Validation and Test Data, Model Performance Testing

Each ML and AI model requires training, validation, and test data. Training data is used to extract information to build a model, which generally performs classification, prediction, or pattern recognition tasks. The test data, on the other hand, remains unseen by the model during training and is used for the unbiased evaluation of model performance. Performance is measured by using model-specific loss functions measuring the error when comparing the predicted and real values or class labels of the samples in the test data, and using receiver operating characteristic (ROC) curves [9] for binary classification. The model performance is generally assessed in an n-fold cross-validation loop, where the available randomized data is split into n equal parts, with $n-1$ parts being used for training, while the remaining one part of the data is used for testing. The model building is performed n times, leaving out a different part of the data for testing each time. The loss (error) function is not the only approach for assessing how well the model is performing for predicting outcomes (e.g., physicochemical parameter) or classification, as stable and similar model parameters between cross-validation loops can also be used as indicators of consistent, well-performing models.

Generally, the model training includes parameter optimization, such as the number of principal components in partial least squares discriminant analysis (PLS-DA) [10], the type of kernel and kernel-specific hyperparameters for support vector machines (SVMs), or network architecture and weight of neurons for DL models (see Sect. 4.3 for details on SVMs and Sect. 4.4 for neural networks). When a model parameter is optimized through the training process, which is the case for most model parameters, an unbiased model performance assessment can be made in a double cross-validation loop. In this approach, the data is split into n parts as in the previous n-fold cross-validation, but the training data is further split into training and test data for model parameter optimization (often referred to as validation data), thus forming an inner and outer loop of testing, where the inner loop handles parameter optimization and the outer loop evaluates model stability and overall performance. This approach ensures that the model with the optimal model parameter from the inner loop is assessed against the test data of the outer loop in an independent and unbiased way.

2.2 Sample and Data Space Coverage

The coverage of data space used to train and test the model performance is of vital importance in determining what an ML/AI model can predict, classify, or recognize. Data space coverage is relevant on several different levels. Firstly, data coverage is required across the sample space, since the model can only assess information related to molecules covered by the training sample space when subsequently analyzing either test or real-world data. If this data space is incomplete, there will be missing information for model training and assessment. Examples of partially covered data spaces include non-complete sample collection from a particular stage of disease (e.g., leaving out samples from primary tumors or controls), missing samples from particular contaminated geographical areas for toxicological studies, or not sampling a specific part or developmental phase of a plant when establishing the metabolite composition of a plant species. The sample space is a particularly challenging space in which to achieve data completeness, because its true boundaries are often unclear, and ethical issues may arise when considering, e.g., the collection of normal patient matching control samples from cancer patients. Next, sample storage and preparation also play a crucial role, as they can affect which portion of the analytes is preserved in the sample and in which state. This analyte portion of the sample that will ultimately be analyzed by the analytical instrumentation platform. For example, it is well known that extraction conditions (used solvent, pH, salt, and additive content) play a crucial role in determining which class of metabolites is kept in the analyzed sample [11]. Similarly, sample collection and storage effects may alter sample characteristics such as clotting time of blood for plasma samples, and repeated freeze-thaw cycles can lead to compound degradation [12]. Next, the parameters of the instrument used to measure the prepared samples determine which types of information the acquired raw data will contain. For mass spectrometry-based proteomics and metabolomics, liquid chromatography (LC)–MS/MS platforms specifically, these parameters include factors such as degradation of compounds in the autosampler, ionization and fragmentation efficiency, stochastic sampling in data-dependent acquisition mode, scanning speed, the dynamic range of the detector, ion trapping, and transition efficiency. All of these parameters influence the quality of the acquired data, and therefore form another crucial data space that must be thoroughly sampled. Furthermore, the set of analytes detected by the analytical instrumentation from all measured samples defines the chemical data space that is available for ML/AI machine learning. Raw experimental data is generally preprocessed, and parts of this process may further alter the data used for ML/AI applications. An example of this from LC–MS/MS would be the centroiding of profile data, which is often used to reduce data volume. This process consists of finding peaks in raw profile-mode mass spectra and, for each peak, extracting and storing only a single intensity at a corresponding representative mass-to-charge ratio (m/z) at the spectral or analysis level. This single delta function is then used to represent the peak instead of including all data points originally recorded, thus reducing the data size.

This multitier data space is therefore enormous, and moreover, determining its boundaries is not a trivial exercise. It is important to assess not only which sample and chemical spaces the data covers, but also what could potentially be missing from the available data, as gaps in the data space coverage can lead to blind spots for the eventual ML/AI model. Domain-specific expertise is needed to assess how well the sample and chemical spaces are covered in the acquired dataset. It should also be determined how well a new part of those spaces is covered with the acquisition of data from new samples, or by changing sample collection, storage, and preparation protocols, or changing parameters, or the type of analytical profiling platform.

When acquiring new data that was not yet represented in the data space originally used to train the model, the partial (e.g., by transfer learning) or complete retraining of the model is required for the model to be able meaningfully engage with this new section of the data space. For example, an ML/AI model performing fragment spectra prediction trained exclusively with non-modified and methionine oxidized peptides is not well suited to performing prediction of spectra from phosphopeptides. Complete or partial retraining of the spectra prediction model using high-quality annotated phosphoproteomics data with known sequences and phosphosite localizations is required to improve the performance of the model [13].

2.3 Noise and Annotation Error of the Training Data

The quality of the training and test data influences the performance assessment of a given model, as the training data is considered, for most cases, to be the gold standard against which validation and test real-world data is compared. If, however, the training and test data contain significant noise and annotation error, this can adversely influence the performance of the trained model. There are some strategies that can be implemented to alleviate these issues. For example, an active learning process can iteratively select and label the most informative samples for training, and in this way help to reduce the impact of noise by favoring information obtained from these most valuable samples. Self-training and co-training can use their own predictions to generate additional pseudo-labels for unlabeled or labeled data, which can redefine labels for noisy or borderline samples. Using L1/L2 regularization [14,15] and dropping inconsistent data from the training dataset helps to avoid a strong reliance of the ML/AI model on a single sample.

Another approach to improve the noise tolerability of an ML/AI model is to include augmented data in the training dataset. The augmented data is obtained from the original data through transformations such as scaling, rotation, cropping, and noise addition, and including it in the training process makes the model more robust against these transformations. Combining predictions and outcomes from multiple ML/AI models into an ensemble model is yet another approach to making predictions more accurate and robust. Outlier detection and using loss functions like the Huber loss function, which is less sensitive to outliers or sample mislabelling compared to the typically used mean squared error, can further enhance model performance and robustness. The use of domain-specific noise models and transfer learning with noisier datasets can also improve model robustness. It is important to note that there are no clear guidelines for the selection of the appropriate approach to use to make models robust against noise and sample mislabelling. These approaches should be used with care after proper assessment of model performance on test data, where the quality of the output can be assessed independently by domain experts.

2.4 Effect of Data Structure and Characteristics

A clear understanding of data structure and data characteristics is essential for the development of robust, interpretable, and generalizable AI/ML models in mass spectrometry–based analytical chemistry. Mass spectrometry (MS) data are inherently complex, multidimensional, and highly dependent on the instrumentation, acquisition strategy, and experimental configuration used to generate them.

Key factors defining MS data structure include the type of mass spectrometer, the acquisition mode, the fragmentation strategy, and the parameterization of data acquisition. Instrument-specific characteristics such as the mass analyzer type (e.g., time-of-flight (TOF), Orbitrap, Fourier transform ion cyclotron resonance (FTICR), quadrupole, three-dimensional Paul traps, linear ion traps, or hybrid configurations) determine the mass accuracy, resolving power, scan speed, and dynamic range. These properties directly influence how spectral features are represented and should be explicitly considered when designing AI/ML models or selecting suitable input representations.

MS platforms support multiple acquisition modes, each generating distinct data structures. In single-stage MS (MS1) acquisition, spectra contain information on intact precursor ions, including isotope distributions, but no direct structural information from fragmentation. In Data-Dependent Acquisition (DDA), MS1 scans are followed by MS/MS (MS2) scans of selected precursor ions, typically the most intense features detected in the MS1 survey scan. While DDA provides relatively comprehensive MS1 coverage, MS/MS data are inherently biased toward high-abundance ions, leading to incomplete and stochastic fragmentation coverage across samples.

To address precursor selection bias, Data-Independent Acquisition (DIA) strategies have been developed. An early implementation is MSe (introduced by Waters), in which alternating low- and high-energy scans fragment all ions across the full *m/z* range. Although this approach removes precursor selection bias, it produces highly convoluted fragment spectra that require sophisticated deconvolution based on chromatographic co-elution and other constraints. More recent DIA methods fragment ions within a series of predefined, sequential isolation windows that together span the entire *m/z* range of interest. These windows are typically wider (e.g., 5–25 Da, often variable) than those used in DDA but much narrower than the full-range fragmentation used in MSe, resulting in data that are less convoluted than MSe but more complex than DDA.

Beyond *m/z* and retention time, modern MS workflows increasingly incorporate additional separation dimensions, most notably ion mobility spectrometry (IMS). IMS separates ions based on molecular shape and charge, enabling enhanced moderate orthogonal separation and reducing spectral complexity. A widely adopted implementation is trapped ion mobility spectrometry (TIMS), as used in Bruker timsTOF platforms, which introduces an additional, orthogonal dimension that must be considered in AI/ML model design.

Targeted acquisition modes further alter the data structure. In multiple reaction monitoring (MRM), predefined precursor–fragment ion transitions are monitored, yielding sparse but highly specific and quantitative data. In parallel reaction monitoring (PRM), selected precursor ions are isolated and fragmented, and full high-resolution

MS/MS spectra are recorded for each target, enabling increased specificity while retaining rich fragment information. These targeted strategies generate fundamentally different data representations compared to discovery-based DDA or DIA workflows.

The selected acquisition mode ultimately determines the dimensionality, sparsity, noise structure, and raw data labeling uncertainty of the resulting dataset, which in turn constrains how the data can be encoded for AI/ML models and which modeling approaches are appropriate. Explicit consideration of these data characteristics—along with use of any preprocessing choices such as centroiding, feature detection, and normalization—is therefore critical for the successful application of AI/ML in MS-based proteomics and metabolomics.

3 Types of Supervised Machine Learning Algorithms

While it is important to curate a training dataset that covers the breadth of the data space required to address a specific problem, perhaps of even greater importance is choosing the right ML/AI algorithm to train, as there are many different types of algorithms. Each algorithm serves to solve a specific type of problem and is generally not applicable to other tasks, and among similar algorithms, each has its individual strengths and weaknesses. This section and the next will describe a selection of commonly used algorithms, provide details on how they work, and highlight the type of problems these methods can be applied to. This section in particular will focus on so-called supervised machine learning algorithms, which use training data with known classifications or values of dependent variables to train a model to predict those classifications or values for unknown data.

3.1 Regression Models

Perhaps the simplest type of ML algorithms, regression models are familiar to all analytical chemists. When generating standard curves, such as for concentration determination of an analyte in an unknown sample with MS signal, the fitting of the experimental data to an equation describing a known, analytical relationship between the measured raw data and the compound concentration is done using regression models. The regression models are usually named after the type of function being fitted to the experimental data, leading to branches of regression such as polynomial regression [14], exponential regression [15], locally weighted scatterplot smoothing (LOWESS) [16–19], and linear regression [20–22], with linear regression, of course, being the simplest form.

There are different ways of generating an optimal model for a given regression problem, but since ML regression is an iterative process, these methods cannot be discussed before an approach to evaluating the goodness of fit of the model has been established. In ML regression, the measure of a model is referred to as a loss function and is typically a variant of the residual sum of squares [23, 24] (i.e., the sum of the squared distances from each data point to the model function). The residual sum of squares (SS_{res}) is calculated by the equation $SS_{res} = \sum_i (y_i - f(x_i))^2$, where y_i is the dependent variable of the *i*th data point (e.g., a quantitative readout of a peak intensity from an LC–MS/MS dataset), and $f(x_i)$ is the output of the model function predicting the expected dependent variable using the independent variables, x_i, of the *i*th data point as input (the model estimate of the peak intensity in the previous example). Using such a function to describe the model error has certain advantages, such as differentiability, which are useful in the optimization of the model. However, as the residual sum of squares is dependent on the sample size, and therefore not a comparable measure across models optimized using sample sets of different sizes, it is often averaged over the number of data points, resulting in the mean square error (MSE) given by the equation $MSE = SS_{res}/N$, where N is the number of samples. In some circumstances, other variations such as the root-mean-square deviation (RMSD or the square root of the MSE) or the coefficient of determination [25], which is given by $R^2 = 1 - \frac{\sum_i (y_i - f(x_i))^2}{\sum_i (y_i - \bar{y})^2}$, where $\bar{y}$ is the mean of the *dependent variable* across the experimental data points, may be more convenient, as they offer improved interpretability. The RMSD is measured in the same units as the dependent variable under consideration, which offers a more intuitive understanding of the size of the error, while the coefficient of determination can be interpreted as the percentage of the variability in the dataset that is explained by the model.

With a loss function of the models established, the optimal model is generated by minimizing that loss function. Under the assumption that the model function, f, is known and differentiable, this is usually done with gradient descent methods [26, 27], where the derivatives of the loss function with respect to the model parameters are calculated and used to compute new updated parameter values. For linear regression using the MSE as a loss function, f is simply a first-order polynomial (i.e., $f(x) = a \cdot x + b$), and the loss function derivatives with respect to parameters a and b become $g_a = \frac{\partial MSE}{\partial a} = -\frac{2}{N} \sum_i x_i \cdot (y_i - a \cdot x_i - b)$ and $g_b = \frac{\partial MSE}{\partial b} = -\frac{2}{N} \sum_i (y_i - a \cdot x_i - b)$, respectively, where N is the sample size. The derivatives, or gradients, are used to compute new values for a and b as $a_{new} = a_{old} - g_a \cdot L$ and $b_{new} = b_{old} - g_b \cdot L$, where L is the step size or learning rate of the algorithm. The process is iterated until a function minimum is reached and a final parameter set is obtained. The process can be improved by implementing, for instance, a conjugate gradient method [28], which uses the history of computed gradients in its determination of subsequent gradients, making the process of convergence to a minimum much faster.

While the model function in a regression problem is typically a known and differentiable function with one or more independent variables, it is worth mentioning that solutions to the optimization problem also exist if the model function is known but not differentiable, in which case algorithms such as the proximal gradient method [29] can be used, or if the model function is unknown outside of its output. In the latter case, various zero-order optimization algorithms, which only require the ability to evaluate the function for a given input set (i.e., the first-order derivative of the optimized function is not needed), can be used. An example of such an algorithm would be golden-section search [30]. For multivariate optimization problems, gradient descent and proximal gradient methods will still work for their respective use cases, but golden-section search is strictly reserved for univariate functions. Fortunately, other zero-order optimisation algorithms, such as pattern search [31] or the downhill simplex method [32], can be applied to such multivariate problems. This can be useful in training neural networks [33–35], where interpretability is generally an issue and gradients are not always easily calculable.

The previously discussed regression models performed value prediction, but regression can also be used to generate a classifier model, i.e., a model that will attempt to predict whether a sample has a given attribute or belongs to a specific class. This is achieved by equating a linear function (potentially in multiple variables) to a logit function, i.e., $\ln\left(\frac{y}{1-y}\right)=\sum_j a_j x_j$, where $\ln\left(\frac{y}{1-y}\right)$ is the predicted probability that the data point belongs to the class of interest, x_j is the feature set of the data point (x_1 being unity), and a_j is the parameter set of the model. This is referred to as logistic regression [36], and unlike linear/polynomial regression, it does not use the MSE as a loss function. Instead, logistic regression uses the logistic loss function, also referred to as the cross-entropy between the predicted and known classifications of the training dataset, to evaluate the model. The logistic loss function is essentially a measure of how often the model predicts a wrong outcome, and is defined as $L_{\overline{a}}=-\frac{1}{N}\sum_{i=1}^{N} y_i\cdot\ln(p_{\overline{a},i})+(1-y_i)\cdot\ln(1-p_{\overline{a},i})$, where N is the sample size, y_i is the known classification of the ith data point, and $p_{\overline{a},i}$ is the probability of y_i being unity as predicted by the model with parameter set $\overline{a}$. This loss function can be optimized using gradient descent methods as described earlier, leading to a classifier with minimal misclassifications. To evaluate the model in a manner more easily interpretable by humans, a confusion matrix can be computed using a test dataset. A confusion matrix measures how many times the model predicted a classification and whether it was correct. So the diagonal would consist of the number of true negatives and true positives, while the off-diagonal entries would show the number of false negatives and false positives classifications, with an optimal model producing a diagonal matrix. From the confusion matrix, performance metrics such as the model accuracy (the ratio of true classification to the total number of samples), the sensitivity (the ratio of true positives to the number of positive samples), and the specificity (the ratio of true negatives to the number of negative samples) can be calculated. These metrics can then be used to evaluate the impact of the model classifying a test data point as either positive or negative on the probability of this classification being accurate through the use of Bayesian inference [37, 38].

3.2 Decision Trees

Another approach to classifying data is through the use of a decision tree [39]. This is an approach that uses known data points to generate a flow chart for the classification of unclassified data points. The flow chart, or decision tree, consists of a series of branch nodes, where the data points are separated based on their value for one specific feature of the data points, and ends in a variety of leaf nodes, which ideally contain only one type of known data points and determine the classification of an unclassified data point that has followed the flow chart to a given leaf node. The general idea of decision trees should thus be familiar to many analytical chemists, but the complex rules that govern the generation of such a tree might not be. In order to determine which feature of the data points should be used to split them at any given branch node, a measure of the "purity" of the nodes, i.e., how consistent the classifications of the contained training data points are, is required. In practice, the chosen measure is typically the so-called Gini impurity [40], $I_G(n)$, which actually measures the diversity of the nodes, and is computed as $I_G(n) = 1 - \sum_{i=1}^{J} \left(\frac{n_i}{N}\right)^2$, where n_i is the number of samples in the training set with classification *i*, *j* is the number of different classification options (i.e., the number of groups for the classification outcome), and *N* is the total number of training data points in the node. The Gini impurity has a range between 0 for perfectly pure nodes (nodes only containing one class of sample) and a maximum value of $1 - \frac{1}{J}$ for nodes with equal populations of all the available classifications (i.e., nodes that contain the same number of data points for every classification group).

To determine the improvement in classification accuracy from splitting the data points based on a given data feature, the Gini impurity is calculated for the parent/branch node and all the child nodes that would be generated by the split, and the child impurities are subtracted from the parent impurity after weighting them based on the proportion of the data points from the parent node each child node contains (see Fig. 3). This measure of how much the total node impurity has improved after the split is referred to as the information gain, and once it has been computed for all hypothetical splits based on each available feature of the data points, the real split is made based on the feature resulting in the highest information gain. This

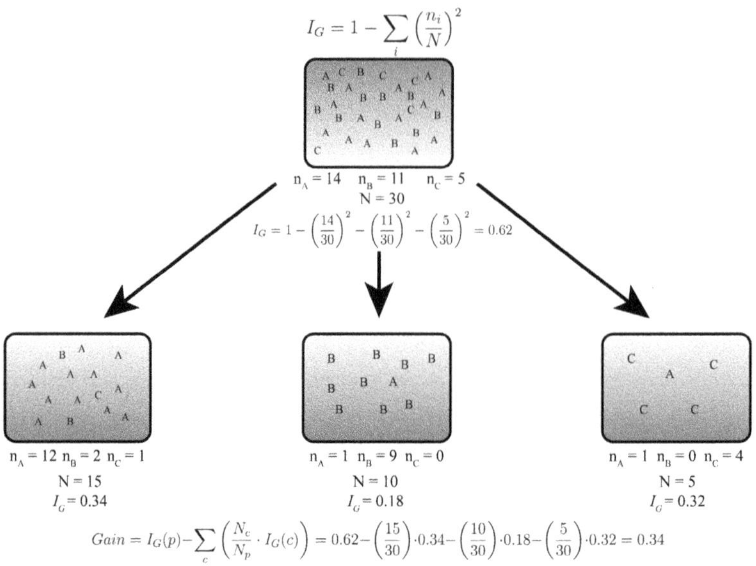

Fig. 3 Determining the information gain from splitting data at a branching node in a decision tree. The Gini impurity of the parent node is 0.62, and the three child nodes have impurities of 0.34, 0.18, and 0.32, respectively. The information gain is calculated by the equation at the bottom to be 0.34

process is repeated until either pure leaf nodes (i.e., nodes with a Gini impurity of 0) are obtained or all the data point features have been used. Note that the order in which the features are used to split the data does not have to be consistent between branches of the tree. As with logistic regression, the quality of a decision tree can be evaluated using a confusion matrix generated from the test dataset.

Even a tree with highly populous, pure leaf nodes will still be prone to occasional misclassification of samples. This is due to the tendency of decision trees to overfit the training dataset [41], which is particularly pronounced with large trees, i.e., trees with many branch nodes. This limitation of decision trees can be mitigated through strategies like pruning [42, 43], where a maximum number of branch nodes is predefined, but can also be circumvented somewhat by building multiple decision trees from random subsets of the training data and using the consensus classification of this so-called random forest [44, 45] as the classification prediction. This process of generating random subsets of the training data is called bagging, and by allowing each data point to be selected more than once, the process can produce random subsets of the same size as the original dataset. Further variation in the random forest training datasets can be obtained by so-called feature bagging [46], where each set of data points uses only a subset of the available data point features to train that particular decision tree.

By using these two types of bagging approaches, sufficient variability can be introduced into the otherwise deterministic generation of multiple decision trees, so that each tree in the random forest has a legitimate chance of coming to different conclusions, thus reducing the risk of the final outcome being overfitted to a particular training dataset.

3.3 K-Nearest Neighbors

Another approach to classifying unknown samples is the K-Nearest Neighbors algorithm [47, 48]. For a data point with N features (i.e., independent variables), this algorithm works by placing the sample and a training dataset of pre-classified samples in an N-dimensional space and calculating the Euclidean distance from the unknown sample to every training sample. The algorithm then selects the k nearest neighboring samples and classifies the unknown sample as the consensus classification of those K training samples.

There are a few considerations to make when generating a K-Nearest Neighbors algorithm. Firstly, since the distances are calculated using the standard Euclidean distance, a feature/dimension with a large dynamic range might dominate the calculated distances, making otherwise key distances in other dimensions irrelevant. To avoid this, each feature of the data points is typically normalized by min–max normalization [49] or standardized using Z-scoring [50], thus ensuring that all features are measured on similar scales [51]. Each of these strategies has strengths and weaknesses [52, 53], with min–max normalization producing truly normalized datasets, where every feature has a dynamic range from 0 to 1, but not handling outliers well, and Z-scoring might be preferable, but if not, min–max normalization is likely the better choice. Secondly, it is necessary to evaluate the accuracy of the model. This is usually done by computing the validation accuracy score of the model using a test dataset containing samples not used for model building and with known classification labels. The validation accuracy score is the percentage of data points the model classified correctly. Thirdly, and perhaps most obviously, is the value of K, which is a model parameter and can be optimized. Fortunately, generating a K-Nearest Neighbors model is not a resource-intensive process, so the simplest solution of trying a large range of options for K and choosing the one that produces the highest validation accuracy score is entirely viable.

4 Types of Unsupervised Machine Learning Algorithms

So far, the algorithms discussed have been based on supervised learning, where training datasets of pre-classified samples (or with known continuous values as dependent variables for the linear regression model) are used to generate a template for how to classify unknown samples. While supervised learning is a powerful tool for situations where it is applicable, it cannot be used to analyze data without having a preexisting, labeled training dataset available to build a model. In the absence of an annotated training dataset, unsupervised learning can be used to find patterns in a given dataset and to identify clusters of similar samples. This process of data clustering can be achieved through the use of different algorithms, of which the most commonly used is called *K*-Means Clustering.

4.1 Clustering Algorithms

The *K*-Means Clustering algorithm [54–57] can almost be considered the inverse of the previously described *K*-Nearest Neighbors algorithm, in that it uses a 1-Nearest Neighbor algorithm to assign each data point in the dataset to one of *K* randomly placed pseudo data points, which have the same independent variables as the real data points, but also have a cluster/classification parameter. For reasons that will become clear shortly, these pseudo data points are referred to as cluster centroids. As this cluster assignment is once again based on the Euclidean distance between two points in an *N*-dimensional space, it is sensitive to dynamic range differences, and often requires normalizing or standardizing the features of the data points as described in the previous section so as to avoid clustering based solely on the features with the broadest ranges [51]. Once all samples have been assigned to a cluster, the mean value of each feature of the data points in a given cluster is computed and used as new feature values for the cluster centroids, which results in the cluster centroids being moved to the centroid position of their respective data point clusters, hence the name. If the cluster centroids move, the cluster assignment of the data points may also change, and so the process can be iterated upon by alternating between assigning samples to clusters and moving the cluster centroids to the mean coordinates of the clustered samples. Convergence is achieved when the cluster centroids no longer move between iterations.

The quality of the produced clusters can be measured by the so-called inertia, which is the within-cluster sum of squared distances, i.e., the sum of the squared distances of each data point in the dataset to the nearest cluster centroid. As with the *K*-Nearest Neighbors algorithm, this quality measure can be used to determine the optimal value of *K*. However, in this case, the inertia needs to be minimized rather than the validation accuracy score being maximized, and a cut-off value must be chosen. This is because the model inertia is a strictly decreasing function of *K* for values of *K* ranging from 1 to the number of data points in the dataset [58]. Obviously, setting the number of clusters equal to the size of the dataset would lead to an inertia of zero, but it would also be a completely useless model, as it fails to group the data points together in any meaningful way. Choosing the best value for *K* is often done by plotting the inertia as a function of *K* and choosing the *K*-value where the descent rate of the graph becomes insignificant. The process could be automated by calculating the inertia of models with an increasing value of *K* until the improvement in the inertia of the $K+1$-clusters model over the K-clusters model drops below a predetermined percentage. While the cluster inertia is a common quality measure for clustering algorithms, it is by no means the only one. A comprehensive review of these various quality measures can be found elsewhere [59].

The K-Means Clustering algorithm, while popular, is not without drawbacks. One problem with the algorithm is that, since the initial cluster centroids are chosen at random, it may occasionally become trapped in a local minimum, where the clusters remain unchanged between iterations, but the inertia has not reached a global minimum [60]. This can be avoided by selecting the initial group of cluster centroids in a nonrandom fashion, such as with the K-Means++ algorithm [61]. Here, the first cluster centroid is selected at random from one of the existing samples, and the remaining cluster centroids are chosen from the other data points using a probability distribution weighted by the distance from each point to the existing cluster centroids, thus making data points far away from the existing cluster centroids more likely to be chosen. Another potential problem with K-Means Clustering is that since it uses the Euclidean distance to assign data points to clusters, the clusters will end up uniform in size, and since real-world clusters may have different variances, this can lead to samples at the edges of a loose cluster being wrongly assigned to a nearby tight cluster. This can be prevented by using an expectation–maximization (EM) algorithm [62, 63], which takes both variances and covariances into account when generating the model clusters, thus allowing the model to accommodate different sizes of real-world clusters at the cost of increased computation time. A third problem with K-Means Clustering, again caused by the use of the Euclidean distance for cluster assignment, is that it does not handle anisotropically shaped clusters well. If, for instance, the real-world clusters are distinct smears across the parameter space, in which the intra-cluster distances may exceed the inter-cluster distances, human intuition would quickly group the data points together from a visual representation, but K-Means Clustering and many other centroid based-clustering algorithms would produce wrong clusters, which would most likely include fractions of multiple different real-world clusters (see Fig. 4). This problem can be circumvented through the use of other clustering algorithms such as Spectral Clustering [64–67], which actually may make use of K-Means Clustering to cluster a proxy dataset derived from the original dataset through the use of similarity graphs and graph Laplacian matrices, or various density-based clustering algorithms such as DBSCAN (Density-Based Spatial Clustering of Applications with Noise) [68] or OPTICS (ordering points to identify the clustering structure) [69].

4.2 Artificial Neurons and Perceptrons

The term "perceptron" has historically been used in several different ways, with some sources using the term to describe the general

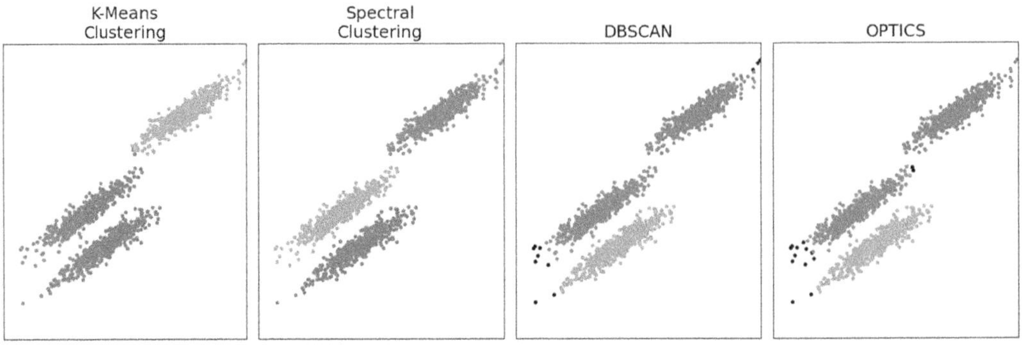

Fig. 4 Scatter plots of clustering methods. Scatter plots of a two-dimensional dataset consisting of anisotropic smears of points with color-coded clusters as determined by *K*-Means Clustering, Spectral Clustering, DBSCAN, and OPTICS. The clustering algorithms were implemented using the scikit-learn Python package [70]

concept of an artificial neuron, others shrinking the definition to a particular type of artificial neuron, and yet others using it to denote the algorithm used to optimize that particular type of artificial neuron. For clarity, this chapter will use the definition referring to a particular type of artificial neuron and refer to the optimisation algorithm as the perceptron algorithm. While perceptrons are significant in that they were the first type of artificial neuron proposed, it is more instructive to first define artificial neurons and then describe the special case of perceptrons.

Artificial neurons are mathematical functions [73], which seek to mimic the operation of biological neurons. Much like real neurons, they take a series of input signals (postsynaptic potentials at neural dendrites), transform those inputs, and if the transformed inputs meet certain criteria, produce a tangible output (axonal action potential). Formally, an artificial neuron takes an input in the form of an m-dimensional vector, $\overline{x}$, with entries x_1 through x_m. The artificial neuron itself consists of $m + 1$ parameters and an activation function. The parameters are m scaling factors and a bias term, sometimes referred to as the scaling factor of a zeroth input $x_0 = 1$. These scaling factors are used to compute the weighted sum of the inputs as $u = \overline{w} \cdot \overline{x} = \sum_{i=0}^{m} w_i x_i = 1 \cdot b + \sum_{i=1}^{m} w_i x_i$, where u is the weighted sum, $\overline{w}$ is the $m + 1$-dimensional vector of the scaling factors, and $b = w_0$ is the bias term. Note that with this notation, the inputs are now $m + 1$-dimensional vectors, which include a constant dimension that produces the bias term as $b = w_0 \cdot x_0 = w_0$. This simplifies the notation, but admittedly can lead to confusion, as the data space is m-dimensional, but the vector representation of each point is $m + 1$-dimensional. The activation function, $\phi(u)$, is a function which takes the weighted sum, u, as input to generate the output, y, of the neuron. The activation function can take various forms and is what determines the type of artificial neuron. Typical activation functions include a pass-through function (the output is simply the input), various sigmoidal functions (which are used as well-behaved, nonlinear activation functions), and the rectifier function (which is a pass-through function if the input is strictly positive and otherwise outputs zero).

For perceptrons [72, 73], the activation function is a so-called Heaviside step function [74], which produces an output of 1 if the input is strictly positive and zero if it is not (some sources use a similar step function, which equals −1 for negative inputs, but this change does not influence the below considerations). Using this activation function turns the perceptron into a linear, binary classification algorithm, which, not unlike the previously described logistic regression algorithm, can determine whether input samples belong to a specified class. This is because the $m + 1$-dimensional neuron parameter vector, $\overline{w}$, defines an m-dimensional subspace, called a hyperplane, as the subspace to which it is a normal vector, and this hyperplane separates the $m + 1$-dimensional space into two sections based on the sign of the weighted sum, u. Given a training dataset of points in the $m + 1$-dimensional space with known classifications, the scaling factors or weights of the perceptron can be adjusted such that the generated hyperplane separates the dataset based on classification, provided the dataset is linearly separable. That is, if a hyperplane exists, which separates the dataset based on classification, it can be represented by a perceptron. The appropriate weights can be found using the perceptron algorithm [75], which iterates over the training dataset and updates the weights vector whenever the perceptron mislabels a sample. The weights are updated based on the equation $\overline{w}(t+1) = \overline{w}(t) + r \cdot (l_j - \phi(u_j(t))) \cdot \overline{x_j}$, where $\overline{w}(t)$ is the weights vector after t iterations, l_j is the label of sample j, $\phi(u_j(t))$ is the perceptron output for sample j using the weights vector after t iterations, $\overline{x_j}$ is the input vector of sample j, and r is the learning rate of the algorithm. The iterative process can continue indefinitely without ever converging if the dataset is not linearly separable, but if it is, the perceptron algorithm is guaranteed to find a valid solution in a finite number of iterations. There is, however, no guarantee that the found solution will be the optimal solution. To handle this complication, algorithms have been further developed to determine the so-called perceptron of optimal stability, which will correspond to the hyperplane that splits the dataset with the maximum margin between the two classes. Examples of algorithms used to determine the perceptron of optimal stability include the Min-Over (minimal overlap) [76] and AdaTron (adaptive perceptron) [77, 78] algorithms.

4.3 Support Vector Machines

The perceptron described above is an excellent approach to binary classification, but it is limited to datasets which are linearly separable. This restriction on the dataset can sometimes cause problems for real-world data, and so various strategies have been developed to handle cases of nonlinearly separable data. These approaches include, among others, perceptrons with errors, which can be implemented as minimal misclassification classifiers [79, 80] or soft margin classifiers [81], and layered networks of perceptron units, such as committee machines [82–84] and parity machines[82, 85, 86].

One highly popular approach is the Support Vector Machine (SVM) [81, 87], which gets around the issue of linear separability by transforming the training dataset through a nonlinear transformation into a space of much higher dimensionality, where the dataset is likely to be linearly separable. It seems reasonable that such a transformation can be found, as the new space can be of arbitrary dimensionality, including infinite, but it is not obvious how to choose the transformation, and the increased dimensionality appears likely to make the classifier prohibitively computationally expensive to train. Fortunately, both of these issues can be circumvented by imposing certain requirements on the transformation function. Specifically, requiring that the transformation satisfies the condition that the dot product between two transformed input vectors must be computable in terms of the input vectors in the original $m + 1$-dimensional space through a so-called kernel function, as $\Psi(\overline{x_i}) \cdot \Psi(\overline{x_j}) = k(\overline{x_i}, \overline{x_j})$, where Ψ is the transform function, $\overline{x_i}$ is the i'th input vector, and k is the kernel function, turns out to be an immensely useful restriction. This is because for a given kernel function, $k(\overline{x_i}, \overline{x_j})$, which satisfies Mercer's condition [88] of being positive semi-definite, a corresponding transformation fulfilling the condition is guaranteed to exist. Furthermore, because the optimization problem in the higher-dimensionality space ends up only featuring the transformation function Ψ in dot products between two transformed input vectors, it can always be substituted by the kernel function. This approach is commonly referred to as the kernel trick [89] and enables the training of the non-linear SVM classifier (i.e., a classifier that can separate complex, curved boundaries in the input data) as the linear perceptron of an optimal stability classifier for a higher-dimensionality, linearly separable proxy dataset without explicit knowledge of the transformation function, Ψ, used to generate that implicit proxy dataset. Which kernel to choose to ensure that the transformed dataset is in fact linearly separable depends on the specific problem, but there are many options to choose from, including polynomial kernels [87, 90], sigmoid kernels [91], Bessel function-based kernels [92], linear spline kernels [93], and various radial basis function kernels [94], and each has merits in certain contexts. If one is unsure which kernel to use, general-purpose kernels like the Gaussian radial basis function kernel are a good place to start, although the choice of kernel function is yet another problem that can be solved through optimization.

4.4 Deep Learning

Deep learning [7] is an area of machine learning which uses multiple instances of the above-described artificial neurons to build networks in an effort to mimic the information processing pathways found in human brains. These Artificial Neural Networks (ANNs) can be structured in many different ways [95], with the most common so-called architecture being the Deep Neural Network (DNN) [96, 97].

Deep Neural Networks (DNN)

Deep Neural Networks (DNNs) consist of a series of fully interconnected layers, each containing a variable number of independent neurons. Each layer uses the outputs of the neurons of the previous layer as inputs, and in turn provides outputs from each neuron, which are used as inputs in the subsequent layer. The first and last layers of the network are called the input layer and the output layer, and they serve the special functions of interacting with the original dataset and outputting the final prediction or effectuating the determined appropriate action, respectively. While these two layers are necessarily limited in their structure and their input/output ranges, the intermediate layers can be structured in myriad different ways, and can vary broadly during the optimization of the network.

Convolutional Neural Network (CNN)

Another common type of ANN is the Convolutional Neural Network (CNN) [98], which is commonly used in image analysis. While DNNs could be used to analyze an image by representing the image as a vector with one entry for each pixel in the image, this is not a good approach to image analysis, as a small change in the input image (say, a feature of the image is moved to a different position) can result in a large and unpredictable change in the vector representation. CNNs instead use convolution layers, which compute the matrix product between a small filter matrix and equally-sized subsections of the input image to generate an activation map corresponding to features that look similar to the filter matrix. By using multiple different filter matrices, edges with different orientations can be detected in the input image and passed to the next layer of the neural network. CNNs also make use of pooling layers, which reduce the dimensions of the image being analyzed by representing multiple pixels with a single value. This could be done by averaging the value of a small square of pixels (say, a 2×2 or 3×3 block) or selecting the maximum value from that small square of pixels. Note that using a 3×3 pooling window does not necessarily mean that the dimensions of the output image are reduced by a factor of three relative to the input image. This is because of a second parameter used in the pooling function called the stride, which determines how many pixels the pooling window is moved between iterations in the mapping process, and thus the factor by which the image dimensions are reduced. Using these convolution and pooling layers, the original image can be reduced to an edge map of much smaller size, which can then be passed to a DNN to produce an output (e.g., a classification of the image).

Recurrent Neural Network (RNN)

Both DNNs and CNNs belong to a type of ANNs collectively referred to as Feedforward Neural Networks (FNNs), because information travels in only one direction through the network. This is in contrast to so-called Recurrent Neural Networks (RNNs) [99], where the information flow is bidirectional in the sense that the output of a given neuron is able to influence the subsequent input of that same neuron. RNNs are uniquely suited to analyzing sequence-style input data (such as DNA or peptide sequences), because this feedback mechanism allows them to "remember" information about previous input, and use this "memory" to add context to future inputs. If, however, the input sequence becomes exceedingly long, these standard RNNs tend to experience training problems called the vanishing and exploding gradient problems, which relate to the gradients used to update the weights of the network, becoming either vanishingly small or stupendously large, and thus severely complicating the training process. Several RNN variants have been developed to circumvent these issues. Specifically, Gated Recurrent Units (GRUs) [100, 101] were developed to handle large gradients by allowing the network to "forget" information, whereas Long Short-Term Memory (LSTM) networks [102] are able to retain information over a large number of iterations. GRUs have been shown to be effective in speech recognition applications [103] and to provide performance comparable to LSTMs for sequence modeling tasks [104]. LSTMs, meanwhile, have found widespread use in a range of different contexts including, but not limited to, drug design [105], prediction of subcellular protein localization [106], control of surgical robots [107], music composition [108], and natural language processing [109, 110], although they have largely been replaced by transformers [111] in the lattermost context.

Model Optimization and Reinforcement Learning

The above are just a few examples of the many different ANN architectures which have been developed, and just as there are many ways to construct an ANN, there are a multitude of diverse approaches to optimizing a network. As mentioned earlier, one option is the downhill simplex optimization, where minimal knowledge about the network is required. All that is needed for this optimization algorithm to work is the ability to modify the internal weights of the network and evaluate the model output, and a loss function. As in standard regression problems, deep learning often uses loss functions like the MSE for value prediction and the logistic loss function for binary classification problems. For multiclass classification problems, loss functions like the categorical cross-entropy [112, 113] can be used. Due to the minimal requirements, such zero-order optimization algorithms can always be implemented, but for well-behaved networks, which use differentiable activation functions, much more efficient options are available. Using the backpropagation algorithm [114–116], the gradient of the loss function with respect to the weights of the network can be calculated in reverse order, starting from the output layer and moving backward through the network, hence the term backpropagation. The algorithm makes use of the chain rule of differentiation [117, 118], which necessitates knowledge of the gradients in one layer to determine the gradients in the previous one. Once the gradients have been computed, the weights can be adjusted using the familiar gradient descent method.

One problem with this approach to training a neural network is that it is computationally expensive. For large networks being trained on a large dataset of high dimensionality, it becomes prohibitively expensive. To combat this problem, stochastic gradient descent [119] has been proposed, in which an approximate gradient for the entire training dataset is calculated based on a random subset of the dataset, thus reducing the computational load of the gradient determination. While this will of course lead to poorer performance in each iteration during the optimization and a slower convergence rate with respect to the number of iterations, the reduction of computation time for each iteration counterbalances the slow convergence to produce an overall faster optimization algorithm for large scale optimization problems [120]. Stochastic gradient descent also has the advantage of enabling the use of the rectifier or Rectified Linear Unit (ReLU) activation function [121], which is one of the most commonly used activation functions today. The ReLU activation function is not fully differentiable, which is required for gradient descent, but since stochastic gradient descent only requires the activation function to be subdifferentiable, this problem can be circumvented. In practice, this is a fairly minor advantage, as most implementations of gradient descent for ANNs simply define the gradient of the ReLU function at the discontinuity to be zero [122], thus allowing it to work with the gradient descent algorithm as well.

The training approaches discussed thus far all have the same significant shortcoming: They optimize the weights in the network, but cannot fundamentally change the architecture of the network. In other words, they require a human to have already designed a network, including selecting the number and types of layers, and the number of neurons in each layer. This shortcoming has led to the development of several so-called genetic algorithms [123], which take inspiration from biological evolution, including implementing processes for mutation, recombination, and selection of candidate networks, to produce optimal models. These algorithms are a type of reinforcement learning, in which the performance of a large number of slightly different (mutated) models is evaluated through the application of a reward function (the model will be given a positive score for desired behavior and a negative score for undesirable behavior).

The highest scoring models (selection) are used to generate a new set of models to test through crossover exchange (recombination) of model components, random alterations (mutation) such as the addition or removal of neurons or layers, and the modification of a number of weights. The process is iterated upon through several "generations" until a satisfactory model is obtained, or it becomes clear from the behavior of the models that the reward function does not in fact reward the expected behaviors, at which point it is necessary to modify the reward function and restart the model optimization.

Many different versions of genetic algorithms have been suggested including GeNeralized Acquisition of Recurrent Links (GNARL) [124], NeuroEvolution of Augmenting Topologies (NEAT) [125] and several variants thereof (rtNEAT [126], HyperNEAT [127], cgNEAT [128], etc.), Evolutionary Acquisition of Neural Topologies (EANT/EANT2) [129, 130], and many others. A more comprehensive history and comparison of these various algorithms can be found elsewhere [131].

Toward Generative AI

The types of ANNs discussed thus far are generally concerned with data analysis to enable data classification, score optimization, or similar goals. However, another type of AI also exists, which can be useful not only in the training of the data classifying ANNs through the implementation of Generative Adversarial Networks (GANs) [132], but also in entirely separate applications such as predictive models. This type of AI is referred to as generative AI. The earliest generative AI models were hidden Markov models (HMMs) [133, 134], which used Markov chains [135] to generate sequential data, and primarily saw use in natural language processing, where they were eventually replaced by *n*-gram models [136]. Like discriminative AI, generative AI has seen rapid development since the advent of neural networks, with neural networks quickly supplanting *n*-gram models [137], and further development leading to generative RNNs [138]. Present-day generative AI, however, is almost exclusively based on a new type of neural network architecture that, while based on similar ideas to LSTM RNNs, does not feature the recurrent units of RNNs, which makes it less time-consuming to train. This architecture is known as the transformer. Generative AI and Large Language Models.

The development of transformer architectures [139] in 2017 led to a new type of AI based on Large Language Models (LLMs) that are trained on large amounts of text and are more directly applicable to specific domain applications with little or no additional training. Transformers rely on the concept of Attention and long-range connections in a neural network rather than a traditional architecture involving recurrent layers (as described above), and they map better to highly parallel GPU computation [140]. This is largely because they are able to process the training data more holistically rather than sequentially, allowing greater parallelism. One application of LLMs is generative AI, in which an input prompt yields a novel response that has been synthesized (generated) based on the corpus of training data. A common example is a chatbot that replies based either solely on the information in the original LLM or in conjunction with domain-specific text, such as a user manual for a product of a company. LLMs led to a rapid growth in generative AI, particularly with the release of ChatGPT from OpenAI in late 2022, followed shortly after by ChatGPT3.5 in April 2023 [141]. LLMs are themselves a subset of a broader set of models called Foundation Models [142], which are similarly powerful and widely applicable but are trained on more general and possibly heterogeneous data sources rather than text and language.

The impact of foundation models and generative AI on science has been significant, particularly in protein structure prediction, where the mapping of the sequence of a protein to its 3D structure fits naturally in the generative AI paradigm. An early application of transformer-based deep learning in protein science was ProteinBERT, which adapted a sequence-based language model architecture and was trained on more than 100 million protein sequences [143]. ProteinBERT demonstrated strong performance in a range of sequence-level prediction tasks, including functional annotation and the prediction of post-translational modification sites. However, the model does not explicitly predict three-dimensional protein structures; instead, it learns rich sequence representations that can support downstream structure-related or functional inference when combined with other models or methods. More recently, AlphaFold2 and AlphaFold3 have demonstrated the ability to predict 3D protein structures with atomic-level accuracy [142, 144].

Protein science fits well in generative AI because many applications involve a well-defined input query consisting of a peptide sequence, with the desired response being predictable based on data from actual protein structures [145, 146]. In contrast, many aspects of analytical chemistry involve a noisy and multidimensional input, such as an LC–MS dataset, where the desired response is the corresponding peptides that produced the dataset. One related example based on Deep Learning, but not foundation models, is Spec2Mol, which maps MS/MS spectra to chemical structures [147]. Currently, there is little published activity in applications of foundation models and generative AI to mass spectrometry and other areas of analytical chemistry, but the potential is there, as long as a model can be trained on a sufficiently large amount of data to provide a useful response to a given query.

Recently, foundation models [140] for mass spectrometry have emerged that can learn directly from large collections of MS/MS data. A prominent example is DreaMS (Deep Representations Empowering the Annotation of Mass Spectra) [148], a transformer-based neural network pretrained in a self-supervised manner on millions of unannotated tandem mass spectra from the GeMS (GNPS Experimental Mass Spectra) dataset [149]. During pretraining, DreaMS learns to reconstruct masked fragment spectral peaks (*m/z*-intensity pairs) and to predict chromatographic retention order among pairs of spectra, enabling the model to acquire rich, structure-related representations of small molecules directly from raw spectral data.

Rather than directly annotating spectra, DreaMS learns dense, fixed-length embeddings of precursor–fragment spectra pairs that capture chemically meaningful features of molecular structure. Once pretrained, DreaMS can be fine-tuned via transfer learning for a variety of downstream spectrum annotation tasks. For example, the model has been adapted to predict: molecular fingerprints (e.g., Molecular ACCess System (MACCS) keys [150, 151] derived from MS/MS spectra), chemical properties (e.g., lipophilicity, polarity, molecular size/shape, Bertz molecular complexity, quantitative estimate of drug-likeness (QED), synthetic accessibility score), spectral similarity, and presence of specific elements, such as fluorine. These predicted features are derived from transformer-based embeddings that encode chemically relevant information from both precursor and fragment ion spectra. In addition to predicting masked fragments, DreaMS also learns chromatographic elution order as part of its self-supervised objective, implicitly capturing aspects of retention behavior that are critical for compound identification and spectral interpretation. This combination of self-supervised learning on unannotated data and transfer learning for downstream tasks positions DreaMS as a foundational model for small-molecule MS/MS interpretation, capable of improving annotation coverage and robustness beyond conventional library-based approaches. Compared with traditional spectral matching methods, DreaMS provides a more flexible and generalizable framework, particularly for previously unseen or poorly characterized compounds.

Overall, DreaMS represents an important step toward foundation models in analytical chemistry, demonstrating how large-scale self-supervised learning can bridge raw mass spectrometry data and chemical structure space. While still an emerging area, such models highlight the potential of generative and foundation-model paradigms to transform the interpretation of complex MS datasets, provided that sufficiently large and diverse training data are available.

5 AI Models for Mass Spectrometry

5.1 Application of ML/AI in Proteomics and Metabolomics Profiling with LC–MS/MS

Today, multiple aspects of the data generated by LC–MS/MS-based proteomics (and metabolomics) can be assessed, processed, and modeled by ML/AI (Fig. 5). This section will list all the areas where ML/AI algorithms were found to either already be gainfully implemented or show great promise for future implementations.

These areas include the prediction of detectable compounds by an LC–MS/MS platform depending on experimental conditions such as LC column stationary phase and eluent composition, pH and salt concentration, the prediction of protein sample digestion output based on the applied protease, the identification of compounds from raw data and using spectral libraries, the detection and quantification of peaks using ML/AI-supported approaches, the identification of metabolites and peptides from data generated through different data acquisition modes, and finally the prediction of physical, chemical, and analytical system characteristics such as compound re ional cross sections. For each presented tool and approach, the purpose and type of the associated ML/AI models will be discussed, and the model performance assessed. As part of a review published in 2020 by Wen et al. [152], the authors produced an extensive list of bioinformatics tools and approaches making use of deep learning models to analyze mass spectrometry-based proteomics data, which can be found on GitHub [153].

5.2 Predicting the Observability of Peptides Prior to MS Analysis

As mentioned previously, protein science is highly compatible with AI tools due to the sequential structure of peptides. In contrast, predicting the observability of metabolites is much more difficult due to the larger chemical data space, the lack of common building blocks, and the unavailability of complete libraries of metabolites expected to appear in complex biological samples. These factors make it impossible for supervised training to consider all potentially present metabolites in a given sample. As a result, the literature lists only one study predicting metabolite levels in cancer cell lines using

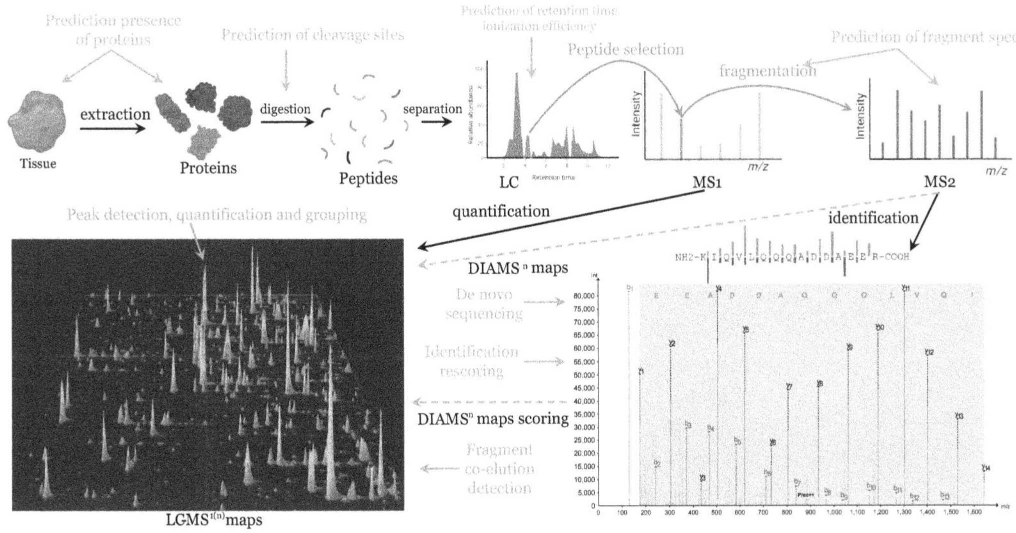

Fig. 5 Common DDA/DIA workflow. This scheme illustrates the common shotgun bottom-up proteomics workflow with data independent (DIA) and data dependent (DDA) acquisition. It shows the main steps (black/red text and arrow) and characteristics that are predicted, rescored, or analyzed with various ML/DL algorithms

metabolite/mRNA correlations [154]. Predicting the observability of peptides, on the other hand, requires only a list of potentially measurable candidate peptides, which can be obtained from a protein sequence database through in silico digestion using the known digestion patterns of a given protease. Peptide observability predictions mainly rely on the amino acid sequence, but may be enhanced by employing the structural, physicochemical, and evolutionary features of the proteins/peptides, which enables the use of ML-based approaches, but at the moment, DL-based approaches outperform ML approaches without relying on this additional information.

In Silico Digestion in Targeted Approaches

The most widely used approach to identify peptides and proteins in LC–MS/MS data is through database (DB) searches. This approach is targeted, i.e., it uses lists of peptides that are supposed to be present in the analyzed samples. The lists are generated by the in silico digestion of species-specific proteome databases using the cleavage rules of the applied protease, as generally established several decades ago using a small cohort of manually observed peptides. With the increasing popularity of MS-based proteomics and data sharing initiatives like PRIDE [155], iPROX [156], jPOST [157] and MassIVE [158] forming the ProteomeXchange Consortium [159], larger volumes of data are becoming publicly available, which allows for the refinement of the peptide cleavage models and improvement of the prediction of peptide products.

Multiple models have now been established for this enzymatic digestion of proteins, which enables the prediction of which peptides will be generated and are therefore potentially detectable. Information on the expected cleavage sites not only limits the search space of possible peptides to be matched against experimental spectra, but can also improve MS-based estimations of protein abundances [160, 161]. Accurate prediction of protease substrates and products facilitates the study of the role of proteases in different diseases and other physiological conditions, leading to a better understanding of their biological functions [162]. In targeted proteomics, a priori knowledge of proteases is particularly important, since the observed masses must be selected in advance. To maximize the information obtained from an experiment, so-called proteotypic peptides and their most intense fragments are usually chosen for observation. Proteotypic peptides are peptides unique to a specific protein (i.e., they can only arise from the digestion of one specific protein in the searched database), but they must also be reliably detectable through MS techniques. The selection of proteotypic peptides, which both ionize and fragment in a predictable fashion and are not lost during sample preparation, is, therefore, a vital aspect of targeted proteomics [161]. More accurate predictions of cleavage sites improve peptide-spectrum matching and enhance protein quantification, especially in targeted proteomics [160, 161].

There is still an unmet need to make these approaches more broadly applicable, which would include the modeling of more enzymes to predict their behavior under different conditions (e.g., different pH, ionic strength, temperature, solvent, and in the presence of cofactors) as reported by Wall et al. and Uliana et al. [163–165]. Furthermore, to understand enzyme cleavage behavior in proteomics, computational models can analyze cleavage sites and their surrounding sequence context using measures such as cleavage and block entropies. These models help distinguish which measures reflect the intrinsic site specificity of the enzyme and which capture the positional preferences of amino acids around the cleavage sites. [164]. This data is useful in training models for peptide product profile predictions, which help tailor multi-protease digestion protocols to particular proteins, classes of protein, or sample types. It should be noted that current protocols use either standard cleavage rules [165] or nonspecific cleavage [164] and use canonical protein databases publicly available from UniProt, which do not account for protein variants and isoforms, and only include the commonly known and frequent posttranslational modifications (PTMs) of proteins. As a result, current models may be biased towards canonical proteins containing only a few PTMs.

Modeling of Peptide Physicochemical Properties in Targeted Approaches

Apart from cleavage site predictions of the used protease, the detectability of peptides depends on their physicochemical properties, which determine their ionization potential and can predict sample loss during sample preparation, in the autosampler, or during chromatography [166, 167]. Several methods have been developed to evaluate the physicochemical properties of peptides for detectability prediction using identified peptide information from public databases or large-scale experiments. For example, Gao et al. [165] published the tool Advanced Proteotypic Peptide Predictor (AP3), which combined the prediction of trypsin peptide products and the detectability of peptides using a decision tree trained on 544 physicochemical properties of amino acids and the list of tryptic peptides with detected/undetected labels. Optimizing the number of peptides potentially present in the sample resulted in more accurately predicted search spaces and improved the identification of peptides and proteins. Fannes et al. [168] used a decision tree considering amino acids around the trypsin cleavage sites defined by the original Keil rules [169] in their tool Cleavage Prediction with Decision Trees (CP-DT) and using peptide identification data from the public PRIDE repository for prediction of trypsin cleavage products. The obtained decision tree showed improvement over the original Keil rules, is interpretable by proteomics experts, and showed good prediction performance for trees trained and tested on different species. The iProt-Sub tool developed by Song et al. [162] predicts the cleavage sites of multiple enzymes from 4 major protease families using 11 different sequence-derived structural and physicochemical properties, such as the local amino acids sequence, solvent accessibility, and native sequence disorders. This information is used to train an SVM predictor, which predicts and ranks potential substrate cleavage sites specific to the studied protease. The model was trained and tested using cleavage products from the MEROPS database [170], which is a collection of the products of enzymatic proteolysis in multiple species. The d::pPop algorithm from Zimmer et al. [171] uses FNNs trained using the peptide intensities and predicted physicochemical properties of observable peptides from yeast and *C. reinhardtii* proteomics datasets to predict the ion intensity ranks of tryptic peptides. The model ranks the peptides according to observability as peptide observability is dependent on the abundance of the proteins, and the accuracy of its predictions was assessed using a QconCAT artificial protein set containing peptides mimicking portions of ten membrane-intrinsic and soluble photosynthesis proteins from *C. reinhardtii*. Sun et al. developed the deep-learning predicting missed cleavages (dpMC) tool, which, as the name indicates, is able to predict missed cleavages and the peptides containing those missed cleavage sites [172]. The tool uses a BiLSTM temporal CNN and encodes the protein sequence input data through a one-hot encoding layer before using this proxy input data to predict expected cleavage and missed cleavage sites for tryptic digestion. The model was trained using publicly available datasets from ProteomeXchange, which were acquired on various different mass analyzers using samples from various species. The output was a series of mass analyzer-specific probabilistic cleavage motifs of cleaved and uncleaved sites, and the list of experimentally detected peptides served as the ground truth for performance assessment. The dpMC algorithm outperformed the DeepDigest tool published by Yang et al. [173] a few months earlier. DeepDigest is built on an LSTM CNN and provides probabilities of cleavage sites for several widely used proteases such as trypsin, ArgC, chymotrypsin, GluC, LysC, AspN, LysN, and LysargiNase. The model was trained using 19 publicly available datasets, which were processed with the Andromeda search engine in MaxQuant. The deep learning models provided better performance compared to interpretable machine learning tools such as logistic regression, SVM, and random forest. Procleave [174] is a tool for predicting protease-specific substrate and cleavage sites based on the primary amino acid sequence and 3D structure of proteins obtained from the PDB database [175]. Structural features up to 4 amino acids away from known cleavage sites were encoded in a conditional random field (a type of undirected graphical model) based on training data consisting of primary amino acid sequences and features representing chemical groups. The training data were obtained from MEROPS, and performance was assessed using ROC curves.

Solutions for Untargeted Approaches

Information on peptide observability (together with digestibility) can also be used in an untargeted context, either for improving scoring, which is of particular interest in applications such as single-cell proteomics where low signal intensity can be an issue, or to reduce the size of the spectral library used during data-independent acquisition (DIA) data processing. Generally, these models are trained on large publicly available datasets using lists of identified peptides. For example, Guruceaga et al. [176] used multiple search engines such as Comet [177], Mascot [178], OMSSA [179], and X!Tandem [180] to analyze data from NCI-60 cell lines comprising 61 shotgun and 9 deep proteome measurements. The author used multiple classifiers and chose a random forest approach to devise a classification model to differentiate the most and least observed peptides based on 550 physicochemical and biochemical properties. Of these properties, 106 nonredundant properties were used in the final peptide prediction model. The model has been used to assess whether proteotypic peptides of missing proteins (proteins that have no evidence at the protein level but ought to be present based on other evidence, such as mRNA expression) can be detected. The previously discussed DeepDigest model [173] was used to reduce the peptide library search space by filtering out fragment spectra corresponding to peptides that have low detectability from the spectral library, which enhanced peptide identification by DIA-NN in DIA LC–MS/MS data [181]. Song et al. [182] developed DbyDeep, which is an innovative end-to-end LSTM model designed to predict peptide detectability in mass spectrometry–based proteomics. The model not only considers the peptide sequence itself but also incorporates its surrounding sequence context, e.g., to confirm the presence of a protease cleavage site, which improves prediction accuracy. DbyDeep was trained on the MassIVE-KB [183] peptide spectral library, which comprises 1,075,832 peptides derived from 19,300 proteins. Direct MS1 response has been predicted by Abdul-Khalek et al. [184] by using an encoder–decoder model with an attention mechanism, where the attention weights correlated with amino acid physicochemical properties. In this work, the authors gained insight into amino acid physicochemical properties governing MS1 response, which the model could later use to predict absolute peptide and protein quantities in label-free LC–MS/MS experiments, thus reducing the need to use stable isotope labeled standards.

All the literature discussed so far has investigated the full proteome. Unfortunately, we did not find a published ML/DL method which predicts the detectability of enriched subsets of the proteome, which could be used for samples, e.g., enriched for phosphorylated peptides. However, prediction models do exist to predict acetylation and phosphorylation sites in proteins. DeepAcet, developed by Wu et al. [185], uses six different encoding schemes and a multilayer perceptron model to perform acetylation site prediction and achieved

84.87% accuracy, 83.46% specificity, 86.28% sensitivity, an area under the curve (AUC) of the ROC curve of 0.8407, and a Matthews Correlation Coefficient (MCC) [186] of 0.6977. For comparison, a model that was randomly guessing would get an AUC under the ROC curve of 0.5, while a perfect model would get a score of 1. Similarly, for the MCC, guessing randomly would result in a score of 0, while a perfect classifier would get a score of 1. PTransIPs [187] is a prediction model for phosphorylation sites, which uses a pretrained language model (PLM). It achieved a quite impressive level of performance (sensitivity and specificity above 80%), yet the generalizability of the model is questionable due to the small size of the employed training set, which included only the protein output of A549 cells infected by COVID-19.

5.3 Prediction of Retention Times and Collisional Cross Sections in LC–MS Systems

Long-Lasting Challenges

Accurate prediction of analyte characteristics such as retention time (RT) and collisional cross section (CCS) can significantly improve the reliability of analyte identification from LC–MS experiments as well as experimental reproducibility. Of the relevant analyte characteristics, RT and CCS prediction are among the most extensively studied, and they have been widely successful applications of ML and DL in LC–MS data analysis [190–192]. Reliable estimations of these properties enhance identification accuracy by reducing ambiguity in candidate matching, especially in complex datasets. It should be noted that the retention time (RT) of an analyte in a chromatographic system depends on multiple factors, including the stationary phase type (e.g., reversed phase, hydrophilic interaction liquid chromatography, (HILIC), mixed-mode, and size exclusion), the mobile phase composition (e.g., acetonitrile, methanol, additives such as trifluoroacetic acid, (TFA) or acetic acid that affect pH, or ion-pairing reagents), the accuracy of solvent mixing and gradient delivery, column temperature, and pH. From this list of variables involved in determining the RT of an analyte, it is clear that the absolute RT value is characteristic of the chromatographic system as much as of the analyte. A well-known example of modeling this complex relationship is the linear solvent strength (LSS) theory, developed by Lloyd R. Snyder and John W. Dolan, which is widely used to describe peptide and metabolite retention in linear-gradient reversed-phase chromatography [193]. Ion mobility characteristics, such as CCS values, meanwhile, are less variable than RT and can more accurately characterize a compound, but their values may still be slightly influenced by external factors such as temperature, pressure, buffer gas, and electrical field characteristics. The strong dependence of retention time values on the chromatographic system being used makes the prediction highly dependent on the training data, which should, therefore, ideally be obtained from the same chromatographic system.

Predictions of RT and CCS are particularly critical in proteomics, where challenges such as spectral complexity, peptide sequence redundancy, and PTMs are common. Subfields like metaproteomics, proteogenomics, immunopeptidomics, and open modification searches benefit greatly from such predictive models. In metabolomics, RT and CCS serve as orthogonal physicochemical features that help distinguish isobaric and structurally similar compounds, including stereoisomers, which are otherwise indistinguishable by mass alone [192].

Beyond enhancing identification, RT prediction supports quality control, enables the deconvolution of chimeric spectra, and plays a crucial role in spectral library construction, particularly in DIA workflows. Moreover, RT models hold promise for targeted proteomics, although systematic studies validating this application are still limited and may require further investigation.

Despite the challenges posed by diverse chromatographic setups and conditions, DL approaches have proven useful for RT prediction in proteomics. DL methods, particularly those employing RNNs, consistently outperform traditional ML models [193]. This is largely due to their ability to capture the sequential nature of peptide sequences and to eliminate the need for manual feature engineering, which may introduce biases stemming from prior assumptions or incomplete domain knowledge. Furthermore, transfer learning has shown promise in generalizing RT predictions to peptides with rare modifications, and accurate adjustment to the chromatographic system has enabled robust performance even in less well-characterized datasets obtained from chromatographic systems different from those used to acquire the training data.

Retention Time Prediction Algorithms

Traditionally, RT prediction was addressed using rule-based or physicochemical descriptor models, but the advent of ML and DL has revolutionized this field by enabling more precise, adaptable, and scalable solutions. This section summarizes the state-of-the-art ML and DL methods for RT prediction, highlighting their methodological principles, training data characteristics, analyte scope, and predictive capabilities.

Traditional ML Approaches Based on Physicochemical Features

Early ML efforts in peptide RT prediction used classical algorithms such as SVMs, trained on in silico calculated features derived from peptide physicochemical properties. Notable examples include the methods by Palmblad et al. [194] and the Elude tool by Moruz et al. [195], which used amino acid-specific properties like hydrophobicity and charge to model peptide chromatographic behavior. These models rely on moderate-sized, well-curated peptide datasets and require careful feature selection, which can introduce bias or limit generalizability to peptides with PTMs or unusual sequences.

Afkham et al. [196] used an ML approach based on Gaussian Process regression [197–199], offering probabilistic predictions and uncertainty estimates. The most comprehensive model for peptide retention time prediction has been developed by Oleg Krokhin and is implemented in the sequence specific retention calculator (SSRCalc) [200]. SSRCalc uses the peptide amino acid sequence, position-specific contributions, the peptide length, and potential chemical modifications-derived characteristics in a multiple linear regression model to predict the RT for a reverse-phase LC system. SSRCalc version 2 [201] improved RT predictions by using positional weighting of amino acids, using larger and more diverse datasets, expanding the gradient system model based on LSS theory, and adding empirical corrections for peptides with multiple charged residues, thus improving predictions for strongly basic or acidic sequences. Although these traditional ML methods demonstrated improvements over rule-based models, their reliance on hand-crafted features and limited ability to model complex sequence dependencies create a niche where modern DL models can improve upon their performance.

Deep Learning Methods for Capturing Complex Sequence Dependencies and Modifications

The introduction of DL architectures was a transformative step in feature prediction, as it enabled leveraging large-scale bottom-up peptide datasets to perform sequence-based modeling without manual feature engineering. DL models such as DeepRT (Ma et al. [202]) and DeepMass (Tiwari et al. [203]) employ convolutional or recurrent ANNs to learn intrinsic peptide sequence patterns influencing RTs. One of the most influential DL frameworks, Prosit by Gessulat et al. [193], integrates GRUs and fully connected layers to predict RTs alongside fragmentation spectra and CCS values. Prosit uses transfer learning, which allows it to adapt to peptides with diverse PTMs, significantly enhancing its applicability in complex proteomics experiments. Similarly, DeepLC by Bouwmeester et al. [204] specializes in predicting RTs for both unmodified and modified peptides, showing robust generalization across different chromatographic setups. More recent tools such as AlphaPeptDeep developed by Zeng et al. [205] extend these capabilities by incorporating multitask learning to predict RT, CCS, and fragment intensities simultaneously. AlphaPeptDeep also employs transfer learning to handle unknown or rare PTMs and integrates DL-based rescoring to improve peptide identification confidence. Wilburn et al. [206] developed the Chronologer tool, which further advances RT prediction for PTM-marked peptides, thus addressing challenges encountered in the expanding landscape of modified proteomes.

Proteoform RT prediction, relevant for top-down proteomics, remains a nascent field due to the scarcity of high-quality, large training datasets. In 2022, Chen et al. [207] assessed several ML and DL approaches for proteoform RT and migration time (MT) prediction for separation by either reversed-phase LC or by capillary electrophoresis (CE). They concluded that, at the time, transfer learning combined with GRUs and fully connected network architectures provided the best RT prediction accuracy for proteoforms by using pretrained models trained on peptide-level data from bottom-up experiments, and that the limited availability of annotated top-down datasets continues to constrain the development of more specialized models.

Specialized Models for Metabolite Retention Time Prediction

While peptide RT prediction dominates proteomics, metabolomics poses unique challenges due to the chemical diversity and variability in chromatographic systems. Several ML and DL approaches have been developed specifically for metabolite RT prediction. For example, PredRet created by Stanstrup et al. [208] uses a monotonically constrained generalized additive model (GAM) [209], which enables model transfer between different LC setups, facilitating RT prediction across laboratories. CALLC from Bouwmeester et al. [204] combined multiple predictive regression models calibrated to specific LC conditions, which improved the accuracy for metabolites under varying experimental parameters. Chen et al. [210] introduced RT-Ensemble Pred, which is a neural network ensemble method designed to predict metabolite RTs across multiple LC–MS platforms, showing enhanced robustness and system transferability. Graph neural networks [211, 212] also show promise in metabolomics, as demonstrated by Kensert et al. [213], who used graph convolutional networks to capture molecular structures more effectively than traditional sequence-based models. A useful resource for ML/DL model training in metabolomics is RepoRT [214], which is a repository of small molecule retention times.

Collisional Cross-Section Prediction

In addition to retention time, CCS has emerged as a valuable compound-specific property, which can be used in analytical chemistry to enhance confidence in molecular identifications. The CCS of an ion is indicative of its effective size and shape and is measured by letting the ion drift through a buffer gas under the influence of an electric field as part of a technique called ion mobility spectrometry (IMS) [215]. This orthogonal, structure-sensitive dimension of separation is particularly useful for resolving isobaric compounds and supporting identifications in both proteomics and metabolomics workflows. While CCS measurement technologies such as trapped ion mobility spectrometry (TIMS) [216], drift tube ion mobility spectrometry (DTIMS) [217], cyclic ion mobility (cIM) [218], and traveling wave ion mobility spectrometry (TWIMS) [219, 220] have become increasingly accessible, accurate computational prediction of CCS remains essential to support spectral annotation and identification, especially in high-throughput applications or for compounds lacking experimental standards.

As with RT prediction, ML and DL approaches have been increasingly employed to model CCS values. These efforts vary by analyte type, training data availability, and algorithmic complexity. For small molecules, CCS prediction is already relatively mature. Traditional ML models trained on curated databases of chemical structures and experimental CCS values have demonstrated robust performance. Early efforts by, e.g., Zhou et al. include algorithms such as MetCCS [221, 222] and LipidCCS [223], which use physicochemical descriptors derived from 2D molecular structures to predict CCS values with good accuracy across metabolite and lipid classes. Similarly, Bijlsma et al. [224] applied ANNs to predict CCS values of pesticides, showing that even relatively shallow networks can capture structural variance effectively when trained on chemically homogeneous subsets. Some efforts have gone beyond classical descriptor-based models. Colby et al. [225], for instance, developed the DarkChem tool, which is based on a CNN that directly learns from SMILES [226–228] representations of molecular structures, employing transfer learning to generalize predictions to novel compounds. Ross et al. [229] explored unsupervised clustering of small molecules based on structural similarity, followed by the training of cluster-specific ML models, which allowed for improved performance in the prediction of CCS values for chemically diverse datasets.

In contrast to small molecules, CCS prediction for peptides is more complex due to the greater conformational flexibility displayed by peptides, the charge state dependency of CCS values, and the limited availability of training data. Early attempts like imPredict by Shah et al. [230] and the work by Wang et al. [231] applied ML models using amino acid descriptors and peptide physicochemical properties as input. While these models provided initial insights, their accuracy and generalizability were limited. More recent studies have adopted DL approaches, particularly RNNs, as they are particularly well suited for modeling sequence-based features. Meier et al. [232] used RNNs to predict CCS values directly from peptide sequences, and Chang et al. [231] further refined these approaches to account for modifications and varying charge states. The IonMob framework developed by Teschner et al. [233] introduced a flexible and extensible platform for CCS prediction based on transformer architectures, which were integrated within a Python environment tailored for peptide data processing. Although performance improvements have been made, the accuracy of peptide CCS prediction remains lower than for small molecules, and most models still require careful calibration or retraining when applied to different instrument platforms or novel peptide types.

Across both analyte classes, the interpretability of ML and especially DL models remains a major concern. While DNNs often outperform classical models, they generally operate as "black boxes," making the reasoning behind their predictions opaque, which can be problematic in regulatory contexts or when analytical decisions must be justified. To address this, recent studies have begun incorporating explainability techniques into DL-based analytical models. For instance, Yeung et al. [234] developed an interpretable DL model combining convolutional and recurrent layers to identify which structural or sequence features most influence RT prediction. Their findings highlighted the importance of CCS and accessible surface area, thus pointing to physically meaningful attributes being key factors, which aligns with existing domain knowledge. Similar strategies are increasingly applied to CCS prediction models, including attention-based architectures and gradient-based attribution methods [235], though their integration into routine workflows remains limited.

Looking forward, the use of ML and DL in predicting analyte-specific physicochemical properties such as RT and CCS continues to grow in importance for analytical chemistry. By modeling these properties directly from chemical or sequence information, it becomes possible to enhance compound identification, improve the quality of spectral libraries, and support DDA targeted and DIA strategies. However, several challenges remain. The variability introduced by different ion mobility platforms and the broad range of experimental conditions employed around the world, as well as the scarcity of high-quality CCS datasets for many compound classes, limit the generalizability of current models. Furthermore, while transfer learning offers a promising solution for adapting RT prediction models to new compound types or conditions, its application in CCS prediction is still emerging and requires more systematic investigation.

The future of predictive modeling in analytical chemistry likely lies in the development of hybrid systems that balance predictive performance with interpretability, leverage both experimental data and domain knowledge, and remain adaptable to diverse analyte types and instrumentation. As high-quality datasets become more readily available and modeling strategies more sophisticated, ML and DL-based CCS prediction is expected to become an integral part of data analysis pipelines in both research and applied settings.

5.4 Quantitative Preprocessing of LC–MS(/MS) Data

LC–MS/MS datasets can be considered multidimensional images composed of many Gaussian-like (or more precisely Cauchy–Lorentzian-like) peaks that frequently overlap and are embedded into electronic noise patterns and may contain chemical noise generated by, e.g., background ions from laboratory air. Quantifying compounds in such data involves multiple sequential steps: Baseline removal, noise filtering, peak detection, peak list filtering, integration of peak areas or volumes, identification of related peaks such as isotopologues forming an isotope cluster, identification of isotope clusters arising from different charge states or adduct forms of the same compounds, and (in DIA workflows) the association of fragment peaks with precursor ions. Additionally, peak alignment across multiple chromatograms and annotation of detected features with compound identities are essential, particularly in large-scale cohort studies or comparative studies. ML and DL methods are increasingly being applied across these steps to improve the accuracy, sensitivity, specificity, and scalability of quantification workflows, while requiring less user input.

There are many diverse approaches to handling each aspect of LC–MS(/MS) data analysis, but they can generally be grouped into a few overlapping categories based on their aim: Raw peak detection and classification, peak integration for precise quantification, compound feature detection, and chromatogram and peak alignment across samples. The most user-friendly frameworks aim for end-to-end, fully automated analysis, combining several steps, and providing comprehensive protein, peptide, or metabolite quantification tables suitable for statistical analysis as output.

Direct Peak Detection from Raw LC–MS Data

Some applications aim for raw peak detection and pattern recognition in high-resolution LC–MS data. These methods detect peaks based on predefined features such as peak width, peak shape, or peak symmetry, which are extracted through traditional signal processing algorithms. These models operate directly on the raw signal intensities as a function of retention time and *m/z*. Examples of such algorithms include threshold avoiding proteomics pipeline (TAPP) [236] and pipelines and systems for threshold-avoiding quantification (PASTAQ) [229], which detect peaks in continuous LC–MS maps as local maxima and determine the peak shape and boundaries using a slope-based algorithm (TAPP) or fast Gaussian peak fitting (PASTAQ) on these peak centers, thus providing parameterization based on metrics such as peak width or resolution, which are well known to analytical chemists, and enable highly accurate quantification. TAPP and PASTAQ do not use intensity thresholds (or use a very permissive one) for peak detection and perform discrimination of signal from noise at the end of the quantification workflow when performing peak matching across multiple chromatograms at the dataset level, which results in an improvement of the effective quantification dynamic range of one to two orders of magnitude.

By applying more general ML and DL directly to the native raw data, other approaches attempt to learn relevant signal patterns or discriminate true analyte peaks from noise and artefacts intrinsically, while avoiding bias introduced by manually defined descriptors. This strategy is particularly attractive for high-resolution MS data, where signal complexity and richness make manual feature engineering both challenging and potentially limiting. However, it also makes it difficult for the user to interpret the results, as human-defined descriptors such as peak width and resolution are easier for analytical chemists to understand.

Woldegebriel et al. [237] developed an FNN that was applied to grid-based representations of raw signal intensity across the retention time and *m/z* dimensions. This ANN estimates the probability of a feature being present at any point in the data space and propagates the data through the downstream analysis pipeline after weighting it by that probability. This approach of letting an ANN operate directly on the raw acquired data enables the model to incorporate various peak shapes in the search space simply by including them in the training dataset, thus eliminating the need to use computationally expensive regression methods with specific mathematical models. Another example was shown by Kantz et al. [238], who developed a CNN tailored to the classification of chromatographic peaks and distinction of true metabolite peaks from false signals. While the CNN demonstrated superior performance over logistic regression, its generalizability can be limited by its dependence on analytical conditions and expert-curated training datasets. Borgsmüller et al. [239] introduced the WiPP algorithm based on an SVM approach that requires users to generate their own training dataset of gas chromatography (GC)–MS peaks classified by quality for supervised learning. Similarly, Risum et al. [240] combined PARAFAC2 [241, 242] with a CNN classifier to model elution profiles in GC–MS experiments as single peaks, shoulder peaks, baseline drift, or other, thus integrating peak shape information into an automated classification workflow. Melnikov et al. [243] proposed the "peakonly" tool, which combines peak detection and integration based on a dual CNN system trained on extracted ion chromatogram (EIC) traces from LC–MS data, which first categorizes the EICs based on chromatographic peak content and shape, and then performs peak integration. This method supports a wide variety of LC–MS platforms and emphasizes shape-based accuracy independent of the peak intensity.

More recently, Gloaguen et al. [244] introduced NeatMS, a CNN model that reduces false positives by classifying peaks into quality tiers and supports transfer learning so that curated models can adapt to new datasets with different peak characteristics. Eilertz et al. [245] built the automRm R package for automated peak detection and integration of targeted multiple reaction monitoring (MRM) LC–MS/MS data. Here, ML models such as random forests, neural networks, SVMs, or XGBoost [246] are trained on 20 chromatographic quality scores such as peak height and width, deviation from the expected RT, and peak Gaussianity. The ground truth required to train the models was obtained by manual peak annotation, which requires domain expertise, but also enables the detection and scoring of slightly atypical peaks by including them in the training data.

Stancliffe et al. [247] presented PeakDetective, a peak curation tool for untargeted metabolomics, which is based on a hybrid semi-supervised DL approach requiring minimal manual annotation of the raw EIC traces. An autoencoder learns latent representations of the EICs, enabling a small FNN to classify features while requiring fewer than a hundred labeled training samples. The fully trained model can perform peak detection, intensity normalization, RT alignment, and automated annotation, thus improving peak curation efficiency in untargeted metabolomics.

Peak Integration for Quantitative Analysis

Integrating feature detection with precise peak integration remains vital for accurate quantification. Wu et al. [248] developed PB-Net, an RNN-based algorithm for automatic chromatographic peak integration within MRM data, which processes RT-dependent intensities for each transition to locate boundaries and quantify peaks. Li et al. [249] extended this to DIA data with their implementation in the MSSort-DIA^{XMBD} tool, which uses CNNs and double-threshold segmentation to classify precursor signals and select high-confidence transitions from OpenSWATH [250] output. Lee et al. [251] presented HeapMS, which combined two pre-trained CNN architectures (CoAtNet [252] and Faster R-CNN [253]) to assist in quality control and boundary detection for MRM chromatograms, flagging potentially problematic peaks for further inspection. Park et al. [254] released DeepMRM, an integrated DL framework combining classification and regression to detect, delineate, and quantify peak groups in parallel PRM, MRM, and DIA data by using two-channel heatmap inputs and subnetworks to eliminate noise and interference in heavy/light peptide comparisons.

Chromatogram and Peak Alignment Across Samples

Another important challenge is chromatogram and peak alignment across multiple samples. Signal alignment algorithms, such as dynamic time warping (DTW) [255], were adopted from speech recognition at an early stage. The DTW algorithm, along with newer ones such as correlation optimized warping (COW) [256] and parametric time warping (PTW) [257], was first implemented using single traces such as base peak or total ion chromatograms and operated on pairs of LC/GC–MS datasets. Later, these methods were extended to operate on selected mass traces that contain significant signal components, using the component detection algorithm (CODA) [258]. This led to combined approaches such as continuous-wavelet transform (CWT)-CODA [259], COW-CODA [260], and DTW-CODA and PTW-CODA [261]. Improvements in the performance of these algorithms were achieved by increasing the inclusion of signals from shared compounds while reducing contributions from noise. One of the most powerful among these improved algorithms is WARP2D [262], which maximizes the overlap of 2D Gaussian peaks between pairs of peak lists from LC–MS(/MS) chromatograms. Other algorithms use locally weighted scatterplot smoothing (LOWESS) to find the optimal alignment path between pairs of chromatograms

[263]. This approach is also implemented in the commercial software Spectronaut [264, 265], which aligns all chromatograms to a measured or theoretical (pseudo) elution profile of iRT peptide standards. These methods generally provide a monotonic retention time correction, ensuring that the relative elution order of compounds is preserved after alignment. This monotonic function can be used to identify outliers, i.e., points that deviate from the monotonic trend (see Fig. 6,)which typically correspond to false identifications or to compounds that are not properly aligned by the dataset-dependent retention time model. Fig. 6 A and B show two examples of retention time alignment plots with different outlier propensities.

Another example is the ChromAlignNet algorithm, which was developed for analysis of GC–MS metabolomics data by Li et al. [266]. Using RNNs and Siamese architecture [267] subnetworks, this method encodes spectra, peak profiles, and chromatogram segments to align identical analytes across samples, even in noisy or shifting chromatographic contexts. Colby et al. [268] developed DEIMoS LC–IMS–MS/MS data, incorporating SVM-based regression and agglomerative clustering to achieve reference-based cross-sample alignment using automatically annotated features from preprocessed reference data, thus enabling robust alignment even across heterogeneous runs.

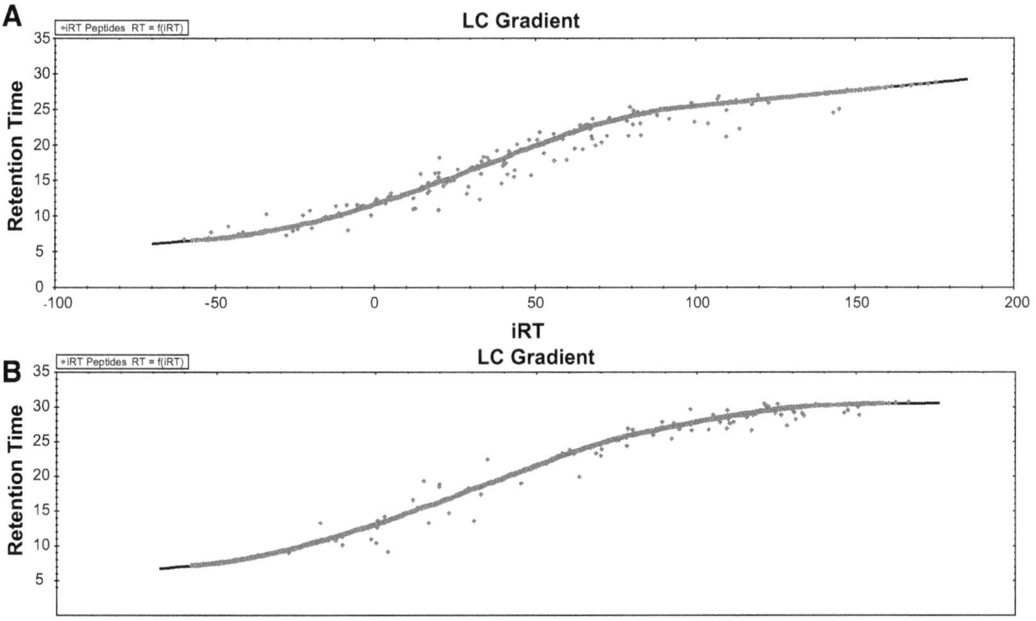

Fig. 6 Retention time alignment scatter plot. This plot shows the alignment of peptides to a theoretical pseudo-iRT elution profile by Spectronaut in two different datasets. Most peptides align along a nonlinear curve, representing well-identified peptides that behave consistently under the retention time alignment algorithm. Peptides scattering around this curve are either misidentified or not properly aligned by the retention time model

Peptide Feature Detection in Proteomics LC–MS/MS Data

Within proteomics, peptide feature detection is essential for accurate identification and quantification. A feature typically refers to a coherent signal in MS1 data that represents a peptide ion, characterized by its mass-to-charge ratio (*m/z*), retention time, and isotopic pattern, and often across multiple charge states. Zohora et al. [269] introduced DeepIso, which combines CNN and RNN architectures to detect discrete isotope features across multiple charge states from raw LC–MS/MS data without requiring any preprocessing. Their follow-up model, PointIso [270], improved on this by using a point cloud segmentation network and adding compatibility with ion mobility data. Ivanov et al. [271] developed DirectMS1 using LightGBM decision tree boosting [272] to score feature matches within a single LC–MS/MS run by comparing experimental MS1 features to theoretical peptide features predicted from an in silico digested protein database. This approach improves peptide feature identification and enables the detection of 2000 proteins in short 5-min gradient DDA LC–MS/MS Orbitrap proteomics data.

Fully Automated and Complete Analysis Pipelines

Finally, some frameworks aim at fully automated and unconstrained analysis of raw high-resolution data, encompassing deisotoping, peak classification, molecular formula prediction, and quantification. Boiko et al. [273] proposed an ML/DL pipeline where gradient boosted decision trees perform deisotoping and an LSTM-based RNN generates molecular formulas from fine isotopic patterns, thereby enabling de novo metabolite detection without the use of standard libraries.

The rich ecosystem of ML and DL tools addresses different phases of the quantification pipeline in proteomics and metabolomics. Some models operate directly on raw data pixels or RT-*m/z* maps, while others leverage processed peak lists or manually defined descriptors. Some emphasize cross-sample alignment and peak quality curation, while others excel at feature detection in complex ion mobility contexts or automated peak integration for targeted assays. Across these approaches, the most advanced models increasingly combine CNNs, RNNs, autoencoders, and transformer-like architectures, often incorporating transfer learning or active learning strategies to maximize performance with minimal annotation.

While these models offer dramatic gains in throughput and accuracy, they also raise challenges regarding interpretability, robustness, and generalizability. Many rely on domain-specific training data, so cross-platform transferability can be limited unless transfer learning is applied. To foster trust and transparency, emerging work emphasizes explainability by visualizing which features or signal patterns drive classification, integrating attention mechanisms, or exposing latent representations in autoencoders, thus enabling the implementation of ML/DL models in clinical or regulatory settings, where decisions must be logically justifiable.

5.5 MS/MS Spectrum Prediction

In the early days of automated large-scale peptide identification from LC–MS/MS data, heuristic rules were employed to predict the most probable fragment ions generated from peptide sequences. These early approaches primarily focused on predicting the *m/z* values of expected fragment ions and were integrated into database search engines and mass sequence tag-based strategies. Examples of database search tools include MSFragger [274], Comet [177], Mascot [178], X!Tandem [180], InsPecT [275], and DirecTag [276]. These tools rely on different scoring models and typically evaluate the error rate of identified spectra, peptides, and proteins using a decoy-based strategy, such as searching against inverted or shuffled peptide sequences. Alternatively, error rates can be estimated by statistical approaches, for instance by fitting mixture models that approximate two distributions: One representing false matches and the other representing true identifications. The use of fragment ion intensity information emerged later with spectral library search strategies, where consensus spectra compiled from multiple high-confidence peptide identifications were aggregated into annotated spectral libraries. An example of such a tool is SpectraST [277], which is part of the Trans-Proteomic Pipeline suite [278]. Since then, the accurate prediction of MS/MS spectra, including both fragment *m/z* values and their relative intensities, has remained a major objective in computational mass spectrometry. Unlike MS1 precursor ion *m/z* values, which can be deterministically calculated from the elemental composition of a molecule, MS2 spectra are far more complex to predict due to intricate and partially stochastic fragmentation pathways, which are influenced by fragmentation methods (e.g., collision-induced dissociation (CID) [279], higher-energy collisional dissociation (HCD) [280], and electron-transfer dissociation (ETD) [281]), collision energies, charge states, and other instrument parameters. Instrumental design also influences the observed spectra. The Athena Ion Processor (AIP) integrated into the timsUltra AIP platform, for instance, optimizes the transmission efficiency of low-mass fragment ions by modifying ion dynamics after the collision cell, thereby enhancing sensitivity to light fragments.

Figure 7 illustrates a triangular scheme representing the peptide identification workflow, encompassing three key components: Fragment spectra prediction from peptide amino acid sequences, raw MS/MS data scoring, and peptide identification by comparison of the raw data and the predicted spectra.

Peptide MS/MS Spectrum Prediction

Early Solutions

A seminal contribution to peptide fragmentation modeling was made by Zhongqi Zhang, who developed comprehensive regression-based models describing peptide MS/MS fragmentation kinetics to improve peptide identification [282]. His initial models addressed unmodified peptides [282, 283], but were later extended to peptides with

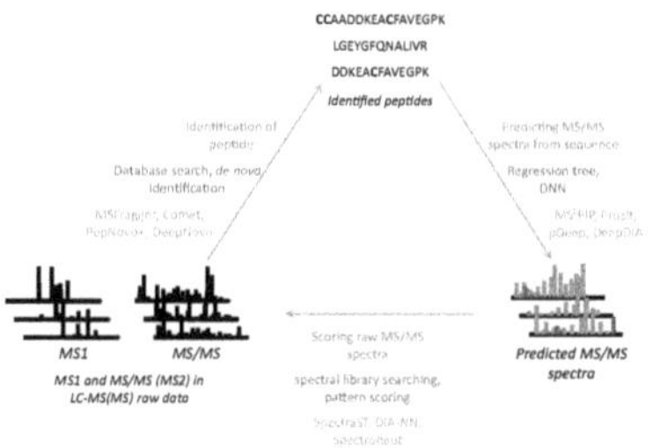

Fig. 7 Three aspects of the peptide identification workflow. This triangular scheme illustrates peptide identification through three interconnected processes: scoring of raw LC–MS/MS data, prediction of MS/MS spectra from peptide amino acid sequences, and evaluation of peptide presence by comparing observed data with the predicted fragment spectra

common and structurally simple PTMs such as phosphorylation and acetylation [284], and subsequently to glycopeptides [285]. These models incorporated key factors, including fragmentation energy, peptide charge state, and amino acid cleavage propensities, to estimate the probabilities of fragment ion formation and fragment peak intensities. The work of Zhang not only enhanced spectrum prediction accuracy but also deepened mechanistic insight into a range of fragmentation processes. Furthermore, the models informed subsequent developments such as the PepNovo+ de novo peptide sequencing tool developed by Ari Frank and colleagues [286–288].

The first ML methods developed for predicting peptide MS/MS spectra leveraged classical algorithms such as SVMs, dynamic Bayesian networks, and decision tree ensembles. As DL frameworks became more accessible, newer approaches began incorporating neural networks to enhance prediction accuracy and scalability. Among the earlier ML-based tools, MS^2PIP stands out for its strong performance. It initially used random forest regressors [289] and later adopted gradient tree boosting models [290], resulting in robust and accurate fragment ion intensity predictions. These models effectively captured the relationship between peptide sequence features and MS/MS signal intensities, and MS^2PIP continues to serve as a reliable and widely adopted tool, even as DL methods have emerged.

Deep Learning Approaches

A major leap in MS/MS spectra prediction using DL was achieved with the introduction of Prosit [193], a DNN trained on millions of spectra from the ProteomeTools dataset [291]. Prosit demonstrated the ability to accurately predict both *b*- and *y*-ion intensities with high generalizability across charge states and fragmentation conditions. Subsequent models, such as the Prosit Transformer [292], used transformer architectures for improved long-range dependency modeling and improved fragment intensity prediction accuracy. Prosit was extended to enable analysis of timsTOF data by Adams et al. [293], while Oktoberfest [294] and MSLibrarian [295] repurposed Prosit for scoring optimisation and spectral library generation.

In parallel, the pDeep family of models introduced the use of BiLSTM RNN architectures to capture sequential amino acid dependencies in peptides. The original pDeep [296] was succeeded by pDeep2 [297], which incorporated transfer learning for adaptation to instrument specifics such as type of instrument and fragmentation type, and allowed for the analysis of TMT labeled data. The latest development introduced pDeep3 [298], which enabled model training from sparse data through a meta-learning framework, which allows rapid adaptation to different instruments or labeling strategies using as few as ~100 peptide-spectrum matches (PSMs). Specialized extensions like pDeepXL [299] allowed for accurate prediction of spectra from cross-linked peptides.

Other DL models introduced diverse architectural innovations. MS2CNN used convolutional layers to capture sequence-level and physicochemical features [300], while DeepDIA focused on DIA-specific spectrum prediction using DNN [301]. Liu et al. [302] developed a model to predict full MS/MS spectra directly from peptide sequences without preselecting ion types. The output includes fragments that do not originate from peptide backbone fragmentation, but which enhance the accuracy of MS/MS spectra prediction [228]. Application-specific models such as DeepSCP [303] and Deephos [304] further extended the applicability of DL-based fragment prediction to single cell proteomics, and to the analysis of PTM-enriched and TMT-labeled proteomics data, respectively. Finally, commercial tools like INFERYS by MSAID [305] have also demonstrated strong performance from proprietary platforms by using instrument-specific neural networks for spectrum prediction and rescoring. INFERYS integrates seamlessly with the Proteome Discoverer environment from Thermo Fisher, where it enhances PSM scoring by incorporating predicted MS/MS spectra as additional features for Percolator-based rescoring. Its underlying model is trained on extensive high-resolution HCD datasets and is optimized for Orbitrap data, where it offers substantial improvements in identification sensitivity, particularly for low-abundance and TMT-labeled samples.

Since high-quality datasets that can provide ground truth for PTM placement, such as datasets generated from synthetic peptides with PTMs at precisely defined amino acid positions, are still not widely available, predictions of PTM site localization often employ alternative strategies. For small chemical modifications such as phosphorylation or acetylation, the general fragmentation pattern of the peptide backbone remains similar to the unmodified form, though phosphorylation in particular can introduce characteristic neutral losses and intensity redistributions in addition to mass shifts of fragment ions. Taking advantage of this fact, Suni et al. [306] developed SimPhospho, which builds on the MSSimulator [307] module from OpenMS [308] to predict spectra for unmodified peptides and then introduces appropriate mass shifts and fragment-intensity adjustments to simulate phosphorylated counterparts with known site localization. More recently, Yang et al. [309] demonstrated that fragment-spectrum prediction followed by phospho-specific adjustments (mass shifts, phosphoric acid neutral loss peaks, and a "budding" strategy) enables accurate spectral matching for site localization of singly and multiply phosphorylated peptides. The authors systematically benchmarked modern fragmentation prediction engines like MS²PIP, Prosit, and DeepMS2, which is the fragment spectra prediction tool for peptides in DeepDIA [301], and pDeep2 for phosphopeptides. They showed that non-modified predictors can be adapted by introducing phospho-specific rules for ion intensities and loss channels, enabling more accurate generation of phosphopeptide spectral libraries. These approaches make it possible to construct large-scale synthetic spectral libraries of phosphopeptides with defined modification sites, which can be used for spectral library searching or as training data for deep learning–based phosphopeptide prediction models.

MS/MS Fragmentation Prediction for Small Molecules and Metabolites

Early Solutions

Unlike peptide-based models, where sequence-based rules can be derived from linear chains of amino acids, small molecule fragmentation prediction poses additional challenges due to the diversity of chemical structures, isomers, and reaction pathways, and the lack of general building blocks such as amino acids for peptides. Rule-based fragmentation systems such as MetFrag [310], MIDAS [311], EPIC (now part of the Sirius tool) [312], MAGMa+ [313], and MS-Finder [314, 315] apply curated heuristics to simulate the cleavage of known bond types, generate fragment candidates, and score them against observed spectra [316]. Though efficient, these systems are limited by their inability to capture less common fragmentation routes.

Combinatorial approaches go a step further by recursively cutting bonds in the molecular graph to enumerate all possible substructures, thereby generating a comprehensive set of potential fragment ions. This strategy can reveal nonintuitive fragmentation patterns, but it is computationally demanding and, without careful scoring or filtering, may suffer from inflated search spaces and high false-positive rates.

Probabilistic generative models like Competitive Fragmentation Modeling (CFM-ID) offer a data-driven alternative by learning fragmentation probabilities from annotated spectra and applying them to candidate structures. CFM-ID [317] has proven effective in ranking potential candidates and is accessible through various web tools and software packages.

Fragmentation tree-based fingerprinting tools such as SIRIUS [318] and CSI:FingerID [319] construct fragmentation trees from experimental spectra and compute molecular fingerprints that are used for database searching. These approaches have achieved high accuracy in challenges like the Critical Assessment of Small Molecule Identification (CASMI) contest and are particularly useful for novel compound identification.

Deep Learning Approaches

DL methods have recently entered the small molecule MS/MS prediction domain with models like MolDiscovery [320], which uses learned fragmentation probabilities from large spectral datasets, models fragment graphs, and matches the learned graphs via scoring functions to yield improved accuracy and speed. Graph-based architectures such as MassFormer [321] employ transformers to model long-range interactions in molecular structures and predict fragmentation across different collision energies. SingleFrag [322] further innovates fragment spectra learning and prediction by working on individual fragment peaks rather than full fragment spectra, thus offering enhanced flexibility and interpretability.

Quantum mechanical modeling represents a fundamentally different approach. Tools like QCEIMS [323] and its extension QCxMS [324] model fragmentation by performing on-the-fly molecular dynamics simulations using force fields parameterized with semi-empirical tight-binding quantum chemical methods such as GFN-xTB [325] and GFN2-xTB [326]. These models offer first-principles insights into bond dissociation and rearrangement mechanisms based on fundamental physical or chemical laws using quantum mechanical calculations. Schnegotzki et al. [327] demonstrated the successful application of QCxMS to CID spectra of drug compounds, while Hecht et al. [328] extended this to electron ionization MS spectra of environmental chemicals analyzed by GC–MS. Despite their accuracy, quantum models remain computationally intensive and are currently more suited for detailed mechanistic studies rather than high-throughput applications.

Together, these diverse approaches reflect the growing maturity of MS/MS spectra prediction. As high-resolution data availability continues to expand across proteomics and metabolomics, the integration of ML/DL-based models with traditional chemistry-aware systems and quantum simulations promises increasingly accurate, interpretable, and generalizable fragment prediction pipelines. These developments will not only improve compound identification rates but also provide a deeper understanding of fragmentation chemistry in analytical contexts.

5.6 Identification of Peptides and Metabolites Directly from Fragment Spectra

Metabolites lacking both reference MS/MS spectra and structural information pose a significant challenge for identification using conventional LC–MS/MS approaches. Even when reference spectra exist, spectral variance across instrument types or collision energy settings can lead to mismatches. A parallel challenge exists in proteomics when dealing with protein variants not present in sequence databases or peptides bearing unanticipated modifications. De novo identification methods address these gaps by inferring molecular structures or peptide sequences directly from the raw LC–MS/MS data, without relying on predefined candidate lists or reference libraries. This approach is inherently more flexible and capable of discovering novel molecules or unannotated variants in untargeted studies.

The employed de novo identification methods differ between the metabolomics and proteomics fields, since the type of analyte (metabolites versus peptides) dictates which ML and DL strategies are applicable. For metabolites, methods typically focus on predicting the molecular structure or class from spectral input without a reference library. For peptides, de novo algorithms reconstruct amino acid sequences directly or via tag-based inference using dynamic programming informed by learned models.

Spectral Clustering for Data Preprocessing

Before implementing peptide-level de novo sequencing, it can be advantageous to perform spectral clustering in order to improve computational efficiency and sensitivity. Tools such as MS-Cluster [329], spectra-cluster [330] and MaRaCluster [331] group redundant spectra, which reduces the search space for downstream de novo tools. These clustering methods rely on similarity metrics and hierarchical or complete-linkage strategies rather than learned models, but dramatically reduce the computational burden. Later developments like msCRUSH [332] and GLEAMS [333] introduced ML and embedding-based similarity scoring to scale clustering up across massive public datasets, thus enabling the identification of frequently occurring but unannotated spectra. MatchMS [334] was developed to clean up spectral libraries by validating structure annotations and applying various filters. MS2Query [335], Spec2Vec [336], and MS2DeepScore [337] apply DL or learned spectral embeddings to find structurally similar spectra beyond exact matches, which improves analog detection in metabolomics and proteomics studies.

De Novo Identification Methods for Metabolomics

Approaches for metabolite identification without reference libraries vary in complexity and predictive aim. DeepMASS [338] employs an FNN to predict candidate structures that best explain an observed MS/MS pattern. Despite training on only 752 compounds, it demonstrated generalization across instrument types by learning spectrum–structure relationships directly. MetFID [339] uses an ANN to predict binary chemical fingerprints, i.e., descriptors indicating the presence or absence of functional groups, from input spectra, enabling annotation of unknown features. SF-Matching by Li et al. [340] takes a structural approach by matching subfragment patterns to chemical motifs via ML models trained on the structural features of model compounds containing chemical structures commonly found in biological contexts, thus allowing inference of chemical identities from unexplained fragments. CANOPUS [341] uses a DNN to assign chemical ontology classes and compound categories to fragment spectra, abstracting beyond exact structures to class-level annotation. Xing et al. [342] presented SteroidXtract, a DL-based pattern recognition tool optimized to rapidly identify steroid-like metabolites, which illustrates the power of domain-specific pattern learning. Each of these methods uses supervised learning, where pretraining is done based on annotated spectra to predict structural information or chemical classes from unknown samples. They each have different strengths and weaknesses, with DeepMASS being able to propose candidate structures but being limited by dataset size, MetFID and CANOPUS prioritizing generalization to unseen structures but with less details (fingerprints or class), and SF-Matching excelling at capturing subtle fragment–structure relationships but depending on extensive subfragment annotation.

De Novo Sequencing Methods for Proteomics

True de novo peptide sequencing methods reconstruct full peptide sequences from MS/MS data without using a protein database, but the approach to achieving this goal varies between models. Novor [343] uses a dynamic programming algorithm, where decision trees are used to build peptides sequentially based on observed fragment intensities and to score peptide candidates. The DeepNovo engine [344] replaced these heuristics with a DL model based on a CNN combined with LSTM units and was trained to predict succeeding amino acid tokens given a fragment spectrum and previous predictions. DeepNovoV2 [345] refined the architecture based on the results of PointNet and changed the training procedure to improve accuracy in noise detection and identification of peptides with PTMs or missed cleavages. DeepNovoV2 was later adapted by Gueto-Tettay et al. [346] to analyze multi-species samples digested with multiple enzymes. The DeepNovo architecture was also adapted by SMSNet [347], which combines de novo predictions with a supplied peptide database, thus improving both sensitivity and accuracy while retaining de novo capabilities. The Kaiko model [348] trained on a vast library of diverse soil microorganisms using DL similarly to DeepNovo, but aimed at greater generalization. Newer successor models such as PointNovo [349] utilized transformers for improved sequence context modeling and automatic adaptation to the resolution of the acquired data. Another recent algorithm is Casanovo [350], which uses transformer neural networks to yield faster and more accurate sequencing and provides excellent performance for tryptic peptides, nonenzymatic peptides, immunopeptides, and metaproteome datasets. PepNet [351] proposed a CNN-based architecture for de novo sequencing, while DePS [352] and InstaNovo [353] explored new network architectures and sequencing strategies. These DL-based de novo tools offer high sensitivity and can accommodate unanticipated modifications, but require large annotated training datasets and suffer from limited interpretability.

Across both metabolomics and proteomics applications, de novo methods use powerful supervised learning frameworks trained on curated spectral libraries or peptide identifications. Metabolite-focused methods are constrained by smaller training sets and the enormous chemical diversity, often limiting identification to the metabolite class or chemical motif fingerprint levels rather than providing exact structures. Peptide de novo sequencing benefits from standardized fragmentation patterns and larger peptide spectral libraries, but still struggles with incomplete fragmentation, noise, chimeric spectra, and rare modifications. The main disadvantages of DL methods in this space include the need for large and diverse training datasets, limited interpretability, and challenges in handling prediction uncertainty, since errors at individual residues can propagate across the peptide sequence, and the models may produce confident but incorrect outputs. Spectral clustering methods serve as preprocessing mechanisms to reduce dataset scale and improve downstream de novo performance due to lower noise content, but they themselves are not de novo identification tools.

The challenge of interpretability is caused by neural network-based methods being opaque in their internal decision logic, so while they excel at prediction, they lead to complications in validation or error analysis. Methods like SMSNet, which constrain outputs with database reconciliation, can offer improved reliability. De novo tools also tend to generate candidate sequences or structures with confidence scores rather than fully certain identifications, which means the output requires manual review or downstream rescoring. As spectral libraries grow and new architectures emerge, these methods are poised to become increasingly practical and indispensable, particularly when integrated into hybrid workflows combining database searches, de novo predictions, and synthetic spectral library expansion such as ProteomeTools.

5.7 Assessing Confidence in Analyte-Spectrum Assignment

Post-identification analysis in LC–MS/MS workflows plays a crucial role in increasing the reliability and quantity of compound identifications, particularly in proteomics. Among these strategies, rescoring tools are frequently employed to enhance discrimination between true and false PSMs. These methods typically utilize statistical or ML frameworks to re-evaluate the initial identification scores by integrating auxiliary information, such as fragmentation spectra characteristics, predicted RTs, precursor ion properties, and co-elution patterns into the scoring function. By learning discriminative features from curated datasets, these tools can significantly improve sensitivity while maintaining rigorous false discovery rate (FDR) control. In recent years, DL-based scoring and confidence assessment approaches have emerged to leverage the wealth of information available in MS data. These methods increasingly appear in both DDA and DIA pipelines and are now also expanding into metabolomics applications.

Early Solutions

One of the earliest and most widely adopted ML-based post-processing tools is Percolator, introduced by Käll et al. [354] in 2007. Percolator uses a semi-supervised learning approach based on an SVM to distinguish correct from incorrect PSMs. It learns discriminative features from decoy and target identifications to adjust PSM confidence scores and control FDR. Percolator has become a standard in proteomics pipelines and is often used in conjunction with various search engines. A refinement of this approach, presented by Fondrie and Noble [355], adapted Percolator to small-scale datasets through static modeling, making it suitable for challenging low-input applications such as single-cell proteomics.

Complementing Percolator, mProphet, introduced by Reiter et al. [356], established a flexible and modular framework for scoring PSMs in targeted proteomics workflows. mProphet applies a mixture model to estimate posterior error probabilities and integrates features such as RT deviation and transition co-elution in MRM assays. Its Python reimplementation, PyProphet [357], has extended these capabilities to DIA workflows.

Deep Learning Approaches

Several newer ML/DL-based tools further enhance rescoring capabilities. DeepRescore by Li et al. [358] applied a DL model to improve immunopeptidomics data analysis. Their approach learns complex fragmentation rules and retention behaviors across large immunopeptidomics datasets to rescore PSMs with higher precision. Similarly, MS2Rescore, introduced by Silva et al. [359] and extended by Declercq et al. [290], combines predicted spectra (from MS^2PIP or Prosit) with RT predictions and search engine features to generate improved rescoring models, effectively integrating multiple orthogonal data dimensions.

The interpretability of rescoring models remains a priority for many researchers. For example, BoltzMatch by Sulimov et al. [360] incorporates chemically explainable fragmentation patterns into an interpretable neural network model, where the pairwise matching of theoretical and experimental peaks guides the scoring. This approach maintains transparency in scoring decisions and aids analytical chemists in understanding the behavior of the model.

Alternative strategies are also gaining attention. Slider, proposed by Kudriavtseva et al. [361], utilizes a DNN trained directly on database search results for scoring. While less frequently cited, its end-to-end DL design represents a shift toward integrated learning pipelines. Another example is the rescoring algorithm of INFERYS [305], which uses predicted fragment intensities to improve PSM confidence scores. Tool independence and cross-platform compatibility are key considerations in developing ML-based rescoring systems. TIDD by Li et al. [362] addresses this challenge by implementing an SVM-based rescoring engine that operates independently of the specific database search tool, thus broadening its applicability across software environments.

The importance of spectral prediction in boosting PSM validation was demonstrated in the inSPIRE pipeline introduced by Cormican et al. [363], which uses Prosit predictions for both spectra and RTs to enhance rescoring. Spectral similarity between predicted and experimental spectra forms a central feature in the learned model and leads to improved confidence scores and identification yield. Integration of predicted precursor properties has also proven beneficial. A generalized precursor prediction model proposed by Scott et al. [364] showed that including precursor ion properties as learned features enhances the separation between correct and incorrect PSMs. Similarly, Percolator validation incorporating RT and spectral predictors was found to significantly improve the detection of variant peptides in a recent study by Skiadopoulou et al. [365].

Recent advancements in Python-based frameworks such as mokapot [366] facilitate the integration of Percolator-like semi-supervised learning into modern pipelines, and offer flexibility in algorithm choice (SVM or random forest) and applicability in complex identification scenarios including open modification searches. A study by Arab et al. [367] demonstrated the integration of mokapot into the ANN-SoLo [368] spectral library search engine, highlighting the benefit of ML-enhanced PSM rescoring in open modification search workflows.

In DIA-specific workflows, the DIA-NN software suite incorporates DL modules based on a fully connected FNN and using 63 different input scores to enhance scoring and discrimination between target and decoy peptides. DIA-NN also supports integration with external DL tools such as Alpha-XIC [369] and Alpha-Tri [370] using DNN to assess peak co-elution and spectral similarity information, respectively. Alpha-XIC scores the co-elution of fragment ion groups, while Alpha-Tri computes a triple-spectrum similarity feature using predicted, measured, and correlation spectra. Both contribute to improved identifications and FDR estimation in DIA workflows.

Although rescoring tools in metabolomics are comparatively underdeveloped, key contributions have emerged. Scheubert et al. [371] introduced the passatutto tool to generate decoy MS/MS spectra for metabolomics FDR estimation. DIAMetAlyzer by Alka et al. [372] adapted the PyProphet workflow to metabolomics and employs linear discriminant analysis (LDA) to build composite scores from peak group features, while using passatutto for decoy spectrum generation. PeakDecoder by Bilbao et al. [373] represents a recent ML-based approach. It uses supervised learning to differentiate between true and false co-elution and co-mobility events in

LC–IMS–MS data, enabling accurate metabolite identification and error rate estimation.

ML and DL-based rescoring approaches represent a powerful addition to proteomics and metabolomics pipelines, enabling more confident and comprehensive compound identification. These tools build on the increasing availability of high-quality training data and predictive models of RTs and spectra. While, to date, proteomics has benefited most from such tools, the metabolomics field is also making great improvements with emerging ML-enhanced workflows. Moving forward, interpretability, tool independence, and cross-platform compatibility will remain key goals for the next generation of confidence assessment methods.

5.8 Practical Considerations for Applying Machine Learning and Deep Learning in Analytical Chemistry

The implementation of ML and DL in proteomics and metabolomics has become increasingly accessible, offering researchers powerful tools to enhance data interpretation, prediction, and discovery. However, integrating these methods into the research of a user requires careful planning and awareness of both their strengths and limitations. A fundamental aspect of the successful application of ML/DL lies in the selection and preparation of training data. The principle of "garbage in, garbage out" remains critically relevant: Poorly curated or biased training datasets can lead to models that perform well during development, but fail under real-world conditions. Researchers should prioritize high-quality data acquisition and apply stringent filtering criteria when building training datasets, particularly in light of the noise and variability inherent in LC–MS/MS datasets.

Equally important is the use of separate validation and test datasets. These datasets ensure that the models are not simply memorizing the training data, but are capable of generalizing to new, unseen samples. Cross-validation and holdout testing are standard practices that provide insight into model robustness and reduce the risk of overfitting, which is a common issue in high-dimensional data environments like mass spectrometry.

Pretrained models, such as MS^2PIP for fragment ion intensity prediction, offer a valuable shortcut for researchers who may not have the computational resources or large annotated datasets required to train models from scratch. However, these models are often developed under specific experimental conditions, and their generalizability may be limited when applied to different instrument types, acquisition settings, or sample types. Custom training or the use of transfer learning using in-house data can yield improved performance, but comes at the cost of additional effort in data collection and model optimisation. Therefore, the decision to use pretrained models versus custom-built solutions should be made with a clear understanding of the experimental context and data characteristics.

Despite the progress in ML/DL applications, several limitations continue to constrain their full potential. First and foremost is the lack of high-quality ground truth datasets for training and benchmarking. In proteomics, the complexity and dynamic range of biological samples make it difficult to define a gold standard to train models against. While synthetic peptide datasets, such as those generated in the ProteomeTools project [291, 374], have proven useful, they are not without limitations. Imperfect synthesis, incomplete fragmentation, and peak broadening may impact the reliability of such data. Additionally, synthetic datasets often fail to capture the complexity of real biological samples, including endogenous modifications, interfering signals, and biological variability.

Recent efforts to develop benchmarking datasets for label-free quantification (LFQ), such as the one published by Puyvelde et al. [375], represent a step forward. Their study introduced a controlled proteome mixture design combining human, yeast, and *E. coli* lysates at known concentration ratios, measured in triplicates on six different LC–MS/MS platforms, providing a realistic reference for assessing quantitative accuracy and reproducibility. Nevertheless, these datasets cover only a narrow range of experimental scenarios, and there is still a pressing need for comprehensive reference sets that incorporate diverse sample types, modification states, and acquisition protocols. A review by Neely et al. [376] offers a practical guide for implementing quality control (QC) in proteomics core facilities, describing a range of off-the-shelf QC materials (e.g., single-protein or peptide mixtures, whole-cell lysates) and user-friendly software tools for monitoring instrument suitability and workflow robustness, and listing resources to find already acquired LC–MS/MS data that can be used for QC purposes. Another article by Neely et al. [377] discusses the use of synthetic LC–MS/MS data as a means to benchmark proteomics workflows and support ML/AI model development. These synthetic datasets are generated in silico using already acquired LC–MS/MS QC measurements, rather than from newly prepared protein or peptide mixtures. The simulated data replicate key experimental features, including RTs, intensities, and peptide detectability, providing realistic variability for algorithm testing. This approach allows for the systematic evaluation of computational pipelines, including feature detection, peptide identification, and LFQ, under controlled and reproducible conditions. Overall, these in silico QC materials represent a valuable resource for developing, training, and validating ML/AI models in proteomics.

A second major challenge is model interpretability. While traditional analytical features, such as retention time, peak width, or resolution, are directly interpretable by chemists, the abstract representations learned by DNN are often opaque. Post-hoc explainability methods, which aim to elucidate what a model has learned after training, are being actively explored [378]. These tools may help bridge the gap between black-box predictions and chemically meaningful insights, but the field is still in its early stages. Ideally, ML/DL algorithms should not only improve identification performance but also contribute to a deeper understanding of the physicochemical properties that drive analytical separation and fragmentation patterns.

Model selection and optimisation also remain nontrivial, especially in the absence of standardized performance metrics or comparison frameworks. As new algorithms continue to emerge at a rapid pace, transparent benchmarking and open data sharing will be essential. We encourage researchers to adhere to the FAIR principles (Findable, Accessible, Interoperable, and Reusable) [379] when sharing data and models, but also follow the FAIR principles for experimental design, coding, and bioinformatics workflow development [380]. Public repositories such as PRIDE and MassIVE should be used for raw data deposition, and accompanying metadata, such as instrument parameters, chromatography conditions, and data processing pipelines, should be made available to facilitate reproducibility and reuse. Furthermore, researchers should strive to incorporate their models into existing end-to-end pipelines and evaluate the potential synergy or performance decline using specifically designed benchmarking datasets such as proteome mixtures or spike compounds absent in the matrix added to biological samples across a large dynamic range [381].

Ultimately, the widespread adoption of ML/DL will only fulfill its transformative promise if it leads to better, faster, and more meaningful science. The excitement around ML and AI should be matched by a rigorous commitment to methodological clarity and chemical interpretability. As the field evolves, it will be critical to continuously evaluate not only the predictive accuracy of these methods, but also their ability to advance our understanding of the underlying chemical and biological processes.

Conflict of Interest

The authors have no conflicts of interest to declare.

References

1. European Commission's High-Level Expert Group on Artificial Intelligence (2018). A definition of AI: main capabilities and scientific disciplines. https://ec.europa.eu/futurium/en/system/files/ged/ai_hleg_definition_of_ai_18_december_1.pdf
2. IEEE Board of Directors (2019). IEEE Position Statement on Artificial Intelligence. https://globalpolicy.ieee.org/wp-content/uploads/2019/06/IEEE18029.pdf
3. Turing AM (1950) I.—Computing machinery and intelligence. Mind LIX, pp 433–460. https://doi.org/10.1093/mind/LIX.236.433
4. Baum ZJ, Yu X, Ayala PY, Zhao Y, Watkins SP, Zhou Q (2021) Artificial intelligence in chemistry: current trends and future directions. J Chem Inf Model 61:3197–3212. https://doi.org/10.1021/acs.jcim.1c00619
5. Domingos P (2012) A few useful things to know about machine learning. Commun ACM 55:78–87. https://doi.org/10.1145/2347736.2347755

6. Laqtib S, Yassini KE, Hasnaoui ML (2019) A deep learning methods for intrusion detection systems based machine learning in MANET. Proc 4th Int Conf Smart City Appl. ACM, Casablanca Morocco, pp 1–8 https://doi.org/10.1145/3368756.3369021
7. LeCun Y, Bengio Y, Hinton G (2015) Deep learning. Nature 521:436–444. https://doi.org/10.1038/nature14539
8. Yue L, Zhang F, Sun R, Sun Y, Yuan C, Zhu Y, Guo T (2020) Generating proteomic big data for precision medicine. Proteomics 20:1900358. https://doi.org/10.1002/pmic.201900358
9. Fawcett T (2006) An introduction to ROC analysis. Pattern Recognit Lett. 27:861–874. https://doi.org/10.1016/j.patrec.2005.10.010
10. Wold S, Sjöström M, Eriksson L (2001) PLS-regression: a basic tool of chemometrics. Chemom Intell Lab Syst 58:109–130. https://doi.org/10.1016/S0169-7439(01)00155-1
11. Andresen C, Boch T, Gegner HM, Mechtel N, Narr A, Birgin E, Rasbach E, Rahbari N, Trumpp A, Poschet G, Hübschmann D (2022) Comparison of extraction methods for intracellular metabolomics of human tissues. Front Mol Biosci 9 https://doi.org/10.3389/fmolb.2022.932261
12. Mitra V, Govorukhina N, Zwanenburg G, Hoefsloot H, Westra I, Smilde A, Reijmers T, Van Der Zee AGJ, Suits F, Bischoff R, Horvatovich P (2016) Identification of analytical factors affecting complex proteomics profiles acquired in a factorial design study with analysis of variance: simultaneous component analysis. Anal Chem 88:4229–4238. https://doi.org/10.1021/acs.analchem.5b03483
13. Zong Y, Wang Y, Yang Y, Zhao D, Wang X, Shen C, Qiao L (2023) DeepFLR facilitates false localization rate control in phosphoproteomics. Nat Commun 14 2269. https://doi.org/10.1038/s41467-023-38035-1
14. Gergonne J-D (1815) Analise. Application de la méthode des moindres quarrés à l'interpolation des suites. Ann Math Pures Appl 6:242–256
15. XXIV (1825) On the nature of the function expressive of the law of human mortality, and on a new mode of determining the value of life contingencies. In a letter to Francis Baily, Esq. F. R. S. &c. Philos Trans R Soc Lond 115:513–583. https://doi.org/10.1098/rstl.1825.0026
16. Savitzky A, Golay MJE (1964) Smoothing and differentiation of data by simplified least squares procedures. Anal Chem 36:1627–1639. https://doi.org/10.1021/ac60214a047
17. Cleveland WS (1979) Robust locally weighted regression and smoothing scatterplots. J Am Stat Assoc 74:829–836. https://doi.org/10.1080/01621459.1979.10481038
18. Cleveland WS (1981) LOWESS: a program for smoothing scatterplots by robust locally weighted regression. Am Stat 35:54. https://doi.org/10.2307/2683591
19. Cleveland WS, Devlin SJ (1988) Locally weighted regression: an approach to regression analysis by local fitting. J Am Stat Assoc 83:596–610. https://doi.org/10.1080/01621459.1988.10478639
20. Galton F (1889) Natural inheritance. Macmillan, London, London, https://galton.org/books/natural-inheritance/pdf/galton-nat-inh-1up-clean.pdf
21. Su X, Yan X, Tsai C (2012) Linear regression. WIREs Comput Stat. 4:275–294. https://doi.org/10.1002/wics.1198
22. Maulud D, Abdulazeez AM (2020) A review on linear regression comprehensive in machine learning. J Appl Sci Technol Trends 1:140–147. https://doi.org/10.38094/jastt1457
23. Legendre AM (1806) Nouvelles méthodes pour la détermination des orbites des comètes: avec un supplément contenant divers perfectionnemens de ces méthodes et leur application aux deux comètes de 1805. Courcier, Paris
24. Gauss CF (1809) Theoria motus corporum coelestium in sectionibus conicis solem ambientium. Sumtibus F. Perthes et I. H. Besser, Hamburg
25. Wright S (1921) Correlation and causation. J Agric Res 20:557–585
26. Cauchy A-L (1847) Méthode générale pour la résolution des systèmes d'équations simultanées. C R L'Acad Sci Paris 25:536–538
27. Debye P (1909) Näherungsformeln für die Zylinderfunktionen für große Werte des Arguments und unbeschränkt veränderliche Werte des Index. Math Ann 67:535–558. https://doi.org/10.1007/BF01450097
28. Hestenes MR, Stiefel E (1952) Methods of conjugate gradients for solving linear systems. J Res Natl Bur Stand 49:409–435
29. Rockafellar RT (1976) Monotone operators and the proximal point algorithm. SIAM J Control Optim 14:877–898. https://doi.org/10.1137/0314056
30. Kiefer J (1953) Sequential minimax search for a maximum. Proc Am Math Soc 4:502–506. https://doi.org/10.1090/S0002-9939-1953-0055639-3
31. Hooke R, Jeeves TA (1961) "Direct search" solution of numerical and statistical problems. J ACM 8:212–229. https://doi.org/10.1145/321062.321069
32. Nelder JA, Mead R (1965) A simplex method for function minimization. Comput J 7:308–313. https://doi.org/10.1093/comjnl/7.4.308

33. Dornier M, Heyd B, Danzart M (1998) Evaluation of the Simplex method for training simple multilayer neural networks. Neural Comput Appl. 7:107–114. https://doi.org/10.1007/BF01414162
34. Dragović S, Onjia A, Bačić G (2006) Simplex optimization of artificial neural networks for the prediction of minimum detectable activity in gamma-ray spectrometry. Nucl Instrum Methods Phys Res Sect A 564:308–314. https://doi.org/10.1016/j.nima.2006.03.047
35. Grošelj N, Zupan J, Magallanes J, Reich S (2008) Optimization of neural networks by SIMPLEX method performed on environmental data. Acta Chim Slov 55(2):286–293
36. Berkson J (1944) Application of the logistic function to bio-assay. J Am Stat Assoc 39:357–365. https://doi.org/10.1080/01621459.1944.10500699
37. Bayes T, LII (1763) An essay towards solving a problem in the doctrine of chances. By the late Rev. Mr. Bayes, F. R. S. communicated by Mr. Price, in a letter to John Canton, A. M. F. R. S. Philos Trans R Soc Lond 53:370–418. https://doi.org/10.1098/rstl.1763.0053
38. Stigler SM (1982) Thomas Bayes's Bayesian Inference. J R Stat Soc Ser Gen 145:250. https://doi.org/10.2307/2981538
39. Breiman, L, Friedman, JH, Olshen, RA, Stone, CJ, 2017. Classification and regression trees. Routledge. https://doi.org/10.1201/9781315139470
40. Ceriani L, Verme P (2012) The origins of the Gini index: extracts from Variabilità e Mutabilità (1912) by Corrado Gini. J Econ Inequal 10:421–443. https://doi.org/10.1007/s10888-011-9188-x
41. Dietterich T (1995) Overfitting and undercomputing in machine learning. ACM Comput Surv 27:326–327. https://doi.org/10.1145/212094.212114
42. Mingers J (1989) An empirical comparison of pruning methods for decision tree induction. Mach Learn 4:227–243. https://doi.org/10.1023/A:1022604100933
43. Bramer M (2002) Using J-pruning to reduce overfitting in classification trees. Knowl.-Based Syst 15:301–308. https://doi.org/10.1016/S0950-7051(01)00163-0
44. Tin Kam Ho (1995) Random decision forests. Proc 3rd Int Conf Doc Anal Recognit, IEEE Comput. Soc. Press, Montreal, QC, Canada, pp 278–282 https://doi.org/10.1109/ICDAR.1995.598994
45. Breiman L (2001) Random Forests. Mach Learn 45:5–32. https://doi.org/10.1023/A:1010933404324
46. Tin Kam Ho (1998) The random subspace method for constructing decision forests. IEEE Trans Pattern Anal Mach Intell 20:832–844. https://doi.org/10.1109/34.709601
47. Fix E, Hodges JL (1989) Discriminatory analysis. nonparametric discrimination: consistency propeties. Int. Stat. Rev 57:238. https://doi.org/10.2307/1403797
48. Cover T, Hart P (1967) Nearest neighbor pattern classification. IEEE Trans Inf Theory 13:21–27. https://doi.org/10.1109/TIT.1967.1053964
49. Han J, Kamber M (2001) Data mining: concepts and techniques. Morgan Kaufmann Publishers
50. Abdi H, Williams L (2010) Normalizing data. Encycl Res Des. SAGE Publications, Inc, pp 935–936 https://doi.org/10.4135/9781412961288.n278pp
51. Hastie T, Friedman J, Tibshirani R (2001) The elements of statistical learning. Springer New York, New York, NY. https://doi.org/10.1007/978-0-387-21606-5
52. Eesa AS, Arabo WKh (2017) A normalization methods for backpropagation: a comparative study. Sci J Univ Zakho 5:319. https://doi.org/10.25271/2017.5.4.381
53. Henderi H (2021) Comparison of min-max normalization and Z-Score normalization in the K-nearest neighbor (kNN) algorithm to test the accuracy of types of breast cancer. IJIIS Int J Inform Inf Syst 4:13–20. https://doi.org/10.47738/ijiis.v4i1.73
54. Steinhaus H (1956) Sur la division des corps matériels en parties. Bull L'Acad Pol Sci 4:801–804
55. Forgy EW (1965) Cluster analysis of multivariate data: efficiency versus interpretability of classifications. Biometrics 21:768–769. https://doi.org/10.2307/2528201
56. MacQueen J et al. (1967) Some methods for classification and analysis of multivariate observations. Proc Fifth Berkeley Symp. Math. Stat. Probab, Oakland, CA, USA, pp 281–297
57. Lloyd S (1982) Least squares quantization in PCM. IEEE Trans Inf Theory 28:129–137. https://doi.org/10.1109/TIT.1982.1056489
58. Tibshirani R, Walther G, Hastie T (2001) Estimating the number of clusters in a data set via the gap statistic. J R Stat Soc Ser B Stat Methodol 63:411–423. https://doi.org/10.1111/1467-9868.00293
59. Milligan GW, Cooper MC (1985) An examination of procedures for determining the number of clusters in a data set. Psychometrika 50:159–179. https://doi.org/10.1007/BF02294245
60. Hartigan JA, Wong MA (1979) Algorithm AS 136: a K-means clustering algorithm. J R Stat Soc Ser C Appl Stat 28:100–108. https://doi.org/10.2307/2346830

61. Arthur D, Vassilvitskii S (2007) K-Means++: the advantages of careful seeding. Proc. eighteenth annu. ACM-SIAM symp. discrete algorithms. Society for Industrial and Applied Mathematics, USA, pp 1027–1035
62. Dempster AP, Laird NM, Rubin DB (1977) Maximum likelihood from incomplete data via the EM algorithm. J R Stat Soc Ser B Methodol 39:1–22. https://doi.org/10.1111/j.2517-6161.1977.tb01600.x
63. Meng X-L, Dyk DVan (1997) The EM algorithm—an Old Folk-song sung to a fast new tune. J R Stat Soc Ser B Stat Methodol 59:511–567. https://doi.org/10.1111/1467-9868.00082
64. Jianbo Shi JMalik (2000) Normalized cuts and image segmentation. IEEE Trans Pattern Anal Mach Intell 22:888–905. https://doi.org/10.1109/34.868688
65. Ng A, Jordan M, Weiss Y (2001) On spectral clustering: analysis and an algorithm. In: Dietterich T, Becker S, Ghahramani Z (eds) Adv. neural inf. process. syst.. MIT Press. https://proceedings.neurips.cc/paper_files/paper/2001/file/801272ee79cfde7fa5960571fee36b9b-Paper.pdf
66. Fiedler M (1972) Algorithms for partitioning graphs and computer logic based on eigenvectors of connection matrices. IBM Tech Discl Bull 15:938–944
67. Donath WE, Hoffman AJ (1973) Lower bounds for the partitioning of graphs. IBM J Res Dev 15:825–830
68. Ester M, Kriegel H-P, Sander J, Xu X, Density-Based A (1996) Algorithm for Discovering Clusters in Large Spatial Databases with Noise. In: Simoudis E, Han J, Fayyad UM (eds) KDD. AAAI Press, pp 226–231https://dblp.org/rec/conf/kdd/EsterKSX96.htmlpp
69. Ankerst M, Breunig MM, Kriegel H-P, Sander J (1999) OPTICS: ordering points to identify the clustering structure. ACM SIGMOD Rec 28:49–60. https://doi.org/10.1145/304181.304187
70. Pedregosa F, Varoquaux G, Gramfort A, Michel V, Thirion B, Grisel O, Blondel M, Prettenhofer P, Weiss R, Dubourg V, Vanderplas J, Passos A, Cournapeau D, Brucher M, Perrot M, Duchesnay É (2011) Scikit-learn: Machine learning in Python. J Mach Learn Res 12:2825–2830
71. Jain, Jianchang Mao AK, Mohiuddin KM (1996) Artificial neural networks: a tutorial. Computer 29:31–44. https://doi.org/10.1109/2.485891
72. McCulloch WS, Pitts W (1943) A logical calculus of the ideas immanent in nervous activity. Bull Math Biophys 5:115–133. https://doi.org/10.1007/BF02478259
73. Rosenblatt F (1958) The perceptron: a probabilistic model for information storage and organization in the brain. Psychol Rev 65:386–408. https://doi.org/10.1037/h0042519
74. Heaviside IIIOliver (1893) On operators in physical mathematics. Part I. Proc R Soc Lond 52:504–529. https://doi.org/10.1098/rspl.1892.0093
75. Rosenblatt F (1963) Principles of neurodynamics. Perceptrons and the theory of brain mechanisms. Am J Psychol 76:705
76. Krauth W, Mezard M (1987) Learning algorithms with optimal stability in neural networks. J Phys Math Gen 20:L745. https://doi.org/10.1088/0305-4470/20/11/013
77. Anlauf JK, Biehl M (1989) The AdaTron: an adaptive perceptron algorithm. Europhys Lett EPL 10:687–692. https://doi.org/10.1209/0295-5075/10/7/014
78. Biehl M, Anlauf JK, Kinzel W (1991) Perceptron learning by constrained optimization: the AdaTron algorithm. In: Pasemann F, Doebner HD (eds) Neurodynamics Proc 9th summer workshop math. phys. Arnold Sommerfeld Inst. Clausthal 1990. World Scientific Publishing, pp 194–210
79. Gallant SI (1990) Perceptron-based learning algorithms. IEEE Trans Neural Netw 1:179–191. https://doi.org/10.1109/72.80230
80. Wendemuth A (1995) Learning the unlearnable. J Phys Math Gen 28:5423–5436. https://doi.org/10.1088/0305-4470/28/18/030
81. Cortes C, Vapnik V (1995) Support-vector networks. Mach Learn 20:273–297. https://doi.org/10.1007/BF00994018
82. Mitchison GJ, Durbin RM (1989) Bounds on the learning capacity of some multi-layer networks. Biol Cybern 60:345–365. https://doi.org/10.1007/BF00204772
83. Schwarze H, Hertz J (1993) Generalization in fully connected committee machines. Europhys Lett EPL 21:785–790. https://doi.org/10.1209/0295-5075/21/7/012
84. Aubin B, Maillard A, Barbier J, Krzakala F, Macris N, Zdeborová L (2019) The committee machine: computational to statistical gaps in learning a two-layers neural network. J Stat Mech 2019:124023. https://doi.org/10.1088/1742-5468/ab43d2
85. Barkai E, Hansel D, Kanter I (1990) Statistical mechanics of a multilayered neural network. Phys Rev Lett 65:2312–2315. https://doi.org/10.1103/PhysRevLett.65.2312
86. Biehl M, Opper M (1991) Tilinglike learning in the parity machine. Phys Rev A 44:6888–6894. https://doi.org/10.1103/PhysRevA.44.6888

87. Boser BE, Guyon IM, Vapnik VN (1992) A training algorithm for optimal margin classifiers. Proc. Fifth Annu. Workshop Comput. Learn. Theory. ACM, Pittsburgh Pennsylvania USA, pp 144–152 https://doi.org/10.1145/130385.130401
88. Mercer J (1909) XVI. Functions of positive and negative type, and their connection the theory of integral equations. Philos. Trans. R. Soc. Lond. 209:415–446. https://doi.org/10.1098/rsta.1909.0016
89. Aizerman MA, Braverman EA, Rozonoer L (1964) Theoretical foundations of the potential function method in pattern recognition learning. In: Automation and remote control, pp 821–837.
90. Zhou D-X, Jetter K (2006) Approximation with polynomial kernels and SVM classifiers. Adv Comput Math 25:323–344. https://doi.org/10.1007/s10444-004-7206-2
91. Vapnik VN (2000) The nature of statistical learning theory. Springer New York, New York, NY. https://doi.org/10.1007/978-1-4757-3264-1
92. Xiang L, Quanyin Z, Liuyang W (2013) Research of Bessel Kernel function of the first kind for support vector regression. Inf Technol J 12:2673–2682. https://doi.org/10.3923/itj.2013.2673.2682
93. Izmailov R, Vapnik V, Vashist A (2013) Multidimensional splines with infinite number of knots as SVM kernels. In: 2013 International Joint Conference on Neural Networks IJCNN, IEEE, Dallas, TX, USA, pp 1–7. https://doi.org/10.1109/IJCNN.2013.6706860.
94. Scholkopf, Kah-Kay Sung B, Burges CJC, Girosi F, Niyogi P, Poggio T, Vapnik V (1997) Comparing support vector machines with Gaussian kernels to radial basis function classifiers. IEEE Trans Signal Process 45:2758–2765. https://doi.org/10.1109/78.650102
95. Basheer IA, Hajmeer M (2000) Artificial neural networks: fundamentals, computing, design, and application. J Microbiol Methods 43:3–31. https://doi.org/10.1016/S0167-7012(00)00201-3
96. Amari S (1967) A theory of adaptive pattern classifiers. IEEE Trans Electron Comput EC-16:299–307. https://doi.org/10.1109/PGEC.1967.264666
97. Sze V, Chen Y-H, Yang T-J, Emer JS (2017) Efficient processing of deep neural networks: a tutorial and survey. Proc. IEEE 105:2295–2329. https://doi.org/10.1109/JPROC.2017.2761740
98. LeCun Y, Boser B, Denker JS, Henderson D, Howard RE, Hubbard W, Jackel LD (1989) Backpropagation applied to handwritten zip code recognition. Neural Comput 1:541–551. https://doi.org/10.1162/neco.1989.1.4.541
99. Rumelhart DE, Hinton GE, Williams RJ (1986) Learning representations by back-propagating errors. Nature 323:533–536. https://doi.org/10.1038/323533a0
100. Cho K, Merriënboer B, Gulcehre C, Bahdanau D, Bougares F, Schwenk H, Bengio Y (2014) Learning Phrase Representations using RNN encoder–decoder for statistical machine translation. In: Moschitti A, Pang B, Daelemans W (Eds.) Proc. 2014 Conf. Empir. Methods Nat. Lang. Process. EMNLP. Association for Computational Linguistics, Doha, Qatar, van1724–1734 https://doi.org/10.3115/v1/D14-1179
101. Cho K, vanMerriënboer B, Bahdanau D, Bengio Y (2014) On the properties of neural machine translation: encoder–decoder approaches. In: Wu D, Carpuat M, Carreras X, Vecchi EM (eds) Proc. SSST-8 Eighth Workshop Syntax Semant. Struct. Stat. Transl. Association for Computational Linguistics, Doha, Qatar, pp 103–111 https://doi.org/10.3115/v1/W14-4012
102. Hochreiter S, Schmidhuber J (1997) Long short-term memory. Neural Comput 9:1735–1780. https://doi.org/10.1162/neco.1997.9.8.1735
103. Ravanelli M, Brakel P, Omologo M, Bengio Y (2018) Light gated recurrent units for speech recognition. IEEE Trans Emerg Top Comput Intell. 2:92–102. https://doi.org/10.1109/TETCI.2017.2762739
104. Chung J, Gulcehre C, Cho K, Bengio Y (2014) Empirical evaluation of gated recurrent neural networks on sequence modeling. https://doi.org/10.48550/ARXIV.1412.3555
105. Gupta A, Müller AT, Huisman BJH, Fuchs JA, Schneider P, Schneider G (2018) Generative recurrent networks for de novo drug design. Mol Inform 37:1700111. https://doi.org/10.1002/minf.201700111
106. Thireou T, Reczko M (2007) Bidirectional long short-term memory networks for predicting the subcellular localization of eukaryotic proteins. IEEE/ACM Trans Comput Biol Bioinform 4:441–446. https://doi.org/10.1109/tcbb.2007.1015
107. Mayer H, Gomez F, Wierstra D, Nagy I, Knoll A, Schmidhuber J (2006) A system for robotic heart surgery that learns to tie knots using recurrent neural networks. In: 2006 IEEERSJ Int Conf Intell Robots Syst, pp 543–548. https://doi.org/10.1109/IROS.2006.282190
108. Eck D, Schmidhuber J (2002) Learning the long-term structure of the blues. In: Dorronsoro JR (ed) Artif. Neural Netw.

— ICANN 2002. Springer Berlin Heidelberg, Berlin, Heidelberg, pp 284–289 https://doi.org/10.1007/3-540-46084-5_47

109. Lakretz Y, Kruszewski G, Desbordes T, Hupkes D, Dehaene S, Baroni M (2019) The emergence of number and syntax units in LSTM language models. In: Burstein J, Doran C, Solorio T (eds) Proc. 2019 Conf. North Am. Chapter Assoc. Comput. Linguist. Hum. Lang. Technol. Vol. 1 Long Short Pap., Association for Computational Linguistics, Minneapolis. pp 11–20 https://doi.org/10.18653/v1/N19-1002
110. Gers FA, Schmidhuber E (2001) LSTM recurrent networks learn simple context-free and context-sensitive languages. IEEE Trans Neural Netw 12:1333–1340. https://doi.org/10.1109/72.963769
111. Wolf T, Debut L, Sanh V, Chaumond J, Delangue C, Moi A, Cistac P, Rault T, Louf R, Funtowicz M, Davison J, Shleifer S, vonPlaten P, Ma C, Jernite Y, Plu J, Xu C, LeScao T, Gugger S, Drame M, Lhoest Q, Rush A (2020) Transformers: state-of-the-art natural language processing. In: Liu Q, Schlangen D (eds) Proc. 2020 Conf. Empir. Methods Nat. Lang. Process. Syst. Demonstr., Association for Computational Linguistics. pp 38–45 https://doi.org/10.18653/v1/2020.emnlp-demos.6pp
112. Shannon CE (1948) A mathematical theory of communication. Bell Syst Tech J 27:379–423. https://doi.org/10.1002/j.1538-7305.1948.tb01338.x
113. Kullback S, Leibler RA (1951) On information and sufficiency. Ann Math Stat 22:79–86. https://doi.org/10.1214/aoms/1177729694
114. Linnainmaa S (1970) The representation of the cumulative rounding error of an algorithm as a Taylor expansion of the local rounding errors. Master's Thesis, University of Helsinki
115. Linnainmaa S (1976) Taylor expansion of the accumulated rounding error. BIT Numer. Math 16:146–160. https://doi.org/10.1007/BF01931367
116. Schmidhuber J (2015) Deep learning in neural networks: an overview. Neural Netw 61:85–117. https://doi.org/10.1016/j.neunet.2014.09.003
117. Leibniz GW (1676) Calculus tangentium differentialis
118. von Leibniz Gottfried Wilhelm, FJames M, JMGerhardt, KI(translator) Child.Eds. (1920) The early mathematical manuscripts of Leibniz. The Open Court Publishing Company, Chicago, London
119. Robbins H, Monro S (1951) A stochastic approximation method. Ann Math Stat 22:400–407. https://doi.org/10.1214/aoms/1177729586
120. Bottou L, Bousquet O (2007) The tradeoffs of large scale learning. Proc. 20th Int. Conf. Neural Inf. Process. Syst., Curran Associates Inc. Red Hook, NY, USA, pp 161–168
121. Fukushima K (1969) Visual feature extraction by a multilayered network of analog threshold elements. IEEE Trans Syst Sci Cybern 5:322–333. https://doi.org/10.1109/TSSC.1969.300225
122. Bertoin D, Bolte J, Gerchinovitz S, Pauwels E (2021) Numerical influence of ReLU'(0) on backpropagation. In: Ranzato M, Beygelzimer A, Dauphin Y, Liang PS, Vaughan JW (eds) Adv. Neural Inf. Process. Syst., Curran Associates, Inc. pp 468–479. https://proceedings.neurips.cc/paper_files/paper/2021/file/043ab21fc5a1607b381ac-3896176dac6-Paper.pdf
123. Holland JH (1975) Adaptation in natural and artificial systems: an introductory analysis with applications to biology, control, and artificial intelligence. University of Michigan Press, Ann Arbor
124. Angeline PJ, Saunders GM, Pollack JB (1994) An evolutionary algorithm that constructs recurrent neural networks. IEEE Trans Neural Netw 5:54–65. https://doi.org/10.1109/72.265960
125. Stanley KO, Miikkulainen R (2002) Evolving neural networks through augmenting topologies. Evol Comput 10:99–127. https://doi.org/10.1162/106365602320169811
126. Stanley KO, Bryant BD, Miikkulainen R (2005) Real-time neuroevolution in the NERO Video Game. IEEE Trans Evol Comput 9:653–668. https://doi.org/10.1109/TEVC.2005.856210
127. Stanley KO, D'Ambrosio DB, Gauci J (2009) A hypercube-based encoding for evolving large-scale neural networks. Artif Life 15:185–212. https://doi.org/10.1162/artl.2009.15.2.15202
128. Hastings EJ, Guha RK, Stanley KO (2009) Automatic content generation in the *Galactic Arms Race* Video Game. IEEE Trans Comput Intell AI Games 1:245–263. https://doi.org/10.1109/TCIAIG.2009.2038365
129. Kassahun Y, Sommer G (2005) Efficient reinforcement learning through evolutionary acquisition of neural topologies. In: Proc. 13th Eur. Symp. Artif. Neural Netw. ESANN 2005, Bruges, Belgium, pp 259–266
130. Siebel NT, Sommer G (2007) Evolutionary reinforcement learning of artificial neural networks. Int J Hybrid Intell Syst 4:171–183. https://doi.org/10.3233/HIS-2007-4304

131. Baldominos A, Saez Y, Isasi P (2020) On the automated, evolutionary design of neural networks: past, present, and future. Neural Comput Appl 32:519–545. https://doi.org/10.1007/s00521-019-04160-6
132. Goodfellow I, Pouget-Abadie J, Mirza M, Xu B, Warde-Farley D, Ozair S, Courville A, Bengio Y (2014) Generative Adversarial Nets. In: Ghahramani Z, Welling M, Cortes C, Lawrence N, Weinberger KQ (eds) Adv. Neural Inf. Process. Syst., Curran Associates, Inc. https://papers.nips.cc/paper_files/paper/2014/file/f033ed80deb0234979a61f95710dbe25-Paper.pdf
133. Knill K, Young S (1997) Hidden Markov Models in speech and language processing, In: S. Young, G. Bloothooft (eds), Corpus-Based Methods in Language and Speech Processing. Springer Netherlands, Dordrecht, pp 27–68
134. Fine S, Singer Y, Tishby N (1998) The hierarchical Hidden Markov model: analysis and applications. Mach Learn 32:41–62. https://doi.org/10.1023/A:1007469218079
135. Markov AA (1906) Rasprostranenie zakona bol'shikh chisel na velichiny, zavisyashchie drug ot druga. Izv Fiz 15:135–156
136. Brown PF, Della Pietra VJ, deSouza PV, Lai JC, Mercer RL (1992) Class-based *n*-gram models of natural language. Comput Linguist 18:467–480
137. Bengio Y, Ducharme R, Vincent P (2000) A neural probabilistic language model. In: Leen T, Dietterich T, Tresp V (eds) Adv. Neural Inf. Process. Syst.. MIT Press, https://proceedings.neurips.cc/paper_files/paper/2000/file/728f206c2a01bf572b5940d7d9a8fa4c-Paper.pdf
138. Mikolov T, Karafiát M, Burget L, Černocký J, Khudanpur S (2010) Recurrent neural network based language model. In: Proc. Interspeech 2010, ISCA, pp 1045–1048
139. Vaswani A, Shazeer N, Parmar N, Uszkoreit J, Jones L, Gomez AN, Kaiser I (2017) Polosukhin, Attention is All You Need. In: Guyon I, Luxburg UV, Bengio S, Wallach H, Fergus R, Vishwanathan S, Garnett R (eds) Adv. Neural Inf. Process. Syst. 30, Curran Associates, Inc.. pp 5998–6008
140. Bommasani R, Hudson DA, Adeli E, Altman R, Arora S, von Arx S, Bernstein MS, Bohg J, Bosselut A, Brunskill E, Brynjolfsson E, Buch S, Card D, Castellon R, Chatterji N, Chen A, Creel K, Davis JQ, Demszky D, Donahue C, Doumbouya M, Durmus E, Ermon S, Etchemendy J, Ethayarajh K, Fei-Fei L, Finn C, Gale T, Gillespie L, Goel K, Goodman N, Grossman S, Guha N, Hashimoto T, Henderson P, Hewitt J, Ho DE, Hong J, Hsu K, Huang J, Icard T, Jain S, Jurafsky D, Kalluri P, Karamcheti S, Keeling G, Khani F, Khattab O, Koh PW, Krass M, Krishna R, Kuditipudi R, Kumar A, Ladhak F, Lee M, Lee T, Leskovec J, Levent I, Li XL, Li X, Ma T, Malik A, Manning CD, Mirchandani S, Mitchell E, Munyikwa Z, Nair S, Narayan A, Narayanan D, Newman B, Nie A, Niebles JC, Nilforoshan H, Nyarko J, Ogut G, Orr L, Papadimitriou I, Park JS, Piech C, Portelance E, Potts C, Raghunathan A, Reich R, Ren H, Rong F, Roohani Y, Ruiz C, Ryan J, Ré C, Sadigh D, Sagawa S, Santhanam K, Shih A, Srinivasan K, Tamkin A, Taori R, Thomas AW, Tramèr F, Wang RE, Wang W, Wu B, Wu J, Wu Y, Xie SM, Yasunaga M, You J, Zaharia M, Zhang M, Zhang T, Zhang X, Zhang Y, Zheng L, Zhou K, Liang P (2022) On the opportunities and risks of foundation models. arXiv:2108.07258. https://doi.org/10.48550/arXiv.2108.07258.
141. OpenAI (2022) Introducing ChatGPT. https://openai.com/blog/chatgpt
142. Yang Z, Zeng X, Zhao Y, Chen R (2023) AlphaFold2 and its applications in the fields of biology and medicine. Signal Transduct Target Ther 8:115. https://doi.org/10.1038/s41392-023-01381-z
143. Brandes N, Ofer D, Peleg Y, Rappoport N, Linial M (2022) ProteinBERT: a universal deep-learning model of protein sequence and function. Bioinformatics 38:2102–2110. https://doi.org/10.1093/bioinformatics/btac020
144. Abramson J, Adler J, Dunger J, Evans R, Green T, Pritzel A, Ronneberger O, Willmore L, Ballard AJ, Bambrick J, Bodenstein SW, Evans DA, Hung C-C, O'Neill M, Reiman D, Tunyasuvunakool K, Wu Z, Žemgulytė A, Arvaniti E, Beattie C, Bertolli O, Bridgland A, Cherepanov A, Congreve M, Cowen-Rivers AI, Cowie A, Figurnov M, Fuchs FB, Gladman H, Jain R, Khan YA, Low CMR, Perlin K, Potapenko A, Savy P, Singh S, Stecula A, Thillaisundaram A, Tong C, Yakneen S, Zhong ED, Zielinski M, Žídek A, Bapst V, Kohli P, Jaderberg M, Hassabis D, Jumper JM (2024) Accurate structure prediction of biomolecular interactions with AlphaFold 3. Nature 630:493–500. https://doi.org/10.1038/s41586-024-07487-w
145. Anon. (2023) AI will transform science—now researchers must tame it. Nature 621:658–658. https://doi.org/10.1038/d41586-023-02988-6
146. Van Noorden R, Perkel JM (2023) AI and science: what 1,600 researchers think. Nature 621:672–675. https://doi.org/10.1038/d41586-023-02980-0

147. Litsa EE, Chenthamarakshan V, Das P, Kavraki LE (2023) An end-to-end deep learning framework for translating mass spectra to de-novo molecules. Commun Chem 6:132. https://doi.org/10.1038/s42004-023-00932-3
148. Bushuiev R, Bushuiev A, Samusevich R, Brungs C, Sivic J, Pluskal T (2025) Self-supervised learning of molecular representations from millions of tandem mass spectra using DreaMS. Nat Biotechnol 1–11. https://doi.org/10.1038/s41587-025-02663-3
149. Roman Bushuiev and pluskal-lab, GeMS (GNPS Experimental Mass Spectra) Dataset (2025). https://huggingface.co/datasets/roman-bushuiev/GeMS
150. Durant JL, Leland BA, Henry DR, Nourse JG (2002) Reoptimization of MDL keys for use in drug discovery. J Chem Inf Comput Sci 42:1273–1280. https://doi.org/10.1021/ci010132r
151. David L, Thakkar A, Mercado R, Engkvist O (2020) Molecular representations in AI-driven drug discovery: a review and practical guide. J Cheminformatics 12:56. https://doi.org/10.1186/s13321-020-00460-5
152. Wen B, Zeng W-F, Liao Y, Shi Z, Savage SR, Jiang W, Zhang B (2020) Deep learning in proteomics. Proteomics 20:1900335. https://doi.org/10.1002/pmic.201900335
153. Deep learning tools in proteomics (2024). https://github.com/bzhanglab/deep_learning_in_proteomics. Accessed August 30, 2025
154. Cavicchioli MV, Santorsola M, Balboni N, Mercatelli D, Giorgi FM (2022) Prediction of metabolic profiles from transcriptomics data in human cancer cell lines. Int J Mol Sci 23:3867. https://doi.org/10.3390/ijms23073867
155. Perez-Riverol Y, Bandla C, Kundu DJ, Kamatchinathan S, Bai J, Hewapathirana S, John NS, Prakash A, Walzer M, Wang S, Vizcaíno JA (2025) The PRIDE database at 20 years: 2025 update. Nucleic Acids Res 53:D543–D553. https://doi.org/10.1093/nar/gkae1011
156. Ma J, Chen T, Wu S, Yang C, Bai M, Shu K, Li K, Zhang G, Jin Z, He F, Hermjakob H, Zhu Y (2019) iProX: an integrated proteome resource. Nucleic Acids Res 47:D1211–D1217. https://doi.org/10.1093/nar/gky869
157. Moriya Y, Kawano S, Okuda S, Watanabe Y, Matsumoto M, Takami T, Kobayashi D, Yamanouchi Y, Araki N, Yoshizawa AC, Tabata T, Iwasaki M, Sugiyama N, Tanaka S, Goto S, Ishihama Y (2019) The jPOST environment: an integrated proteomics data repository and database. Nucleic Acids Res 47:D1218–D1224. https://doi.org/10.1093/nar/gky899
158. Choi M, Carver J, Chiva C, Tzouros M, Huang T, Tsai T-H, Pullman B, Bernhardt OM, Hüttenhain R, Teo GC, Perez-Riverol Y, Muntel J, Müller M, Goetze S, Pavlou M, Verschueren E, Wollscheid B, Nesvizhskii AI, Reiter L, Dunkley T, Sabidó E, Bandeira N, Vitek O (2020) MassIVE.quant: a community resource of quantitative mass spectrometry–based proteomics datasets. Nat Methods 17:981–984. https://doi.org/10.1038/s41592-020-0955-0
159. Deutsch EW, Bandeira N, Perez-Riverol Y, Sharma V, Carver JJ, Mendoza L, Kundu DJ, Wang S, Bandla C, Kamatchinathan S, Hewapathirana S, Pullman BS, Wertz J, Sun Z, Kawano S, Okuda S, Watanabe Y, MacLean B, MacCoss MJ, Zhu Y, Ishihama Y, Vizcaíno JA (2023) The ProteomeXchange consortium at 10 years: 2023 update. Nucleic Acids Res 51:D1539–D1548. https://doi.org/10.1093/nar/gkac1040
160. Ishihama Y, Oda Y, Tabata T, Sato T, Nagasu T, Rappsilber J, Mann M (2005) Exponentially modified protein abundance index (emPAI) for estimation of absolute protein amount in proteomics by the number of sequenced peptides per protein * S. Mol Cell Proteomics 4:1265–1272. https://doi.org/10.1074/mcp.M500061-MCP200
161. Reker D, Malmström L (2012) Bioinformatic challenges in targeted proteomics. J Proteome Res 11:4393–4402. https://doi.org/10.1021/pr300276f
162. Song J, Wang Y, Li F, Akutsu T, Rawlings ND, Webb GI, Chou K-C (2019) iProt-Sub: a comprehensive package for accurately mapping and predicting protease-specific substrates and cleavage sites. Brief Bioinform 20:638–658. https://doi.org/10.1093/bib/bby028
163. Wall MJ, Crowell AMJ, Simms GA, Liu F, Doucette AA (2011) Implications of partial tryptic digestion in organic–aqueous solvent systems for bottom-up proteome analysis. Anal Chim Acta 703:194–203. https://doi.org/10.1016/j.aca.2011.07.025
164. Uliana F, Vizovišek M, Acquasaliente L, Ciuffa R, Fossati A, Frommelt F, Goetze S, Wollscheid B, Gstaiger M, De Filippis V, auf dem Keller U, Aebersold R (2021) Mapping specificity, cleavage entropy, allosteric changes and substrates of blood proteases in a high-throughput screen. Nat Commun 12:1693. https://doi.org/10.1038/s41467-021-21754-8
165. Gao Z, Chang C, Yang J, Zhu Y, Fu Y (2019) AP3: an advanced proteotypic peptide predictor for targeted proteomics by incorporating peptide digestibility. Anal Chem 91:8705–8711. https://doi.org/10.1021/acs.analchem.9b02520

166. Van Midwoud PM, Rieux L, Bischoff R, Verpoorte E, Niederländer HAG (2007) Improvement of recovery and repeatability in liquid chromatography–mass spectrometry analysis of peptides. J Proteome Res 6:781–791. https://doi.org/10.1021/pr0604099
167. Stejskal K, Potěšil D, Zdráhal Z (2013) Suppression of peptide sample losses in autosampler vials. J Proteome Res 12:3057–3062. https://doi.org/10.1021/pr400183v
168. Fannes T, Vandermarliere E, Schietgat L, Degroeve S, Martens L, Ramon J (2013) Predicting tryptic cleavage from proteomics data using decision tree ensembles. J Proteome Res 12:2253–2259. https://doi.org/10.1021/pr4001114
169. Keil B (1992) Essential substrate residues for action of endopeptidases. Specif Proteolysis. Springer Berlin Heidelberg, Berlin, Heidelberg, 43–228 https://doi.org/10.1007/978-3-642-48380-6_5pp
170. Rawlings ND, Barrett AJ, Thomas PD, Huang X, Bateman A, Finn RD (2018) The MEROPS database of proteolytic enzymes, their substrates and inhibitors in 2017 and a comparison with peptidases in the PANTHER database. Nucleic Acids Res 46:D624–D632. https://doi.org/10.1093/nar/gkx1134
171. Zimmer D, Schneider K, Sommer F, Schroda M, Mühlhaus T (2018) Artificial intelligence understands peptide observability and assists with absolute protein quantification. Front Plant Sci 9 https://doi.org/10.3389/fpls.2018.01559
172. Sun B, Smialowski P, Straub T, Imhof A (2021) Investigation and highly accurate prediction of missed tryptic cleavages by deep learning. J Proteome Res 20:3749–3757. https://doi.org/10.1021/acs.jproteome.1c00346
173. Yang J, Gao Z, Ren X, Sheng J, Xu P, Chang C, Fu Y (2021) DeepDigest: prediction of protein proteolytic digestion with Deep Learning. Anal Chem 93:6094–6103. https://doi.org/10.1021/acs.analchem.0c04704
174. Li F, Leier A, Liu Q, Wang Y, Xiang D, Akutsu T, Webb GI, Smith AI, Marquez-Lago T, Li J, Song J (2020) Procleave: predicting protease-specific substrate cleavage sites by combining sequence and structural information. Genomics Proteomics Bioinform 18:52–64. https://doi.org/10.1016/j.gpb.2019.08.002
175. Berman HM, Westbrook J, Feng Z, Gilliland G, Bhat TN, Weissig H, Shindyalov IN, Bourne PE (2000) The Protein Data Bank. Nucleic Acids Res 28:235–242. https://doi.org/10.1093/nar/28.1.235
176. Guruceaga E, Garin-Muga A, Prieto G, Bejarano B, Marcilla M, Marín-Vicente C, Perez-Riverol Y, Casal JI, Vizcaíno JA, Corrales FJ, Segura V (2017) Enhanced missing proteins detection in NCI60 cell lines using an integrative search engine approach. J Proteome Res 16:4374–4390. https://doi.org/10.1021/acs.jproteome.7b00388
177. Eng JK, Jahan TA, Hoopmann MR (2013) Comet: an open-source MS/MS sequence database search tool. Proteomics 13:22–24. https://doi.org/10.1002/pmic.201200439
178. Perkins DN, Pappin DJC, Creasy DM, Cottrell JS (1999) Probability-based protein identification by searching sequence databases using mass spectrometry data. Electrophoresis 20:3551–3567. https://doi.org/10.1002/(SICI)1522-2683(19991201)20:18%253C3551::AID-ELPS3551%253E3.0.CO;2-2
179. Geer LY, Markey SP, Kowalak JA, Wagner L, Xu M, Maynard DM, Yang X, Shi W, Bryant SH (2004) Open mass spectrometry search algorithm. J Proteome Res 3:958–964. https://doi.org/10.1021/pr0499491
180. Fenyö D, Beavis RC (2003) A method for assessing the statistical significance of mass spectrometry-based protein identifications using general scoring schemes. Anal Chem 75:768–774. https://doi.org/10.1021/ac0258709
181. Demichev V, Messner CB, Vernardis SI, Lilley KS, Ralser M (2020) DIA-NN: neural networks and interference correction enable deep proteome coverage in high throughput. Nat Methods 17:41–44. https://doi.org/10.1038/s41592-019-0638-x
182. Son J, Na S, Paek E (2023) DbyDeep: exploration of MS-detectable peptides via deep learning. Anal Chem 95:11193–11200. https://doi.org/10.1021/acs.analchem.3c00460
183. Wang M, Wang J, Carver J, Pullman BS, Cha SW, Bandeira N (2018) Assembling the community-scale discoverable human proteome. Cell Syst 7:412–421.e5. https://doi.org/10.1016/j.cels.2018.08.004
184. Abdul-Khalek N, Wimmer R, Overgaard MT, Gregersen Echers S (2023) Insight on physicochemical properties governing peptide MS1 response in HPLC-ESI-MS/MS: a deep learning approach. Comput Struct Biotechnol J 21:3715–3727. https://doi.org/10.1016/j.csbj.2023.07.027
185. Wu M, Yang Y, Wang H, Xu Y (2019) A deep learning method to more accurately recall known lysine acetylation sites. BMC Bioinformatics 20:49. https://doi.org/10.1186/s12859-019-2632-9
186. Matthews BW (1975) Comparison of the predicted and observed secondary structure of T4 phage lysozyme. Biochim Biophys

Acta BBA 405:442–451. https://doi.org/10.1016/0005-2795(75)90109-9

187. Xu Z, Zhong H, He B, Wang X, Lu T (2024) PTransIPs: identification of phosphorylation sites enhanced by protein PLM embeddings. IEEE J Biomed Health Inform 28:3762–3771. https://doi.org/10.1109/JBHI.2024.3377362
188. Osipenko S, Nikolaev E, Kostyukevich Y, Osipenko S, Nikolaev E, Kostyukevich Y (2022) Retention time prediction with message-passing neural networks. Separations 9. https://doi.org/10.3390/separations9100291
189. Osipenko S, Bashkirova I, Sosnin S, Kovaleva O, Fedorov M, Nikolaev E, Kostyukevich Y (2020) Machine learning to predict retention time of small molecules in nano-HPLC. Anal Bioanal Chem 412:7767–7776. https://doi.org/10.1007/s00216-020-02905-0
190. Li X, Wang H, Jiang M, Ding M, Xu X, Xu B, Zou Y, Yu Y, Yang W, Li X, Wang H, Jiang M, Ding M, Xu X, Xu B, Zou Y, Yu Y, Yang W (2023) Collision cross section prediction based on machine learning. Molecules 28. https://doi.org/10.3390/molecules28104050
191. Snyder LR, Dolan JW (2006) High-Performance Gradient Elution. The Practical Application of the Linear-Solvent-Strength Model. 1st edn. Wiley, https://doi.org/10.1002/0470055529
192. Teschner D, Gomez-Zepeda D, Declercq A, Łącki MK, Avci S, Bob K, Distler U, Michna T, Martens L, Tenzer S, Hildebrandt A (2023) Ionmob: a Python package for prediction of peptide collisional cross-section values. Bioinformatics 39:btad486. https://doi.org/10.1093/bioinformatics/btad486
193. Gessulat S, Schmidt T, Zolg DP, Samaras P, Schnatbaum K, Zerweck J, Knaute T, Rechenberger J, Delanghe B, Huhmer A, Reimer U, Ehrlich H-C, Aiche S, Kuster B, Wilhelm M (2019) Prosit: proteome-wide prediction of peptide tandem mass spectra by deep learning. Nat Methods 16:509–518. https://doi.org/10.1038/s41592-019-0426-7
194. Palmblad M, Ramström M, Markides KE, Håkansson P, Bergquist J (2002) Prediction of chromatographic retention and protein identification in liquid chromatography/mass spectrometry. Anal Chem 74:5826–5830. https://doi.org/10.1021/ac0256890
195. Moruz L, Staes A, Foster JM, Hatzou M, Timmerman E, Martens L, Käll L (2012) Chromatographic retention time prediction for posttranslationally modified peptides. Proteomics 12:1151–1159. https://doi.org/10.1002/pmic.201100386
196. Maboudi Afkham H, Qiu X, The M, Käll L (2017) Uncertainty estimation of predictions of peptides' chromatographic retention times in shotgun proteomics. Bioinformatics 33:508–513. https://doi.org/10.1093/bioinformatics/btw619
197. Krige DG (1951) A statistical approach to some mine valuation and allied problems on the witwatersrand, M.Sc. thesis. University of the Witwatersrand, Faculty of Engineering. http://hdl.handle.net/10539/17975
198. Matheron G (1963) Principles of geostatistics. Econ Geol 58:1246–1266. https://doi.org/10.2113/gsecongeo.58.8.1246
199. Seeger M (2004) Gaussian processes for machine learning. Int J Neural Syst 14:69–106. https://doi.org/10.1142/S0129065704001899
200. Krokhin OV, Craig R, Spicer V, Ens W, Standing KG, Beavis RC, Wilkins JA (2004) An Improved model for prediction of retention times of tryptic peptides in ion pair reversed-phase HPLC: its application to protein peptide mapping by off-line HPLC-MALDI MS. Mol Cell Proteomics 3:908–919. https://doi.org/10.1074/mcp.M400031-MCP200
201. Krokhin OV, Spicer V (2010) Predicting peptide retention times for proteomics. Curr Protoc Bioinforma 13:14. https://doi.org/10.1002/0471250953.bi1314s31
202. Ma C, Ren Y, Yang J, Ren Z, Yang H, Liu S (2018) Improved peptide retention time prediction in liquid chromatography through deep learning. Anal Chem 90:10881–10888. https://doi.org/10.1021/acs.analchem.8b02386
203. Tiwary S, Levy R, Gutenbrunner P, Salinas Soto F, Palaniappan KK, Deming L, Berndl M, Brant A, Cimermancic P, Cox J (2019) High-quality MS/MS spectrum prediction for data-dependent and data-independent acquisition data analysis. Nat Methods 16:519–525. https://doi.org/10.1038/s41592-019-0427-6
204. Bouwmeester R, Martens L, Degroeve S (2020) Generalized calibration across liquid chromatography setups for generic prediction of small-molecule retention times. Anal Chem 92:6571–6578. https://doi.org/10.1021/acs.analchem.0c00233
205. Zeng W-F, Zhou X-X, Willems S, Ammar C, Wahle M, Bludau I, Voytik E, Strauss MT, Mann M (2022) AlphaPeptDeep: a modular deep learning framework to predict peptide properties for proteomics. Nat Commun 13:7238. https://doi.org/10.1038/s41467-022-34904-3
206. Wilburn DB, Shannon AE, Spicer V, Richards AL, Yeung D, Swaney DL, Krokhin OV, Searle BC (2023) Deep learning from harmonized peptide libraries enables

retention time prediction of diverse post translational modifications. https://doi.org/10.1101/2023.05.30.542978

207. Chen W, McCool EN, Sun L, Zang Y, Ning X, Liu X (2022) Evaluation of machine learning models for proteoform retention and migration time prediction in top-down mass spectrometry. J Proteome Res 21:1736–1747. https://doi.org/10.1021/acs.jproteome.2c00124
208. Stanstrup J, Neumann S, Vrhovšek U (2015) PredRet: prediction of retention time by direct mapping between multiple chromatographic systems. Anal Chem 87:9421–9428. https://doi.org/10.1021/acs.analchem.5b02287
209. Hastie T, Tibshirani R (1986) Generalized additive models. Stat Sci 1:297–310. https://doi.org/10.1214/ss/1177013604
210. Chen B, Wang C, Fu Z, Yu H, Liu E, Gao X, Li J, Han L (2023) RT-ensemble pred: a tool for retention time prediction of metabolites on different LC-MS systems. J Chromatogr A 1707:464304. https://doi.org/10.1016/j.chroma.2023.464304
211. Scarselli F, Gori M, Tsoi AC, Hagenbuchner M, Monfardini G (2009) The graph neural network model. IEEE Trans Neural Netw 20:61–80. https://doi.org/10.1109/TNN.2008.2005605
212. Micheli A (2009) Neural network for graphs: a contextual constructive approach. IEEE Trans Neural Netw 20:498–511. https://doi.org/10.1109/TNN.2008.2010350
213. Kensert A, Bouwmeester R, Efthymiadis K, Van Broeck P, Desmet G, Cabooter D (2021) Graph convolutional networks for improved prediction and interpretability of chromatographic retention data. Anal Chem 93:15633–15641. https://doi.org/10.1021/acs.analchem.1c02988
214. Kretschmer F, Harrieder E-M, Hoffmann MA, Böcker S, Witting M (2024) RepoRT: a comprehensive repository for small molecule retention times. Nat Methods 21:153–155. https://doi.org/10.1038/s41592-023-02143-z
215. McDaniel EW, Martin DW, Barnes WS (1962) Drift tube-mass spectrometer for studies of low-energy ion-molecule reactions. Rev Sci Instrum 33:2–7. https://doi.org/10.1063/1.1717656
216. Hernandez DR, DeBord JD, Ridgeway ME, Kaplan DA, Park MA, Fernandez-Lima F (2014) Ion dynamics in a trapped ion mobility spectrometer. Analyst 139:1913–1921. https://doi.org/10.1039/C3AN02174B
217. Hoaglund CS, Valentine SJ, Sporleder CR, Reilly JP, Clemmer DE (1998) Three-dimensional ion mobility/TOFMS analysis of electrosprayed biomolecules. Anal Chem 70:2236–2242. https://doi.org/10.1021/ac980059c
218. Gabelica V, Marklund EJ (2019) A cyclic ion mobility-mass spectrometry system. Anal Chem 91:10116–10123. https://doi.org/10.1021/acs.analchem.9b01838
219. Gabelica V, Marklund E (2018) Fundamentals of ion mobility spectrometry. Curr Opin Chem Biol 42:51–59. https://doi.org/10.1016/j.cbpa.2017.10.022
220. Giles K, Pringle SD, Worthington KR, Little D, Wildgoose JL (2004) Applications of a travelling wave-based radio-frequency-only stacked ring ion guide. Rapid Commun Mass Spectrom 18:2401–2404. https://doi.org/10.1002/rcm.1641
221. Zhou Z, Shen X, Tu J, Zhu Z-J (2016) Large-scale prediction of collision cross-section values for metabolites in ion mobility-mass spectrometry. Anal Chem 88:11270–11277. https://doi.org/10.1021/acs.analchem.6b03091
222. Zhou Z, Xiong X, Zhu Z-J (2017) MetCCS predictor: a web server for predicting collision cross-section values of metabolites in ion mobility-mass spectrometry based metabolomics. Bioinformatics 33:2235–2237. https://doi.org/10.1093/bioinformatics/btx140
223. Zhou Z, Tu J, Xiong X, Shen X, Zhu Z-J (2017) LipidCCS: prediction of collision cross-section values for lipids with high precision to support ion mobility-mass spectrometry-based lipidomics. Anal Chem 89:9559–9566. https://doi.org/10.1021/acs.analchem.7b02625
224. Bijlsma L, Bade R, Plass V, Ibáñez M, Mollerup C, Sancho JV, Hernández F (2017) Prediction of collision cross-section values for small molecules: application to pesticide residue analysis. Anal Chem 89:6814–6821. https://doi.org/10.1021/acs.analchem.7b00741
225. Colby SM, Nuñez JR, Hodas NO, Corley CD, Renslow RR (2020) Deep learning to generate in silico chemical property libraries and candidate molecules for small molecule identification in complex samples. Anal Chem 92:1720–1729. https://doi.org/10.1021/acs.analchem.9b02348
226. Weininger D (1988) SMILES, a chemical language and information system. 1. Introduction to methodology and encoding rules. J Chem Inf Comput Sci 28:31–36. https://doi.org/10.1021/ci00057a005
227. Weininger D, Weininger A, Weininger JL (1989) SMILES. 2. Algorithm for generation of unique SMILES notation. J Chem Inf Comput Sci. 29:97–101. https://doi.org/10.1021/ci00062a008

228. Weininger D (1990) SMILES. 3. DEPICT. Graphical depiction of chemical structures. J Chem Inf Comput Sci 30:237–243. https://doi.org/10.1021/ci00067a005
229. Ross DH, Cho JH, Xu L (2020) Breaking down structural diversity for comprehensive prediction of ion-neutral collision cross sections. Anal Chem 92:4548–4557. https://doi.org/10.1021/acs.analchem.9b05772
230. Shah AR, Agarwal K, Baker ES, Singhal M, Mayampurath AM, Ibrahim YM, Kangas LJ, Monroe ME, Zhao R, Belov ME, Anderson GA, Smith RD (2010) Machine learning based prediction for peptide drift times in ion mobility spectrometry. Bioinformatics 26:1601–1607. https://doi.org/10.1093/bioinformatics/btq245
231. Wang B, Zhang J, Chen P, Ji Z, Deng S, Li C (2013) Prediction of peptide drift time in ion mobility mass spectrometry from sequence-based features. BMC Bioinformatics 14: S9. https://doi.org/10.1186/1471-2105-14-S8-S9
232. Meier F, Köhler ND, Brunner A-D, Wanka J-MH, Voytik E, Strauss MT, Theis FJ, Mann M (2021) Deep learning the collisional cross sections of the peptide universe from a million experimental values. Nat Commun 12:1185. https://doi.org/10.1038/s41467-021-21352-8
233. Chang C-H, Yeung D, Spicer V, Ogata K, Krokhin O, Ishihama Y (2021) Sequence-specific model for predicting peptide collision cross section values in proteomic ion mobility spectrometry. J Proteome Res 20:3600–3610. https://doi.org/10.1021/acs.jproteome.1c00185
234. Yeung D, Spicer V, Zahedi RP, Krokhin O (2023) Exploring the variable space of shallow machine learning models for reversed-phase retention time prediction. Comput Struct Biotechnol J 21:2446. https://doi.org/10.1016/j.csbj.2023.02.047
235. Wang C, Yuan C, Wang Y, Shi Y, Zhang T, Patti GJ (2024) Predicting collision cross-section values for small molecules through chemical class-based multimodal graph attention network. J Chem Inf Model 64:6305–6315. https://doi.org/10.1021/acs.jcim.3c01934
236. Suits F, Hoekman B, Rosenling T, Bischoff R, Horvatovich P (2011) Threshold-avoiding proteomics pipeline. Anal Chem 83:7786–7794. https://doi.org/10.1021/ac201332j
237. Woldegebriel M, Derks E (2017) Artificial neural network for probabilistic feature recognition in liquid chromatography coupled to high-resolution mass spectrometry. Anal Chem 89:1212–1221. https://doi.org/10.1021/acs.analchem.6b03678
238. Kantz ED, Tiwari S, Watrous JD, Cheng S, Jain M (2019) Deep neural networks for classification of LC-MS spectral peaks. Anal Chem 91:12407–12413. https://doi.org/10.1021/acs.analchem.9b02983
239. Borgsmüller N, Gloaguen Y, Opialla T, Blanc E, Sicard E, Royer A-L, Le Bizec B, Durand S, Migné C, Pétéra M, Pujos-Guillot E, Giacomoni F, Guitton Y, Beule D, Kirwan J (2019) WiPP: workflow for improved peak picking for gas chromatography-mass spectrometry (GC-MS) data. Metabolites 9:171. https://doi.org/10.3390/metabo9090171
240. Risum AB, Bro R (2019) Using deep learning to evaluate peaks in chromatographic data. Talanta 204:255–260. https://doi.org/10.1016/j.talanta.2019.05.053
241. Harshman RA (1972) PARAFAC2: mathematical and technical notes, pp 30–44
242. Harshman RA (1970) Foundations of the PARAFAC procedure: models and conditions for an "explanatory" multi-modal factor analysis. University of California, Los Angeles. https://www.psychology.uwo.ca/faculty/harshman/wpppfac0.pdf
243. Melnikov AD, Tsentalovich YP, Yanshole VV (2020) Deep learning for the precise peak detection in high-resolution LC–MS data. Anal Chem 92:588–592. https://doi.org/10.1021/acs.analchem.9b04811
244. Gloaguen Y, Kirwan JA, Beule D (2022) Deep learning-assisted peak curation for large-scale LC-MS metabolomics. Anal Chem 94:4930–4937. https://doi.org/10.1021/acs.analchem.1c02220
245. Eilertz D, Mitterer M, Buescher JM (2022) automRm: an R package for fully automatic LC-QQQ-MS data preprocessing powered by machine learning. Anal Chem 94:6163–6171. https://doi.org/10.1021/acs.analchem.1c05224
246. Chen T, Guestrin C (2016) XGBoost: a scalable tree boosting system. Proc. 22nd ACM SIGKDD Int. Conf. Knowl. Discov. Data Min. ACM, San Francisco, CA, USA, pp 785–794 https://doi.org/10.1145/2939672.2939785
247. Stancliffe E, Patti GJ (2023) PeakDetective: a semisupervised deep learning-based approach for peak curation in untargeted metabolomics. Anal Chem 95:9397–9403. https://doi.org/10.1021/acs.analchem.3c00764
248. Wu Z, Serie D, Xu G, Zou J (2020) PB-Net: automatic peak integration by sequential deep learning for multiple reaction monitoring. J Proteomics 223:103820. https://doi.org/10.1016/j.jprot.2020.103820
249. Li Y, He Q, Guo H, Zhong C-Q, Li X, Li Y, Han J, Shuai J (2022) MSSort-DIAXMBD:

a deep learning classification tool of the peptide precursors quantified by OpenSWATH. J Proteomics 259:104542. https://doi.org/10.1016/j.jprot.2022.104542
250. Röst HL, Rosenberger G, Navarro P, Gillet L, Miladinović SM, Schubert OT, Wolski W, Collins BC, Malmström J, Malmström L, Aebersold R (2014) OpenSWATH enables automated, targeted analysis of data-independent acquisition MS data. Nat Biotechnol 32:219–223. https://doi.org/10.1038/nbt.2841
251. Lee C-C, Lin Y-C, Pan TY, Yang CH, Li P-H, Chen SY, Gao JJ, Yang C, Chu LJ, Huang P-J, Yeh Y-M, Tang P, Chang Y-S, Yu J-S, Hsiao Y-C (2023) HeapMS: an automatic peak-picking pipeline for targeted proteomic data powered by 2D heatmap transformation and convolutional neural networks. Anal Chem 95:15486–15496. https://doi.org/10.1021/acs.analchem.3c01011
252. Dai Z, Liu H, Le QV, Tan M (2021) CoAtNet: marrying convolution and attention for all data sizes. Adv. Neural Inf. Process. Syst. NeurIPS. 3965–3977 https://doi.org/10.48550/arXiv.2106.04803
253. Ren S, He K, Girshick RB, Sun J (2017) Faster R-CNN: towards real-time object detection with region proposal networks. IEEE Trans Pattern Anal Mach Intell 39:1137–1149. https://doi.org/10.1109/TPAMI.2016.2577031
254. Park J, Wilkins C, Avtonomov D, Hong J, Back S, Kim H, Shulman N, MacLean BX, Lee S-W, Kim S (2023) Targeted proteomics data interpretation with DeepMRM. Cell Rep Methods 3:100521. https://doi.org/10.1016/j.crmeth.2023.100521
255. Sakoe H, Chiba S (1978) Dynamic programming algorithm optimization for spoken word recognition. IEEE Trans Acoust Speech Signal Process 26:43–49. https://doi.org/10.1109/TASSP.1978.1163055
256. Nielsen N-PV, Carstensen JM, Smedsgaard J (1998) Aligning of single and multiple wavelength chromatographic profiles for chemometric data analysis using correlation optimised warping. J Chromatogr A 805:17–35. https://doi.org/10.1016/S0021-9673(98)00021-1
257. Eilers PHC (2004) Parametric time warping. Anal Chem 76:404–411. https://doi.org/10.1021/ac034800e
258. Windig W, Phalp JM, Payne AW (1996) A noise and background reduction method for component detection in liquid chromatography/mass spectrometry. Anal Chem 68:3602–3606. https://doi.org/10.1021/ac960435y
259. Du P, Kibbe WA, Lin SM (2006) Improved peak detection in mass spectrum by incorporating continuous wavelet transform-based pattern matching. Bioinformatics (Oxford, England) 22:2059–2065. https://doi.org/10.1093/bioinformatics/btl355
260. Christin C, Smilde AK, Hoefsloot HCJ, Suits F, Bischoff R, Horvatovich PL (2008) Optimized time alignment algorithm for LC-MS data: correlation optimized warping using component detection algorithm-selected mass chromatograms. Anal Chem 80:7012–7021. https://doi.org/10.1021/ac800920h
261. Christin C, Hoefsloot HCJ, Smilde AK, Suits F, Bischoff R, Horvatovich PL (2010) Time alignment algorithms based on selected mass traces for complex LC-MS data. J Proteome Res 9:1483–1495. https://doi.org/10.1021/pr9010124
262. Suits F, Lepre J, Du P, Bischoff R, Horvatovich P (2008) Two-dimensional method for time aligning liquid chromatography–mass spectrometry data. Anal Chem 80:3095–3104. https://doi.org/10.1021/ac702267h
263. Podwojski K, Fritsch A, Chamrad DC, Paul W, Sitek B, Stühler K, Mutzel P, Stephan C, Meyer HE, Urfer W, Ickstadt K, Rahnenführer J (2009) Retention time alignment algorithms for LC/MS data must consider non-linear shifts. Bioinformatics 25:758–764. https://doi.org/10.1093/bioinformatics/btp052
264. Bruderer R, Bernhardt OM, Gandhi T, Miladinović SM, Cheng L-Y, Messner S, Ehrenberger T, Zanotelli V, Butscheid Y, Escher C, Vitek O, Rinner O, Reiter L (2015) Extending the limits of quantitative proteome profiling with data-independent acquisition and application to acetaminophen-treated three-dimensional liver microtissues*[S]. Mol Cell Proteomics 14:1400–1410. https://doi.org/10.1074/mcp.M114.044305
265. Bruderer R, Bernhardt OM, Gandhi T, Xuan Y, Sondermann J, Schmidt M, Gomez-Varela D, Reiter L (2017) Optimization of experimental parameters in data-independent mass spectrometry significantly increases depth and reproducibility of results. Mol Cell Proteomics 16:2296–2309. https://doi.org/10.1074/mcp.RA117.000314
266. Li M, Wang XR (2019) Peak alignment of gas chromatography–mass spectrometry data with deep learning. J Chromatogr A 1604:460476. https://doi.org/10.1016/j.chroma.2019.460476
267. Bromley J, Guyon I, LeCun Y, Säckinger E, Shah R (1994) Signature verification using a "siamese" time delay neural network. In: Cowan JD, Tesauro G (eds) Adv. Neural

Inf. Process. Syst. NIPS 1993. Morgan Kaufmann, pp 737–744. https://proceedings.neurips.cc/paper_files/paper/1993/file/288cc0ff022877bd3df94bc9360b9c5d-Paper.pdf

268. Colby SM, Chang CH, Bade JL, Nunez JR, Blumer MR, Orton DJ, Bloodsworth KJ, Nakayasu ES, Smith RD, Ibrahim YM, Renslow RS, Metz TO (2022) DEIMoS: an open-source tool for processing high-dimensional mass spectrometry data. Anal Chem 94:6130–6138. https://doi.org/10.1021/acs.analchem.1c05017
269. Zohora FT, Rahman MZ, Tran NH, Xin L, Shan B, Li M (2019) DeepIso: a deep learning model for peptide feature detection from LC-MS map. Sci Rep 9:17168. https://doi.org/10.1038/s41598-019-52954-4
270. Zohora FT, Rahman MZ, Tran NH, Xin L, Shan B, Li M (2021) Deep neural network for detecting arbitrary precision peptide features through attention based segmentation. Sci Rep 11:18249. https://doi.org/10.1038/s41598-021-97669-7
271. Ivanov MV, Bubis JA, Gorshkov V, Abdrakhimov DA, Kjeldsen F, Gorshkov MV (2021) Boosting MS1-only proteomics with machine learning allows 2000 protein identifications in single-shot human proteome analysis using 5 min HPLC gradient. J Proteome Res 20:1864–1873. https://doi.org/10.1021/acs.jproteome.0c00863
272. Ke G, Meng Q, Finley T, Wang T, Chen W, Ma W, Ye Q, Liu T-Y (2017) LightGBM: a highly efficient gradient boosting decision tree. Adv Neural Inf Process Syst NeurIPS. 3149–3157. https://proceedings.neurips.cc/paper_files/paper/2017/file/6449f44a102fde848669bdd9eb6b76fa-Paper.pdf
273. Boiko DA, Kozlov KS, Burykina JV, Ilyushenkova VV, Ananikov VP (2022) Fully automated unconstrained analysis of high-resolution mass spectrometry data with machine learning. J Am Chem Soc 144:14590–14606. https://doi.org/10.1021/jacs.2c03631
274. Kong AT, Leprevost FV, Avtonomov DM, Mellacheruvu D, Nesvizhskii AI (2017) MSFragger: ultrafast and comprehensive peptide identification in mass spectrometry–based proteomics. Nat Methods 14:513–520. https://doi.org/10.1038/nmeth.4256
275. Tanner S, Shu H, Frank A, Wang L, Zandi E, Mumby M, Pevzner PA, Bafna V (2005) InsPecT: identification of posttranslationally modified peptides from tandem mass spectra. Anal Chem 77:4626–4639. https://doi.org/10.1021/ac050102d
276. Tabb DL, Ma Z, Martin DB, Ham A-JL, Chambers MC (2008) DirecTag: accurate sequence tags from peptide MS/MS through statistical scoring. J Proteome Res 7:3838–3846. https://doi.org/10.1021/pr800154p
277. Lam H, Deutsch EW, Eddes JS, Eng JK, King N, Stein SE, Aebersold R (2007) Development and validation of a spectral library searching method for peptide identification from MS/MS. Proteomics 7:655–667. https://doi.org/10.1002/pmic.200600625
278. Deutsch EW, Mendoza L, Shteynberg DD, Hoopmann MR, Sun Z, Eng JK, Moritz RL (2023) Trans-proteomic pipeline: robust mass spectrometry-based proteomics data analysis suite. J Proteome Res 22:615–624. https://doi.org/10.1021/acs.jproteome.2c00624
279. Loo JA, Udseth HR, Smith RD, Futrell JH (1988) Collisional effects on the charge distribution of ions from large molecules, formed by electrospray-ionization mass spectrometry. Rapid Commun Mass Spectrom 2:207–210. https://doi.org/10.1002/rcm.1290021006
280. Olsen JV, Macek B, Lange O, Makarov A, Horning S, Mann M (2007) Higher-energy C-trap dissociation for peptide modification analysis. Nat Methods 4:709–712. https://doi.org/10.1038/nmeth1060
281. Zubarev RA, Kelleher NL, McLafferty FW (1998) Electron capture dissociation of multiply charged protein cations. A nonergodic process. J Am Chem Soc 120:3265–3266. https://doi.org/10.1021/ja973478k
282. Zhang Z (2004) Prediction of low-energy collision-induced dissociation spectra of peptides. Anal Chem 76:3908–3922. https://doi.org/10.1021/ac049951b
283. Zhang Z (2005) Prediction of low-energy collision-induced dissociation spectra of peptides with three or more charges. Anal Chem 77:6364–6373. https://doi.org/10.1021/ac050857k
284. Zhang Z (2011) Prediction of collision-induced-dissociation spectra of peptides with post-translational or process-induced modifications. Anal Chem 83:8642–8651. https://doi.org/10.1021/ac2020917
285. Zhang Z, Shah B (2010) Prediction of collision-induced dissociation spectra of common N-glycopeptides for glycoform identification. Anal Chem 82:10194–10202. https://doi.org/10.1021/ac102359u
286. Frank AM, Ranking-Based Scoring A (2009) Function for peptide-spectrum matches. J Proteome Res 8:2241–2252. https://doi.org/10.1021/pr800678b
287. Frank AM (2009) Predicting intensity ranks of peptide fragment ions. J Proteome Res 8:2226–2240. https://doi.org/10.1021/pr800677f

288. Frank A, Pevzner P (2005) PepNovo: de novo peptide sequencing via probabilistic network modeling. Anal Chem 77:964–973. https://doi.org/10.1021/ac048788h
289. Degroeve S, Martens L (2013) MS2PIP: a tool for MS/MS peak intensity prediction. Bioinformatics (Oxford, England) 29:3199–3203. https://doi.org/10.1093/bioinformatics/btt544
290. Declercq A, Bouwmeester R, Chiva C, Sabidó E, Hirschler A, Carapito C, Martens L, Degroeve S, Gabriels R (2023) Updated MS²PIP web server supports cutting-edge proteomics applications. Nucleic Acids Res 51:W338–W342. https://doi.org/10.1093/nar/gkad335
291. Zolg DP, Wilhelm M, Schnatbaum K, Zerweck J, Knaute T, Delanghe B, Bailey DJ, Gessulat S, Ehrlich H-C, Weininger M, Yu P, Schlegl J, Kramer K, Schmidt T, Kusebauch U, Deutsch EW, Aebersold R, Moritz RL, Wenschuh H, Moehring T, Aiche S, Huhmer A, Reimer U, Kuster B (2017) Building ProteomeTools based on a complete synthetic human proteome. Nat Methods 14:259–262. https://doi.org/10.1038/nmeth.4153
292. Ekvall M, Truong P, Gabriel W, Wilhelm M, Käll L (2022) Prosit transformer: a transformer for prediction of MS2 spectrum intensities. J Proteome Res 21:1359–1364. https://doi.org/10.1021/acs.jproteome.1c00870
293. Adams C, Gabriel W, Laukens K, Picciani M, Wilhelm M, Bittremieux W, Boonen K (2024) Fragment ion intensity prediction improves the identification rate of non-tryptic peptides in timsTOF. Nat Commun 15:3956. https://doi.org/10.1038/s41467-024-48322-0
294. Picciani M, Gabriel W, Giurcoiu V-G, Shouman O, Hamood F, Lautenbacher L, Jensen CB, Müller J, Kalhor M, Soleymaniniya A, Kuster B, The M, Wilhelm M (2024) Oktoberfest: open-source spectral library generation and rescoring pipeline based on Prosit. Proteomics 24:2300112. https://doi.org/10.1002/pmic.202300112
295. Isaksson M, Karlsson C, Laurell T, Kirkeby A, Heusel M (2022) MSLibrarian: optimized predicted spectral libraries for data-independent acquisition proteomics. J Proteome Res 21:535–546. https://doi.org/10.1021/acs.jproteome.1c00796
296. Zhou X-X, Zeng W-F, Chi H, Luo C, Liu C, Zhan J, He S-M, Zhang Z (2017) pDeep: predicting MS/MS spectra of peptides with deep learning. Anal Chem 89:12690–12697. https://doi.org/10.1021/acs.analchem.7b02566
297. Zeng W-F, Zhou X-X, Zhou W-J, Chi H, Zhan J, He S-M (2019) MS/MS spectrum prediction for modified peptides using pDeep2 trained by transfer learning. Anal Chem 91:9724–9731. https://doi.org/10.1021/acs.analchem.9b01262
298. Tarn C, Zeng W-F (2021) pDeep3: toward more accurate spectrum prediction with fast few-shot learning. Anal Chem 93:5815–5822. https://doi.org/10.1021/acs.analchem.0c05427
299. Chen Z-L, Mao P-Z, Zeng W-F, Chi H, He S-M (2021) pDeepXL: MS/MS spectrum prediction for cross-linked peptide pairs by deep learning. J Proteome Res 20:2570–2582. https://doi.org/10.1021/acs.jproteome.0c01004
300. Lin Y-M, Chen C-T, Chang J-M (2019) MS2CNN: predicting MS/MS spectrum based on protein sequence using deep convolutional neural networks. BMC Genomics 20: 906. https://doi.org/10.1186/s12864-019-6297-6
301. Yang Y, Liu X, Shen C, Lin Y, Yang P, Qiao L (2020) In silico spectral libraries by deep learning facilitate data-independent acquisition proteomics. Nat Commun 11: 146. https://doi.org/10.1038/s41467-019-13866-z
302. Liu K, Li S, Wang L, Ye Y, Tang H (2020) Full-spectrum prediction of peptides tandem mass spectra using deep neural network. Anal Chem 92:4275–4283. https://doi.org/10.1021/acs.analchem.9b04867
303. Wang B, Wang Y, Chen Y, Gao M, Ren J, Guo Y, Situ C, Qi Y, Zhu H, Li Y, Guo X (2022) DeepSCP: utilizing deep learning to boost single-cell proteome coverage. Brief Bioinform 23:bbac214. https://doi.org/10.1093/bib/bbac214
304. Na S, Choi H, Paek E (2022) Deephos: predicted spectral database search for TMT-labeled phosphopeptides and its false discovery rate estimation. Bioinformatics 38:2980–2987. https://doi.org/10.1093/bioinformatics/btac280
305. Zolg DP, Gessulat S, Paschke C, Graber M, Rathke-Kuhnert M, Seefried F, Fitzemeier K, Berg F, Lopez-Ferrer D, Horn D, Henrich C, Huhmer A, Delanghe B, Frejno M (2025) INFERYS rescoring: Boosting peptide identifications and scoring confidence of database search results. Rapid Commun Mass Spectrom 39:e9128. https://doi.org/10.1002/rcm.9128
306. Suni V, Suomi T, Tsubosaka T, Imanishi SY, Elo LL, Corthals GL (2018) SimPhospho: a software tool enabling confident phosphosite assignment. Bioinformatics 34:2690–2692. https://doi.org/10.1093/bioinformatics/bty151

307. Bielow C, Aiche S, Andreotti S, Reinert K (2011) MSSimulator: simulation of mass spectrometry data. J Proteome Res 10:2922–2930. https://doi.org/10.1021/pr200155f
308. Röst HL, Sachsenberg T, Aiche S, Bielow C, Weisser H, Aicheler F, Andreotti S, Ehrlich H-C, Gutenbrunner P, Kenar E, Liang X, Nahnsen S, Nilse L, Pfeuffer J, Rosenberger G, Rurik M, Schmitt U, Veit J, Walzer M, Wojnar D, Wolski WE, Schilling O, Choudhary JS, Malmström L, Aebersold R, Reinert K, Kohlbacher O (2016) OpenMS: a flexible open-source software platform for mass spectrometry data analysis. Nat Methods 13:741–748. https://doi.org/10.1038/nmeth.3959
309. Yang Y, Horvatovich P, Qiao L (2021) Fragment mass spectrum prediction facilitates site localization of phosphorylation. J Proteome Res 20:634–644. https://doi.org/10.1021/acs.jproteome.0c00580
310. Ruttkies C, Schymanski EL, Wolf S, Hollender J, Neumann S (2016) MetFrag relaunched: incorporating strategies beyond in silico fragmentation. J Cheminformatics 8: 3. https://doi.org/10.1186/s13321-016-0115-9
311. Wang Y, Kora G, Bowen BP, Pan C (2014) MIDAS: a database-searching algorithm for metabolite identification in metabolomics. Anal Chem 86:9496–9503. https://doi.org/10.1021/ac5014783
312. Kind T, Fiehn O (2007) Seven Golden Rules for heuristic filtering of molecular formulas obtained by accurate mass spectrometry. BMC Bioinformatics 8: 105. https://doi.org/10.1186/1471-2105-8-105
313. Verdegem D, Lambrechts D, Carmeliet P, Ghesquière B (2016) Improved metabolite identification with MIDAS and MAGMa through MS/MS spectral dataset-driven parameter optimization. Metabolomics 12:98. https://doi.org/10.1007/s11306-016-1036-3
314. Lai Z, Tsugawa H, Wohlgemuth G, Mehta S, Mueller M, Zheng Y, Ogiwara A, Meissen J, Showalter M, Takeuchi K, Kind T, Beal P, Arita M, Fiehn O (2018) Identifying metabolites by integrating metabolome databases with mass spectrometry cheminformatics. Nat Methods 15:53–56. https://doi.org/10.1038/nmeth.4512
315. Tsugawa H, Kind T, Nakabayashi R, Yukihira D, Tanaka W, Cajka T, Saito K, Fiehn O, Arita M (2016) Hydrogen rearrangement rules: computational MS/MS fragmentation and structure elucidation using MS-FINDER software. Anal Chem 88:7946–7958. https://doi.org/10.1021/acs.analchem.6b00770
316. Hufsky F, Scheubert K, Böcker S (2014) Computational mass spectrometry for small-molecule fragmentation. TrAC Trends Anal Chem 53:41–48. https://doi.org/10.1016/j.trac.2013.09.008
317. Allen F, Greiner R, Wishart D (2015) Competitive fragmentation modeling of ESI-MS/MS spectra for putative metabolite identification. Metabolomics 11:98–110. https://doi.org/10.1007/s11306-014-0676-4
318. Dührkop K, Fleischauer M, Ludwig M, Aksenov AA, Melnik AV, Meusel M, Dorrestein PC, Rousu J, Böcker S (2019) SIRIUS 4: a rapid tool for turning tandem mass spectra into metabolite structure information. Nat Methods 16:299–302. https://doi.org/10.1038/s41592-019-0344-8
319. Dührkop K, Shen H, Meusel M, Rousu J, Böcker S (2015) Searching molecular structure databases with tandem mass spectra using CSI:FingerID. Proc Natl Acad Sci USA 112:12580–12585. https://doi.org/10.1073/pnas.1509788112
320. Cao L, Guler M, Tagirdzhanov A, Lee Y-Y, Gurevich A, Mohimani H (2021) MolDiscovery: learning mass spectrometry fragmentation of small molecules. Nat Commun 12:3718. https://doi.org/10.1038/s41467-021-23986-0
321. Young A, Röst H, Wang B (2024) Tandem mass spectrum prediction for small molecules using graph transformers. Nat Mach Intell 6:404–416. https://doi.org/10.1038/s42256-024-00816-8
322. Pérez-Ribera M, Faizan-Khan M, Giné R, Badia JM, Junza A, Yanes O, Sales-Pardo M, Guimerà R (2025) SingleFrag: a deep learning tool for MS/MS fragment and spectral prediction and metabolite annotation. Brief Bioinform 26:bbaf333. https://doi.org/10.1093/bib/bbaf333
323. Grimme S (2013) Towards first principles calculation of electron impact mass spectra of molecules. Angew Chem, Int Ed 52:6306–6312. https://doi.org/10.1002/anie.201300158
324. Koopman J, Grimme S (2021) From QCEIMS to QCxMS: a tool to routinely calculate CID mass spectra using molecular dynamics. J Am Soc Mass Spectrom 32:1735–1751. https://doi.org/10.1021/jasms.1c00098
325. Grimme S, Bannwarth C, Shushkov P, Robust A (2017) Accurate tight-binding quantum chemical method for structures, vibrational frequencies, and noncovalent interactions of large molecular systems parametrized for all spd-block elements (Z = 1–86). J Chem Theory Comput 13:1989–2009. https://doi.org/10.1021/acs.jctc.7b00118
326. Bannwarth C, Ehlert S, Grimme S (2019) GFN2-xTB—an accurate and broadly

parametrized self-consistent tight-binding quantum chemical method with multipole electrostatics and density-dependent dispersion contributions. J Chem Theory Comput 15:1652–1671. https://doi.org/10.1021/acs.jctc.8b01176
327. Schnegotzki R, Koopman J, Grimme S, Süssmuth RD (2022) Quantum chemistry-based molecular dynamics simulations as a tool for the assignment of ESI-MS/MS spectra of drug molecules. Chem – Eur J 28:e202200318. https://doi.org/10.1002/chem.202200318
328. Hecht H, Rojas WY, Ahmad Z, Křenek A, Klánová J, Price EJ (2024) Quantum chemistry-based prediction of electron ionization mass spectra for environmental chemicals. Anal Chem 96:13652–13662. https://doi.org/10.1021/acs.analchem.4c02589
329. Frank AM, Bandeira N, Shen Z, Tanner S, Briggs SP, Smith RD, Pevzner PA (2008) Clustering millions of tandem mass spectra. J Proteome Res 7:113–122. https://doi.org/10.1021/pr070361e
330. Griss J, Perez-Riverol Y, Lewis S, Tabb DL, Dianes JA, del-Toro N, Rurik M, Walzer MW, Kohlbacher O, Hermjakob H, Wang R, Vizcaíno JA (2016) Recognizing millions of consistently unidentified spectra across hundreds of shotgun proteomics datasets. Nat Methods 13:651–656. https://doi.org/10.1038/nmeth.3902
331. The M, Käll L (2016) MaRaCluster: a fragment rarity metric for clustering fragment spectra in shotgun proteomics. J Proteome Res 15:713–720. https://doi.org/10.1021/acs.jproteome.5b00749
332. Wang L, Li S, Tang H (2019) msCRUSH: fast tandem mass spectral clustering using locality sensitive hashing. J Proteome Res 18:147–158. https://doi.org/10.1021/acs.jproteome.8b00448
333. Bittremieux W, May DH, Bilmes J, Noble WS (2022) A learned embedding for efficient joint analysis of millions of mass spectra. Nat Methods 19:675–678. https://doi.org/10.1038/s41592-022-01496-1
334. de Jonge NF, Hecht H, Strobel M, Wang M, van der Hooft JJJ, Huber F (2024) Reproducible MS/MS library cleaning pipeline in matchms. J Cheminformatics 16:88. https://doi.org/10.1186/s13321-024-00878-1
335. de Jonge NF, Louwen JJR, Chekmeneva E, Camuzeaux S, Vermeir FJ, Jansen RS, Huber F, van der Hooft JJJ (2023) MS2Query: reliable and scalable MS2 mass spectra-based analogue search. Nat Commun 14:1752. https://doi.org/10.1038/s41467-023-37446-4
336. Huber F, Ridder L, Verhoeven S, Spaaks JH, Diblen F, Rogers S, van der Hooft JJJ (2021) Spec2Vec: Improved mass spectral similarity scoring through learning of structural relationships. PLoS Comput Biol 17:e1008724. https://doi.org/10.1371/journal.pcbi.1008724
337. Huber F, van der Burg S, van der Hooft JJJ, Ridder L (2021) MS2DeepScore: a novel deep learning similarity measure to compare tandem mass spectra. J Cheminform 13:84. https://doi.org/10.1186/s13321-021-00558-4
338. Ji H, Xu Y, Lu H, Zhang Z (2019) Deep MS/MS-aided structural-similarity scoring for unknown metabolite identification. Anal Chem 91:5629–5637. https://doi.org/10.1021/acs.analchem.8b05405
339. Fan Z, Alley A, Ghaffari K, Ressom HW (2020) MetFID: artificial neural network-based compound fingerprint prediction for metabolite annotation. Metabolomics Off J Metabolomic Soc 16:104. https://doi.org/10.1007/s11306-020-01726-7
340. Li Y, Kuhn M, Gavin A-C, Bork P (2020) Identification of metabolites from tandem mass spectra with a machine learning approach utilizing structural features. Bioinformatics 36:1213–1218. https://doi.org/10.1093/bioinformatics/btz736
341. Dührkop K, Nothias L-F, Fleischauer M, Reher R, Ludwig M, Hoffmann MA, Petras D, Gerwick WH, Rousu J, Dorrestein PC, Böcker S (2021) Systematic classification of unknown metabolites using high-resolution fragmentation mass spectra. Nat Biotechnol 39:462–471. https://doi.org/10.1038/s41587-020-0740-8
342. Xing S, Jiao Y, Salehzadeh M, Soma KK, Huan T (2021) SteroidXtract: deep learning-based pattern recognition enables comprehensive and rapid extraction of steroid-like metabolic features for automated biology-driven metabolomics. Anal Chem 93:5735–5743. https://doi.org/10.1021/acs.analchem.0c04834
343. Ma B (2015) Novor: real-time peptide de novo sequencing software. J Am Soc Mass Spectrom 26:1885–1894. https://doi.org/10.1007/s13361-015-1204-0
344. Tran NH, Zhang X, Xin L, Shan B, Li M (2017) De novo peptide sequencing by deep learning. Proc Natl Acad Sci USA 114:8247–8252. https://doi.org/10.1073/pnas.1705691114
345. Qiao R, Tran NH, Xin L, Shan B, Li M, Ghodsi A (2019) DeepNovoV2: better de novo peptide sequencing with deep learning. https://doi.org/10.48550/arXiv.1904.08514
346. Gueto-Tettay C, Tang D, Happonen L, Heusel M, Khakzad H, Malmström J, Malmström L

(2023) Multienzyme deep learning models improve peptide de novo sequencing by mass spectrometry proteomics. PLoS Comput Biol 19:e1010457. https://doi.org/10.1371/journal.pcbi.1010457

347. Karunratanakul K, Tang H-Y, Speicher DW, Chuangsuwanich E, Sriswasdi S (2019) Uncovering thousands of new peptides with sequence-mask-search hybrid de novo peptide sequencing framework. Mol Cell Proteomics 18:2478–2491. https://doi.org/10.1074/mcp.TIR119.001656
348. Lee J-Y, Mitchell HD, Burnet MC, Wu R, Jenson SC, Merkley ED, Nakayasu ES, Nicora CD, Jansson JK, Burnum-Johnson KE, Payne SH (2022) Uncovering hidden members and functions of the soil microbiome using de novo metaproteomics. J Proteome Res 21:2023–2035. https://doi.org/10.1021/acs.jproteome.2c00334
349. Qiao R, Tran NH, Xin L, Chen X, Li M, Shan B, Ghodsi A (2021) Computationally instrument-resolution-independent de novo peptide sequencing for high-resolution devices. Nat Mach Intell 3:420–425. https://doi.org/10.1038/s42256-021-00304-3
350. Yilmaz M, Fondrie WE, Bittremieux W, Melendez CF, Nelson R, Ananth V, Oh S, Noble WS (2024) Sequence-to-sequence translation from mass spectra to peptides with a transformer model. Nat Commun 15:6427. https://doi.org/10.1038/s41467-024-49731-x
351. Liu K, Ye Y, Li S, Tang H (2023) Accurate de novo peptide sequencing using fully convolutional neural networks. Nat Commun 14:7974. https://doi.org/10.1038/s41467-023-43010-x
352. Ge C, Lu Y, Qu J, Xie L, Wang F, Zhang H, Kong R, Chang S (2022) DePS: an improved deep learning model for de novo peptide sequencing. https://doi.org/10.48550/arXiv.2203.08820
353. Eloff K, Kalogeropoulos K, Mabona A, Morell O, Catzel R, Rivera-de-Torre E, Berg Jespersen J, Williams W, van Beljouw SPB, Skwark MJ, Laustsen AH, Brouns SJJ, Ljungars A, Schoof EM, Van Goey J, auf dem Keller U, Beguir K, Lopez Carranza N, Jenkins TP (2025) InstaNovo enables diffusion-powered de novo peptide sequencing in large-scale proteomics experiments. Nat Mach Intell 7:565–579. https://doi.org/10.1038/s42256-025-01019-5
354. Käll L, Canterbury JD, Weston J, Noble WS, MacCoss MJ (2007) Semi-supervised learning for peptide identification from shotgun proteomics datasets. Nat Methods 4:923–925. https://doi.org/10.1038/nmeth1113
355. Fondrie WE, Noble WS (2020) Machine learning strategy that leverages large data sets to boost statistical power in small-scale experiments. J Proteome Res 19:1267–1274. https://doi.org/10.1021/acs.jproteome.9b00780
356. Reiter L, Rinner O, Picotti P, Hüttenhain R, Beck M, Brusniak M-Y, Hengartner MO, Aebersold R (2011) mProphet: automated data processing and statistical validation for large-scale SRM experiments. Nat Methods 8:430–435. https://doi.org/10.1038/nmeth.1584
357. Teleman J, Röst HL, Rosenberger G, Schmitt U, Malmström L, Malmström J, Levander F (2015) DIANA—algorithmic improvements for analysis of data-independent acquisition MS data. Bioinformatics (Oxford, England) 31:555–562. https://doi.org/10.1093/bioinformatics/btu686
358. Li K, Jain A, Malovannaya A, Wen B, Zhang B (2020) DeepRescore: leveraging deep learning to improve peptide identification in immunopeptidomics. Proteomics 20:e1900334. https://doi.org/10.1002/pmic.201900334
359. Silva ASC, Bouwmeester R, Martens L, Degroeve S (2019) Accurate peptide fragmentation predictions allow data driven approaches to replace and improve upon proteomics search engine scoring functions. Bioinformatics (Oxford, England) 35:5243–5248. https://doi.org/10.1093/bioinformatics/btz383
360. Sulimov P, Voronkova A, Kertész-Farkas A (2020) Annotation of tandem mass spectrometry data using stochastic neural networks in shotgun proteomics. Bioinformatics 36:3781–3787. https://doi.org/10.1093/bioinformatics/btaa206
361. Kudriavtseva P, Kashkinov M, Kertész-Farkas A (2021) Deep convolutional neural networks help scoring tandem mass spectrometry data in database-searching approaches. J Proteome Res 20:4708–4717. https://doi.org/10.1021/acs.jproteome.1c00315
362. Li H, Na S, Hwang K-B, Paek E (2022) TIDD: tool-independent and data-dependent machine learning for peptide identification. BMC Bioinformatics 23: 109. https://doi.org/10.1186/s12859-022-04640-y
363. Cormican JA, Horokhovskyi Y, Soh WT, Mishto M, Liepe J (2022) inSPIRE: AN Open-source Tool for Increased Mass Spectrometry Identification Rates Using Prosit Spectral Prediction. Mol Cell Proteomics 21:100432. https://doi.org/10.1016/j.mcpro.2022.100432
364. Scott AM, Karlsson C, Mohanty T, Hartman E, Vaara ST, Linder A, Malmström J,

Malmström L (2023) Generalized precursor prediction boosts identification rates and accuracy in mass spectrometry based proteomics. Commun Biol 6:628. https://doi.org/10.1038/s42003-023-04977-x
365. Skiadopoulou D, Vašíček J, Kuznetsova K, Bouyssié D, Käll L, Vaudel M (2023) Retention time and fragmentation predictors increase confidence in identification of common variant peptides. J Proteome Res 22:3190–3199. https://doi.org/10.1021/acs.jproteome.3c00243
366. Fondrie WE, Noble WS (2021) mokapot: Fast and flexible semisupervised learning for peptide detection. J Proteome Res 20:1966–1971. https://doi.org/10.1021/acs.jproteome.0c01010
367. Arab I, Fondrie WE, Laukens K, Bittremieux W (2023) Semisupervised machine learning for sensitive open modification spectral library searching. J Proteome Res 22:585–593. https://doi.org/10.1021/acs.jproteome.2c00616
368. Bittremieux W, Meysman P, Noble WS, Laukens K (2018) Fast open modification spectral library searching through approximate nearest neighbor indexing. J Proteome Res 17:3463–3474. https://doi.org/10.1021/acs.jproteome.8b00359
369. Song J, Yu C (2021) Alpha-XIC: a deep neural network for scoring the coelution of peak groups improves peptide identification by data-independent acquisition mass spectrometry. Bioinformatics 38:38–43. https://doi.org/10.1093/bioinformatics/btab544
370. Song J, Yu C (2022) Alpha-Tri: a deep neural network for scoring the similarity between predicted and measured spectra improves peptide identification of DIA data. Bioinformatics 38:1525–1531. https://doi.org/10.1093/bioinformatics/btab878
371. Scheubert K, Hufsky F, Petras D, Wang M, Nothias L-F, Dührkop K, Bandeira N, Dorrestein PC, Böcker S (2017) Significance estimation for large scale metabolomics annotations by spectral matching. Nat Commun 8:1494. https://doi.org/10.1038/s41467-017-01318-5
372. Alka O, Shanthamoorthy P, Witting M, Kleigrewe K, Kohlbacher O, Röst HL (2022) DIAMetAlyzer allows automated false-discovery rate-controlled analysis for data-independent acquisition in metabolomics. Nat Commun 13:1347. https://doi.org/10.1038/s41467-022-29006-z
373. Bilbao A, Munoz N, Kim J, Orton DJ, Gao Y, Poorey K, Pomraning KR, Weitz K, Burnet M, Nicora CD, Wilton R, Deng S, Dai Z, Oksen E, Gee A, Fasani RA, Tsalenko A, Tanjore D, Gardner J, Smith RD, Michener JK, Gladden JM, Baker ES, Petzold CJ, Kim Y-M, Apffel A, Magnuson JK, Burnum-Johnson KE (2023) PeakDecoder enables machine learning-based metabolite annotation and accurate profiling in multidimensional mass spectrometry measurements. Nat Commun 14:2461. https://doi.org/10.1038/s41467-023-37031-9
374. Zolg DP, Wilhelm M, Schmidt T, Médard G, Zerweck J, Knaute T, Wenschuh H, Reimer U, Schnatbaum K, Kuster B (2018) ProteomeTools: systematic characterization of 21 post-translational protein modifications by liquid chromatography tandem mass spectrometry (LC-MS/MS) using synthetic peptides. Mol Cell Proteomics 17:1850–1863. https://doi.org/10.1074/mcp.TIR118.000783
375. Van Puyvelde B, Daled S, Willems S, Gabriels R, Gonzalez de Peredo A, Chaoui K, Mouton-Barbosa E, Bouyssié D, Boonen K, Hughes CJ, Gethings LA, Perez-Riverol Y, Bloomfield N, Tate S, Schiltz O, Martens L, Deforce D, Dhaenens M (2022) A comprehensive LFQ benchmark dataset on modern day acquisition strategies in proteomics. Sci Data 9:126. https://doi.org/10.1038/s41597-022-01216-6
376. Neely BA, Perez-Riverol Y, Palmblad M (2024) Quality control in the mass spectrometry proteomics core: a practical primer. J Biomol Tech 35 https://doi.org/10.7171/3fc1f5fe.42308a9a
377. Neely BA, Dorfer V, Martens L, Bludau I, Bouwmeester R, Degroeve S, Deutsch EW, Gessulat S, Käll L, Palczynski P, Payne SH, Rehfeldt TG, Schmidt T, Schwämmle V, Uszkoreit J, Vizcaíno JA, Wilhelm M, Palmblad M (2023) Toward an integrated machine learning model of a proteomics experiment. J Proteome Res 22:681–696. https://doi.org/10.1021/acs.jproteome.2c00711
378. Samek W, Montavon G, Lapuschkin S, Anders CJ, Müller K-R (2021) Explaining deep neural networks and beyond: a review of methods and applications. Proc IEEE 109:247–278. https://doi.org/10.1109/JPROC.2021.3060483
379. Wilkinson MD, Dumontier M, Aalbersberg IjJ, Appleton G, Axton M, Baak A, Blomberg N, Boiten J-W, da Silva Santos LB, Bourne PE, Bouwman J, Brookes AJ, Clark T, Crosas M, Dillo I, Dumon O, Edmunds S, Evelo CT, Finkers R, Gonzalez-Beltran A, Gray AJG, Groth P, Goble C, Grethe JS, Heringa J, Hoen PAC't, Hooft R, Kuhn T, Kok R, Kok J, Lusher SJ, Martone ME, Mons A, Packer AL, Persson B, Rocca-Serra P, Roos M, van Schaik R, Sansone S-A, Schultes E, Sengstag

T, Slater T, Strawn G, Swertz MA, Thompson M, van der Lei J, van Mulligen E, Velterop J, Waagmeester A, Wittenburg P, Wolstencroft K, Zhao J, Mons B (2016) The FAIR Guiding Principles for scientific data management and stewardship. Sci Data 3:160018. https://doi.org/10.1038/sdata.2016.18

380. de Visser C, Johansson LF, Kulkarni P, Mei H, Neerincx P, Joeri van der Velde K, Horvatovich P, van Gool AJ, Swertz MA, Hoen PAC't, Niehues A (2023) Ten quick tips for building FAIR workflows. PLoS Comput Biol 19:e1011369. https://doi.org/10.1371/journal.pcbi.1011369

381. Hoekman B, Breitling R, Suits F, Bischoff R, Horvatovich P (2012) msCompare: a framework for quantitative analysis of label-free LC-MS data for comparative candidate biomarker studies. Mol Cell Proteomics MCP 11:M111.015974. https://doi.org/10.1074/mcp.M111.015974

Chapter 14

Annotating Tandem Mass Spectra in Metabolomics with DreaMS and MassSpecGym

Filip Jozefov, Roman Bushuiev, Anton Bushuiev, Téo Hebra, Tito Damiani, Josef Šivic and Tomáš Pluskal

Abstract

We present a simple, practical workflow for annotating tandem mass spectrometry (MS/MS) data of small molecules using advanced machine learning and a reference library. The discovery and structural characterization of small molecules is central to biology, chemistry, and medicine. MS/MS has become the leading technology for this task, yet the majority of spectra from untargeted metabolomics remain unannotated, limiting our understanding of the underlying chemical diversity. Artificial intelligence (AI) is rapidly reshaping many areas of science, and mass spectrometry is no exception. In this chapter, we introduce a practical workflow for annotating MS/MS data using two complementary resources: DreaMS, a foundation model for mass spectrometry, and MassSpecGym, a large reference library and benchmarking platform. DreaMS embeds mass spectra into a representation space where structural similarity is preserved, and MassSpecGym provides anchor points for retrieving identical or related molecules. Together, they form an ecosystem that enables efficient annotation of experimental MS/MS data. We demonstrate this approach on alkaloid data from *Piper fimbriulatum*, showing how DreaMS can retrieve exact matches or place unannotated spectra alongside structurally related compounds. All steps are provided as easy-to-follow protocols, designed to run on standard hardware and requiring no modifications to existing metabolomics pipelines. Readers will learn how to integrate AI tools into their own workflows for compound annotation, molecular networking, and discovery. Overall, this chapter aims to empower biologists and chemists to harness machine learning for metabolomics, transforming unannotated spectra into interpretable chemical knowledge.

Keywords Mass spectrometry, Foundation model, Mass spectra embeddings, Retrieval, benchmark

1 Introduction

The identification of small molecules from complex biological samples is central to several research fields such as natural products discovery [1], environmental monitoring [2], and clinical diagnostics [3]. Among analytical techniques, tandem mass spectrometry (MS/MS) is unmatched in its ability to provide structural information at the single compound resolution. In a typical liquid chromatography

Frédérique Lisacek (ed.), *Bioinformatics for Mass Spectrometry*, Methods in Molecular Biology, vol. 3071,
https://doi.org/10.1007/978-1-0716-5519-1_14,

tandem mass spectrometry (LC–MS/MS) experiment, thousands of precursor ions are fragmented to generate MS/MS spectra, each representing a partial fingerprint of an underlying molecular structure.

Despite important technological advances over the past decade, the majority of MS/MS data from untargeted metabolomics remains unannotated. In most studies, fewer than 10% of spectra can be linked to known molecules using spectral libraries or state-of-the-art computational tools [4]. This vast reservoir of unassigned spectra, often referred to as *metabolic dark matter* [5], represents an immense but largely untapped source of biochemical knowledge. Manual interpretation, which involves expert inspection of fragmentation patterns to infer substructures, is infeasible at modern data scales, where public repositories contain hundreds of millions of spectra [6].

Machine learning offers a route to overcome this bottleneck by automatically learning patterns in MS/MS data that relate to molecular structures. However, progress in this field has been constrained by two key limitations. First, supervised machine-learning approaches depend heavily on reference spectral libraries, which are curated collections of MS/MS spectra acquired from pure analytical standards, with each spectrum representing a ground truth fingerprint of a known compound. Well-established libraries such as MoNA [7] or NIST [8] remain limited in both size and chemical coverage. Second, there has been no widely adopted benchmark dataset or protocol to enable fair, reproducible comparison of competing methods, a factor that has slowed the translation of advances in deep learning into mass spectrometry applications. To address these gaps, we have recently proposed two complementary resources: DreaMS [9] and MassSpecGym [10]. Together, they form a unified ecosystem: DreaMS provides structure-aware spectrum embeddings that enable efficient similarity search, while MassSpecGym supplies a reference library that supports the elucidation of unknown compounds in experimental samples.

DreaMS (deep representations empowering the annotation of mass spectra) is a transformer-based *foundation model* for MS/MS spectra (Fig. 1a). Its design addresses a long-standing challenge: while millions of spectra are freely available in public repositories such as GNPS [6], most lack annotation linking them to specific molecular structures and could not previously be used to train supervised models. DreaMS overcomes this limitation through self-supervised learning, which allows it to extract patterns directly from raw spectra without human-provided labels. The pretraining task, similar to the approach used in language models such as ChatGPT [11], involves reconstructing masked peaks and predicting the relative retention order of compounds in chromatography. By leveraging this strategy, DreaMS can learn from vast collections of unlabeled spectra, providing an unprecedented scale of training data. After pretraining, DreaMS takes an input mass spectrum and produces an embedding, a

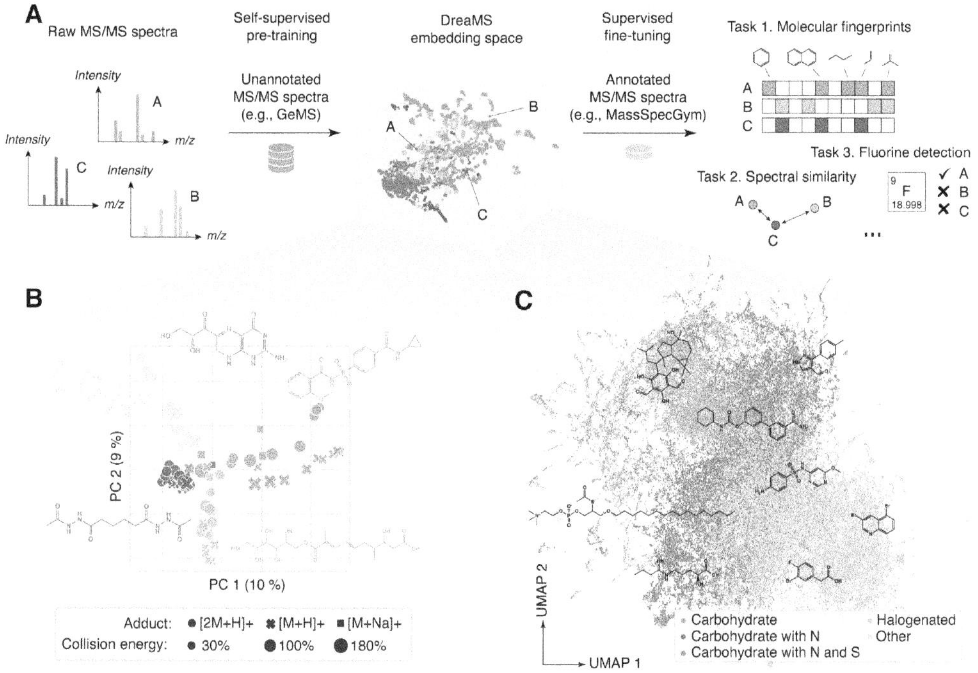

Fig. 1 (**A**) We proposed the DreaMS neural network, which learns molecular representations directly from raw, unannotated mass spectra using self-supervised learning. After pretraining, DreaMS can be fine-tuned via supervised transfer learning to solve a wide range of spectrum annotation tasks, leveraging spectral libraries and other sources of annotated data. (**B**) Applying PCA to selected precursor embeddings shows that DreaMS representations form linearly separable clusters that reflect molecular structures, while remaining robust to variations in ionization adducts and normalized collision energies. (**C**) A UMAP projection of DreaMS embeddings further illustrates how the learned representation space is organized according to molecular formulas-Figures reproduced from Bushuiev et al. [9]

high-dimensional numerical representation that places the spectrum in a space where distances reflect similarities and differences in the chemical and structural features of the underlying molecules (Fig. 1B, C). Once trained in this way, the model can be fine-tuned for specific downstream tasks such as elemental composition prediction, spectral similarity search, or molecular fingerprint prediction.

MassSpecGym is a large-scale, high-quality benchmark specifically created for machine-learning–driven analysis of MS/MS data. It comprises over 231,000 labeled spectra from approximately 29,000 unique molecules. All spectra are drawn from open resources, including MoNA, MassBank [12], GNPS [6], and the MS^{n}Lib [13], and are accessible in both conventional mass spectrometry format and machine-learning–ready representations.

For experimental scientists, the value of MassSpecGym lies in providing a reference library that supports practical tasks such as compound identification: when a new mass spectrum is measured in an experiment, it can be compared against MassSpecGym to find the most similar entries and thus suggest the identity of the compound. For the machine-learning community, MassSpecGym formalizes three essential challenges in computational mass spectrometry—de novo molecular structure generation, molecule retrieval, and spectrum simulation (Fig. 2). This is useful not only for discovering entirely new chemical entities, but also for reliably identifying already known or related compounds. To ensure that evaluation reflects true generalization rather than memorization, MassSpecGym [10] uses a maximum common edge substructure (MCES) [14]-based splitting strategy that prevents structurally similar molecules from appearing in both training and test sets.

When combined with DreaMS, this benchmark forms a powerful ecosystem for both model development and evaluation. The DreaMS foundation model provides structure-aware embeddings that enable efficient similarity search across vast spectral repositories, while MassSpecGym offers curated data, standardized tasks, and rigorous metrics needed for reproducible comparison of methods. Together, they establish a foundation for scalable, benchmark-driven development of new models addressing diverse open challenges in metabolomics. In the remainder of this chapter, we guide the reader through a practical workflow that illustrates how to prepare MS/MS data for embedding-based analysis, perform quality control and deduplication, annotate our data with DreaMS through molecule retrieval using MassSpecGym, and perform molecular networking to further propagate annotation and uncover molecular relationships within the sample. By following these tutorials, researchers will be able to adapt DreaMS to their own datasets, evaluate results, and integrate these approaches into a broader metabolomics pipeline.

Fig. 2 MassSpecGym provides three challenges for benchmarking the discovery and identification of new molecules from MS/MS spectra. De novo molecular structure generation proposes complete structures for previously unannotated spectra, supporting the discovery of new molecules and the reliable identification of known compounds. Molecule retrieval ranks candidate molecules from a reference database for a given spectrum, enabling routine compound identification in biological, environmental, and clinical samples. Spectrum simulation predicts MS/MS spectra from molecular structures, which facilitates in silico expansion of spectral libraries and improves retrieval while deepening understanding of fragmentation. Together, these challenges abstract the process of scientific discovery from biological and environmental samples into well-defined machine-learning problemsFigure reproduced from Bushuiev et al. [10]

2 Materials

To ensure that readers can readily follow and reproduce the procedures described in this book chapter, all data, models, and software dependencies are provided in accessible formats. The complete protocol is designed to run on a standard desktop or laptop computer without the need for high-performance computing resources. All necessary assets are available through public repositories, and the total disk space required does not exceed 10 GB.

- Data and code availability
  All datasets, pretrained model weights, and example notebooks used in this chapter are publicly available. The DreaMS foundation model, MassSpecGym reference library, and *Piper fimbriulatum* alkaloid dataset are provided via the project's GitHub repository and Hugging Face Hub. The repository includes step-by-step Jupyter notebooks, installation instructions, and all scripts required to reproduce the analyses presented in this chapter.
- Software requirements
  To run the protocol, users need a Python environment with Jupyter Notebook for executing the provided tutorials. Interactive tutorials are provided as a collection of Jupyter notebooks that guide the user step-by-step through data preprocessing, model application, and result interpretation. Each notebook contains explanatory text alongside executable code cells, closely mirroring the workflow described in this protocol.
  For a consistent and seamless setup, we recommend installing the Conda [15] environment manager. With Conda in place, all additional software dependencies can be installed automatically using Conda and pip. Detailed installation instructions are available in our *GitHub repository* (https://github.com/pluskal-lab/DreaMS_MIMB) [16] and further in the text below.
- Data resources
  The primary dataset used for library matching and benchmark experiments is the MassSpecGym reference collection. The benchmark file MassSpecGym.mgf can be downloaded from Hugging Face (https://huggingface.co/datasets/roman-bushuiev/MassSpecGym/tree/main/data/auxiliary) [17].
  For demonstration purposes, we also include a dataset of 23 newly reported alkaloids from *Piper fimbriulatum*, characterized by Damiani et al. [18]. This dataset, comprising both annotated and unannotated spectra, is bundled directly with the GitHub repository [16] and requires no separate download.
- Model weights
  The DreaMS foundation model requires two sets of pretrained weights: embedding_model.ckpt and ssl_model.ckpt. These files

are available from *Hugging Face* (https://huggingface.co/roman-bushuiev/DreaMS/tree/main) [19] and should be placed in the model_checkpoints directory of the working environment.

- Code and tutorials
  The core protocol, including all tutorials, scripts, and installation guides, is available on *DreaMS_MIMB GitHub* [16]. This repository provides step-by-step instructions for installing the software stack, running the notebooks, and obtaining results.
- Hardware requirements
  All experiments can be performed on computers running macOS, Linux, or Windows, with a minimum of 12 GB RAM recommended. GPU acceleration is not required for the standard experiments and library matching; all workflow should complete within minutes on typical hardware. While fine-tuning the DreaMS model requires more substantial computational resources (*see* **Notes 8 and 14**), we provide pretrained fine-tuned models for common tasks and demonstrate how users can adapt DreaMS for their own needs.
- Web interface
  For easier access to DreaMS without installation and without requiring necessary computational infrastructure, we also provide a web interface on Hugging Face Spaces [20]. This interface allows users to compute and download DreaMS embeddings from their input datasets (e.g., .mgf or .mzMLmzML files), as well as to automatically run DreaMS library matching against MassSpecGym.

3 Methods

3.1 Environment Setup and Notebook Guide

In this section, we set up the computational environment, introduce the project structure, and outline the use of Jupyter notebooks. The first step is to choose a working directory on your local machine. All code execution and data storage will take place here (*see* **Notes 16**). For example, resources may be stored at:

```
/Users/Myself/Code/
```

Open a terminal and navigate to the selected directory:

```
cd /Users/Myself/Code/
```

Confirm the location with:

```
pwd
```

which should return /Users/Myself/Code/

With the working directory in place, download the project resources from *GitHub*. Verify that Git is installed by running:

```
git --version
```

If installed, this returns a version string (e.g., git version 2.39.1). If not, follow the installation instructions in [21]. To clone the repository, execute:

```
git clone https://github.com/pluskal-lab/DreaMS_
MIMB.git
```

This creates a new folder named DreaMS_MIMB in your directory. Move into it with:

```
cd DreaMS_MIMB
```

Next, create the software environment. The project uses the Conda package manager [15]. Check for Conda with:

```
conda--version
```

If Conda is not installed, follow the instructions in [22]. To install the required dependencies, run:

```
bash install_env.sh
```

This process may take several minutes. When complete, activate the environment:

```
conda activate dreams_mimb
```

Download the necessary datasets by running:

```
python
scripts/download_assets.py
```

This will download all necessary data and place it in the correct directories.

To launch the JupyterLab interface, execute:

```
jupyter lab
```

If a browser window does not open automatically, copy the link displayed in the terminal and paste it into your browser. The root directory should be /Users/Myself/Code/DreaMS_MIMB.

For a more detailed, step-by-step explanation of the installation process, including troubleshooting tips and background on how each step works, please refer to the introductory page of the project (README file). There you will also find alternative procedures in case any step fails.

Once the repository is cloned and the environment installed, your working directory should follow the structure below:

```
├── benchmark
├── data
│   ├── Piper_data_smiles.csv
│   ├── Piper_sirius_all annotated.mgf
│   ├── Piper_sirius_matched_annotated.mgf
│   ├── massspecgym
│   │   ├── MassSpecGym.mgf
│   ├── model_checkpoints
│   │   ├── embedding_model.ckpt
│   │   └── ssl_model.ckpt
├── notebooks
│   ├── 0_notebook_tutorial.ipynb
│   ├── 1_data_preparation.ipynb
│   ├── 2_deduplication.ipynb
│   ├── 3_library_matching.ipynb
│   ├── 4_molecular_networking.ipynb
├── LICENSE
├── configs
├── README.md
├── environment.yml
├── paths.py
├── scripts
└── setup.py
```

With the environment ready and files in place, we can move on to the first tutorial, 0_notebook_tutorial.ipynb, which introduces the Jupyter notebook interface and demonstrates how to execute code interactively. We will begin with an overview of how to navigate and run notebooks, and then proceed to investigate the dataset used in all subsequent tutorials.

We will use high-resolution LC–MS/MS data from *Piper fimbriulatum*, a myrmecophytic plant native to Central America that we recently showed produces alkaloids [23]. The dataset originates from an untargeted metabolomics study [23] that profiled extracts from leaves, stems, and roots, applying feature-based molecular networking [24], in silico annotation, and spectral dereplication to map the plant's chemical space. This analysis revealed 23 structurally confirmed alkaloids spanning five distinct scaffold classes: piperidines, benzylisoquinolines, aporphines, piperolactams, and a novel seco-benzylisoquinoline named fimbriulatumine, as well as numerous additional MS/MS spectra strongly predicted to be alkaloids but lacking definitive structural assignments. Many of these compounds had never been reported in the *Piperaceae* family. The data thus comprise two complementary subsets: (1) spectra for confirmed alkaloid structures, validated by analytical standards or NMR and (2) spectra for alkaloids annotated through computational predictions and network context.

3.2 Data Cleaning with DreaMS Preprocessors

In this tutorial, we begin preparing the mass spectrometry data that the DreaMS model will use to compute spectral embeddings for library retrieval and additional downstream tasks described later in this chapter. The first steps involve loading the raw spectra, converting them into an efficient storage format, and ensuring they meet the model's requirements. The process is demonstrated in 1_data_preparation.ipynb and uses the MSData class, the central interface in DreaMS for loading, exploring, filtering, and saving spectral data. MSData works with common formats such as .mzML and .mgf. On the first load, it automatically detects the file type, parses the contents, and converts them into a single, easy-to-use object. At the same time, it generates a cached HDF5 file, a binary format optimized for fast, large-scale access. This conversion happens transparently: a .hdf5 file appears alongside the original file, and subsequent loads use this cached version, avoiding repeated parsing of large text files (*see* **Notes 3 and 10**).

The HDF5 format used by DreaMS stores MS/MS data as a set of tensors—one for the spectra themselves and others for metadata such as precursor m/z, charge, SMILES strings, etc. This design offers three key advantages: (1) datasets containing millions of spectra can be read efficiently from disk in chunks or spectrum-by-spectrum without loading everything into memory; (2) the tensor structure integrates directly with machine-learning frameworks such as

PyTorch and scikit-learn; and (3) advanced compression reduces file size to about a third of the original while retaining all essential information. Although HDF5 can store full LC–MS/MS runs, including MS1 scans, here we focus exclusively on MS/MS data.

Loading a file with MSData is straightforward, even when working with datasets that use non-standard metadata naming. For example, the MassSpecGym.mgf file in our data/massspecgym/ folder stores the precursor *m/z* under the field name PRECURSOR_MZ. In such cases, MSData allows you to explicitly map these names to the expected ones using parameters like prec_mz_col or spec_col (*see* **Notes 1**). This ensures that the essential attributes for the DreaMS model, such as the spectrum itself and the precursor *m/z*, are consistently available for downstream prediction.

```
from dreams.utils.data import MSData
msdata = MSData.from_mgf(
PROJECT_ROOT / "data" / "Piper_sirius_all_
annotated.mgf", prec_mz_col="PRECURSOR_MZ"
)
```

Once converted, all subsequent loads will use this cached .hdf5 file, eliminating the need to repeatedly parse the original .mgf.

The MSData class is also designed to handle very large datasets efficiently, without the need to load their entire contents into memory. An example is the GeMS-A10 dataset, introduced in the DreaMS paper, which can be downloaded using the script python scripts/download_gems.py. The resulting file, GeMS_A10.hdf5, occupies about 14 GB of disk (compared to an uncompressed size of roughly 47 GB) and contains over 23 million spectra, far more than can fit in the RAM of an average laptop. By opening it with in_mem=False, as in

```
msdata = MSData.from_hdf5(pth, in_mem=False)
```

no spectra or metadata are loaded initially; instead, the data is read on demand in small chunks. This makes it possible to inspect, filter, and process even multi-gigabyte datasets without exhausting system memory.

Once your data is loaded into an MSData object, you can begin exploring both the spectra and their associated metadata. A convenient way to inspect a single entry is by using the.at() method, which returns not only the numerical spectrum and its metadata but also, when SMILES structures are available, a visual display of the molecule alongside its MS/MS spectrum (Fig. 3).

```
msdata.at(333)
```

To work with all spectra at once, you can retrieve them as a single tensor:

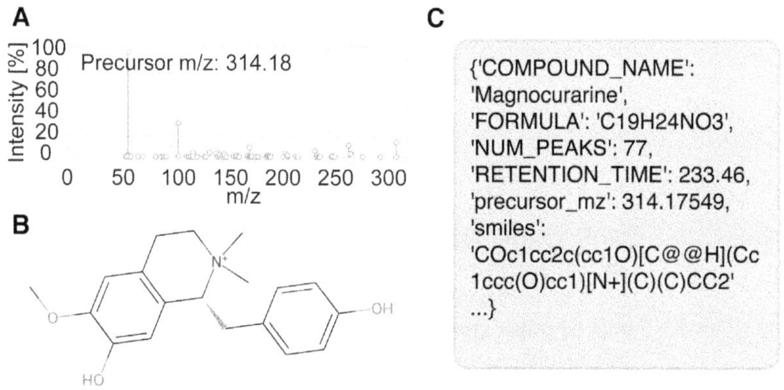

Fig. 3 Example visualization of an annotated alkaloid spectrum using the DreaMS MSData.at() method. The figure shows the MS/MS fragmentation spectrum of *magnocurarine*, one of the alkaloids discovered in *Piper fimbriulatum* [23]. When a valid SMILES structure is available, the tool displays the molecular structure alongside the experimental spectrum (panels **A** and **B**). If no SMILES are present, only the spectrum and associated metadata (panels **A** and **C**) are shown

```
spectra = msdata.get_spectra()
print(f'"spectra" object is a {type(spectra)}
tensor of shape {spectra.shape} containing
{spectra.dtype} elements.')
```

The “spectra” object is a ⟨class “numpy.ndarray”⟩ tensor of shape (213548, 2, 128) containing float64 elements. The three dimensions correspond to: (i) the total number of spectra (213,548), (ii) two rows holding the m/z and intensity values, and (iii) up to 128 entries for each, representing the individual peaks in the spectrum. This fixed length is achieved during the conversion by either padding shorter spectra with zeros or trimming longer ones to keep only the most intense peaks (*see* **Notes 5**).

To retrieve all columns parsed from the input file, one can use the .columns() method (Fig. 4).

```
msdata.columns()
```

To retrieve a column, one can either use the.get_values() method.

```
msdata.get_values('FORMULA')
```

To retrieve specific elements, one can pass indices as a second argument to get_values or index the column obtained with square brackets.

```
msdata.get_values('FORMULA', [0, 1, 2])
```

For a more advanced workflow, the MSData class also supports converting data into external formats such as matchms

```
msdata.columns()
```

```
['COMPOUND_NAME', 'FORMULA',
'INCHI', 'INCHIKEY', 'MS_LEVEL',
'NUM_PEAKS',
'RETENTION_TIME', 'SCANS',
'adduct', 'charge', 'precursor_mz',
'smiles', 'spectrum']
```

Fig. 4 Each MSData object provides an application programming interface (API) [25] that allows users to interact with and explore the loaded mass spectrometry data. One particularly useful feature is the .columns() method, which prints a list of all available attributes for the loaded spectra. This is especially helpful when working with datasets that use non-standard column names, as it enables users to correctly map the relevant attributes, such as precursor *m/z* and spectrum data, to the crucial fields in the load_data object

[26] or merging multiple .hdf5 files into a single dataset (*see* **Notes 11 and 12**).

After loading and inspecting your dataset, you may find that it contains spectra of varying quality, some of which are unsuitable for training or inference with DreaMS. To ensure that our data matches the quality of the spectra used to train the model, we apply a series of MS/MS single-spectrum quality metrics, as described in Fig 2B, C of the DreaMS paper [9]. These metrics filter out low-quality spectra as well as very large-molecule entries.

For this purpose, we have prepared a dedicated script [27] that performs all cleaning steps, producing a dataset that adheres to the same filtering rules as the DreaMS training set. All criteria are enforced as single-spectrum quality control checks. Because the current DreaMS release was trained on and supports only singly charged spectra acquired in positive ionization mode (*see* **Notes 2**), we first restrict the dataset to spectra meeting these conditions. We then retain only tandem MS/MS spectra with precursor *m/z* values up to 1000 Da, fragment ions limited to $m/z \leq 1000$, and between 3 and 128 peaks per spectrum (after padding or truncation, as described above). In addition, the overall intensity range of each spectrum must span at least 20% of the normalized relative intensity scale; that is, the difference between the weakest and strongest signals must cover at least one-fifth of the available range and at least three peaks above 10% relative intensity. The following code demonstrates how to apply these quality control steps (Fig. 5):

```
from dreams.utils.dformats import DataFormatA
spectra = msdata['spectrum']
prec_mzs = msdata['precursor_mz']
dformat = DataFormatA()
```

```
All checks passed                      4758
Number of high intensity peaks >= 3     954
Intensity amplitude >= 20.0             942
Number of peaks between 3 and 128        61
m/z range <= 1000.0                       5
Name: count, dtype: int64
```

Fig. 5 Quality control and filtering of MS/MS spectra using DataFormatA rules. We started with 6732 spectra and applied all predefined quality filters, reducing the dataset to 4758 high-quality spectra. The listing shows how many spectra failed each condition. For example, 954 spectra were removed because they contained fewer than three high-intensity peaks. While we highlight the four most common filtering conditions, several additional criteria are also enforced but not shown here. For a complete list of rules and their implementation, see the source code [28]

```
quality_lvls = [dformat.val_spec(s, p, return_
problems=True) for s, p in zip(spectra, prec_mzs)]
pd.Series(quality_lvls).value_counts()
```

This summary shows the number of spectra passing all checks and the reasons for any failures (*see* **Notes 13**). Finally, select only the high-quality spectra and save them into a new HDF5 file:

```
from dreams.utils.io import append_to_stem
hq_pth = append_to_stem(in_pth, 'high_quality').
with_suffix('.hdf5')
msdata.form_subset(
idx=np.where(np.array(quality_lvls) == 'All checks
passed')[0]
out_pth=hq_pth
)
msdata_hq = MSData.load(hq_pth)
```

The result is a clean, high-quality.hdf5 file (*see* **Notes 9 and 15**), now ready for downstream embedding, clustering, or any other analysis in the next tutorial.

3.3 Deduplication via Locality-Sensitive Hashing

High-resolution LC–MS/MS datasets often contain many spectra that correspond to the same or nearly identical compounds. Such redundancy can arise from repeated measurements, isotopic variants, stereoisomers, or spectra collected under slightly different collision energies or instrument conditions. For downstream machine-learning applications and molecular networking, removing these duplicates is essential: it not only reduces computational overhead, but also prevents statistical bias introduced by over-represented molecules. In this tutorial, we work in 2_deduplication.ipynb, where this step is carried out using Locality Sensitive Hashing (LSH) [29, 30], a technique that makes it possible to group highly similar mass spectra together without the need for exhaustive pairwise comparison.

Locality Sensitive Hashing works by transforming each spectrum into a compact fingerprint in such a way that similar spectra are very likely to share the same fingerprint, or "hash," while dissimilar spectra almost always receive different hashes. This is achieved by first binning the m/z axis to produce a uniform vector representation of each spectrum and then projecting these vectors onto a set of random hyperplanes. For each hyperplane, a bit records on which side of the plane the spectrum lies; concatenating these bits yields the final LSH signature. Spectra that produce identical hashes are assigned to the same bucket, meaning they are considered near-duplicates. Unlike conventional hashing, which tries to avoid collisions, LSH is designed so that collisions occur preferentially among similar items, dramatically reducing the amount of comparison required. In the context of large-scale mass spectrometry, this means millions of spectra can be clustered into groups of near-identical fragmentation patterns in linear time, and can therefore be performed on a laptop.

In practice, computing the LSH signatures in DreaMS can be done in just a few lines of code.

```
from dreams.utils.spectra import
BatchedPeakListRandomProjection
lsh_projector = BatchedPeakListRandomProjection(bin_
step=0.5, n_hyperplanes=50)
lshs = lsh_projector.compute(msdata.get_spectra())
```

Here, bin_step defines the m/z bin width used to discretize spectra before projection, while n_hyperplanes controls the resolution of the hashing. Larger values generally yield higher precision but may fragment clusters, while smaller values group more spectra at the cost of introducing occasional false positives. The optimal choice depends on the desired trade-off between the strictness of deduplication and tolerance for minor spectral variation.

For example, if the goal of clustering is to only group near-duplicate spectra, n_hyperplanes = 128 is a good choice. Conversely, if the goal is to cluster structural analogs, n_hyperplanes = 20 may be a better option. The second parameter, bin_step, does not significantly influence the clustering.

When applied to a real dataset, each spectrum receives a hash value that identifies its cluster. Once the LSH signatures are computed, the distribution of hash frequencies provides an immediate picture of redundancy in the dataset. Large clusters may contain dozens or hundreds of spectra, while many others consist of only a single spectrum. In our experiment, we set bin_step to 0.5 and n_hyperplanes to 50, which led to the largest cluster containing 88 spectra. Inspecting a cluster in detail confirms whether the method is

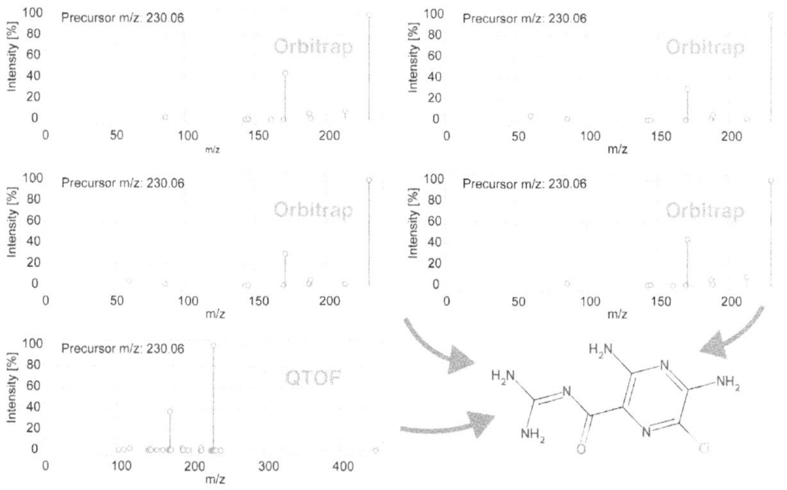

Fig. 6 Locality-sensitive hashing applied for the deduplication of MS/MS spectra. As an example, we visualize a cluster of five spectra corresponding to the same compound with a monoisotopic mass of 230.06 Da. The LSH algorithm groups these spectra together despite differences in acquisition conditions, such as measurements on QTOF and Orbitrap instruments. The method demonstrates strong robustness to experimental noise, particularly visible at *m/z* >400, where additional signals appear in some spectra but do not affect clustering accuracy

grouping truly identical patterns. For example, to examine a cluster of exactly five spectra:

In this example (Fig. 6), all five spectra correspond to the same compound measured under different MS conditions, such as QTOF and Orbitrap instruments. The LSH algorithm is robust to noise, which is evident from extra signals above 230 Da present in some spectra, and all spectra in the cluster display very high modified cosine similarities to one another.

An important property of LSH in DreaMS is that it is deliberately optimized for precision. Clusters are highly pure, grouping together spectra that are genuinely very similar. This tends to result in lower recall: while true duplicates are rarely missed, spectra with more divergent patterns, even if they originate from the same compound, may be split into separate clusters (Fig. 7). This behavior is advantageous when the goal is strict deduplication rather than broad grouping.

We are now investigating spectra, which are from the same compound, and in this example, we picked molecules that have 46 reference spectra. And then for each cluster, we take all spectra and compute pairwise similarity to all spectra in the other cluster, including itself.

In the heatmap above, each black-framed cell corresponds to a separate LSH cluster for the same compound from MassSpecGym, with colors indicating pairwise modified cosine similarity. Not all highly similar spectra are connected, reflecting the low recall of the

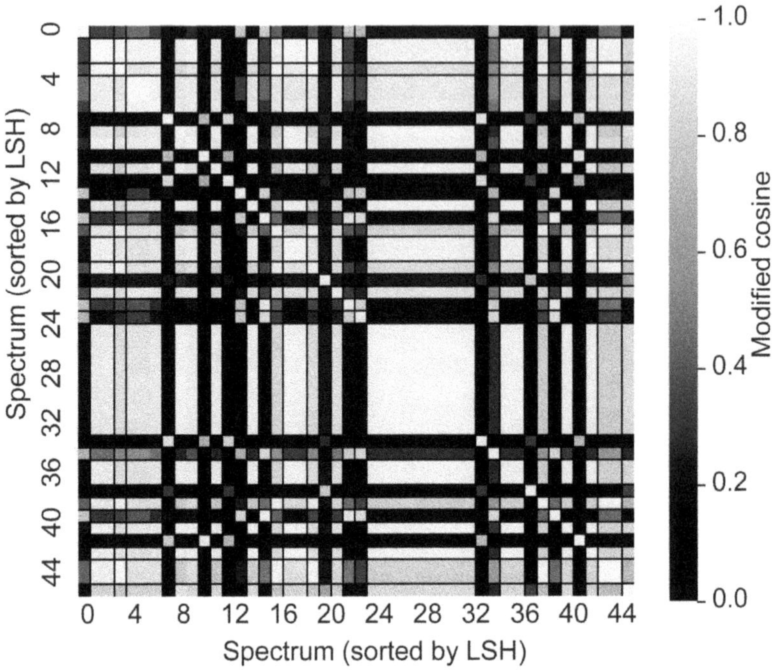

Fig. 7 Visualization of LSH clustering fragmentation for a single compound. In this example, we examine a molecule represented by 46 reference spectra. The heatmap shows pairwise modified cosine similarity between all clusters where this molecule appears, with the diagonal blocks representing similarities within each cluster. These intra-cluster similarities are consistently high, demonstrating the high precision of LSH. However, many off-diagonal similarities are also high, meaning that spectra from the same compound are often separated into multiple clusters, which highlights the algorithm's low recall by design

method, but all intra-cluster similarities are extremely high, underscoring its precision.

The deduplication process is completed by retaining only a single representative spectrum from each cluster. In practice, this involves selecting the first occurrence of each unique hash and discarding the rest. The resulting subset preserves the chemical diversity of the original dataset while removing redundant entries, substantially reducing the total number of spectra. In our experiment, this process reduced the dataset from 4758 to 2799 spectra. This compact, non-redundant set can then be saved as a new HDF5 file and serves as the streamlined input for downstream workflows such as spectral library searching or molecular annotation.

3.4 Embedding and Retrieval

In the earlier tutorials, we focused on preparing our experimental mass spectrometry data, ensuring that each spectrum was carefully processed, quality-checked, and free from redundancy. With this foundation in place, we can now turn to one of the most powerful applications of our workflow: using the DreaMS model for library matching. This step, implemented in 3_library_matching.ipynb, takes our measured spectra and searches for their most similar counterparts within a large, curated reference collection, linking unknown experimental data to known molecular structures.

For this purpose, we will work with MassSpecGym [10], the largest open-access spectral library containing hundreds of thousands of high-quality MS/MS spectra, each annotated with its corresponding molecular structure. By comparing our query spectra to this resource, we can assign chemical identities to compounds we have measured, or, when an exact match is not present, discover structural analogs and related scaffolds that offer valuable chemical clues.

To begin, we load both the MassSpecGym library and our prepared query dataset into MSData objects:

```
from dreams.utils.data import MSData
query_h5 = PROJECT_ROOT / "data" / "query_spectra.
hdf5"
mgf_path = PROJECT_ROOT / "data" / "massspecgym" /
"MassSpecGym.mgf"

msdata_lib = MSData.load(mgf_path, in_mem=True,
prec_mz_col="PRECURSOR_MZ")
msdata_q = MSData.load(query_h5, in_mem=True)
```

The DreaMS model enables this comparison through its embedding space: a learned, 1024-dimensional representation where spectra that arise from similar chemical structures are positioned close together. We first compute DreaMS embeddings for both the library and our query spectra. This step transforms each MS/MS spectrum into a numerical fingerprint optimized for structural similarity search (*see* **Notes 4 and 6**):

```
from dreams.api import dreams_embeddings
embs_lib = dreams_embeddings(msdata_lib)
embs_q = dreams_embeddings(msdata_q)
```

Once both sets of embeddings are in the same space, we can efficiently measure their similarity. Here we use cosine similarity, which quantifies how close two vectors are in orientation, irrespective of their absolute magnitude (ranges from −1 to 1). In the context of embeddings, a higher cosine similarity means the underlying molecules are more likely to be structurally similar:

```
from sklearn.metrics.pairwise import
cosine_similarity
sims = cosine_similarity(embs_q, embs_lib)
```

By ranking the MassSpecGym entries according to their similarity score for each query spectrum, we can retrieve the most likely candidates for identification. High-scoring matches (e.g., >0.9) often correspond to identical or near-identical molecules, while moderate scores (e.g., >0.7) may point toward chemical relatives that share key substructures, functional groups, or core scaffolds. Even in cases where no exact match exists, this approach allows us to generate well-founded hypotheses about the structural class or biosynthetic origin of unknown compounds.

When examining results from the dataset used in this tutorial, the top-ranked matches align closely with its expected chemical profile. All ten highest-scoring spectra correspond to alkaloid structures, as would be expected in a plant metabolomics dataset. For these, the DreaMS similarities exceeded 0.97, while the modified cosine scores on raw spectra were above 0.96, confirming a strong agreement between the embedding-based and peak-based matching approaches. Some of the top hits also correspond to piperine, a compound characteristic of *Piper* species.

However, perfect matches are not always attainable, as the coverage of even the largest public libraries remains incomplete. In such cases, DreaMS similarity provides a way to extract structural hypotheses without a definitive identification. To illustrate, we examine the query spectrum in our dataset with the lowest top-1 DreaMS similarity. Although the best match achieved a similarity of only 0.39 and a modified cosine score of 0.02, inspection of the top five hits revealed a consistent structural motif: every candidate contained at least one benzene ring, and four of them carried a chlorine atom (Fig. 8).

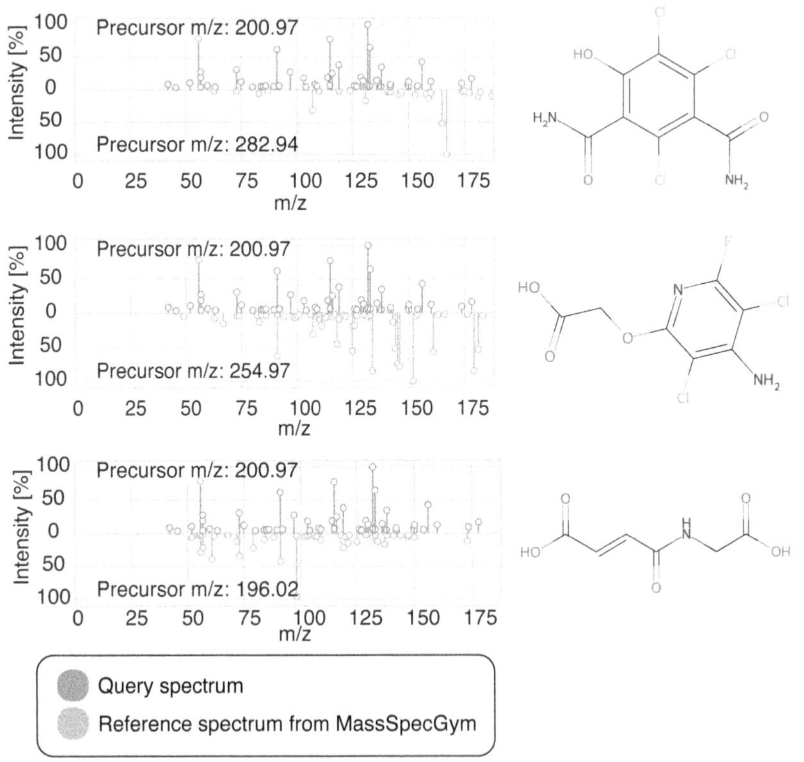

Fig. 8 Example of a query spectrum with low DreaMS similarity (≈0.39) and an even lower modified cosine score (≈0.02). Despite the weak spectral alignment, the top-matched reference spectra consistently share key structural features: four out of five contain a benzene ring, and also include chlorine atoms

This example highlights the interpretive value of the embedding space in low-confidence scenarios. Even when the query cannot be directly annotated with a known molecule, the clustering of chemically related structures among the top hits allows us to form a hypothesis about the underlying scaffold.

Because this dataset was partly annotated using reference standards, we can directly evaluate the accuracy of DreaMS library matching against known identities. Out of the top ten matches, seven are exact matches to the annotated molecules, and another shows a Tanimoto similarity greater than 0.9. Visual inspection of the corresponding molecular graphs confirms that this near-match differs only in benzene ring bond rotation or bond stereochemistry (Fig. 9). These results demonstrate that DreaMS, in combination with MassSpecGym, delivers both high precision and meaningful chemical interpretability, even when the spectral library does not contain a perfect reference.

Finally, obtaining DreaMS embeddings and running library matching against MassSpecGym can also be done using the DreaMS web interface hosted on Hugging Face Spaces [20] (Fig. 10). The interface requires only a single input: an .mgf or .mzML file of the dataset of interest. In our case, we can upload the same example *Piper fimbriulatum* dataset, click the "Run DreaMS" button, and then download a.csv file containing the predicted embeddings and library hits. The results are also visualized directly in the browser.

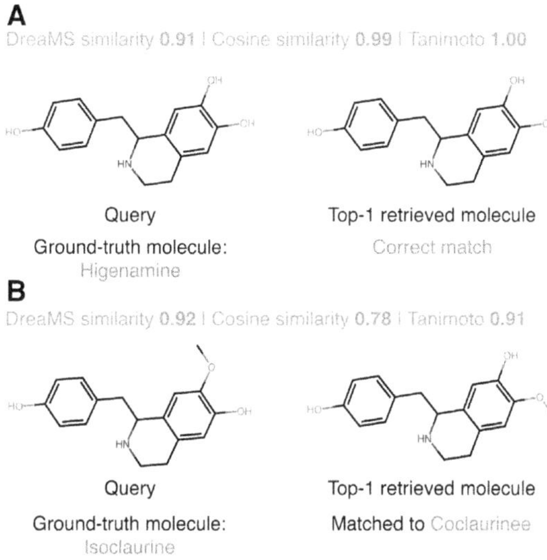

Fig. 9 To validate DreaMS performance on annotated molecules, we examine examples from the partially annotated Piper dataset where the ground truth structures are known. (A) Higenamine is present in MassSpecGym and is correctly matched by DreaMS as the top-1 hit. (B) Isococlaurine is absent from the reference library, yet DreaMS retrieves coclaurine as the top-ranked match. The two molecules are positional isomers, differing only in the O-methyl position on the benzylisoquinoline scaffold while sharing the same secondary-amine core. DreaMS achieves a high similarity score (0.92) despite the absence of an exact reference.

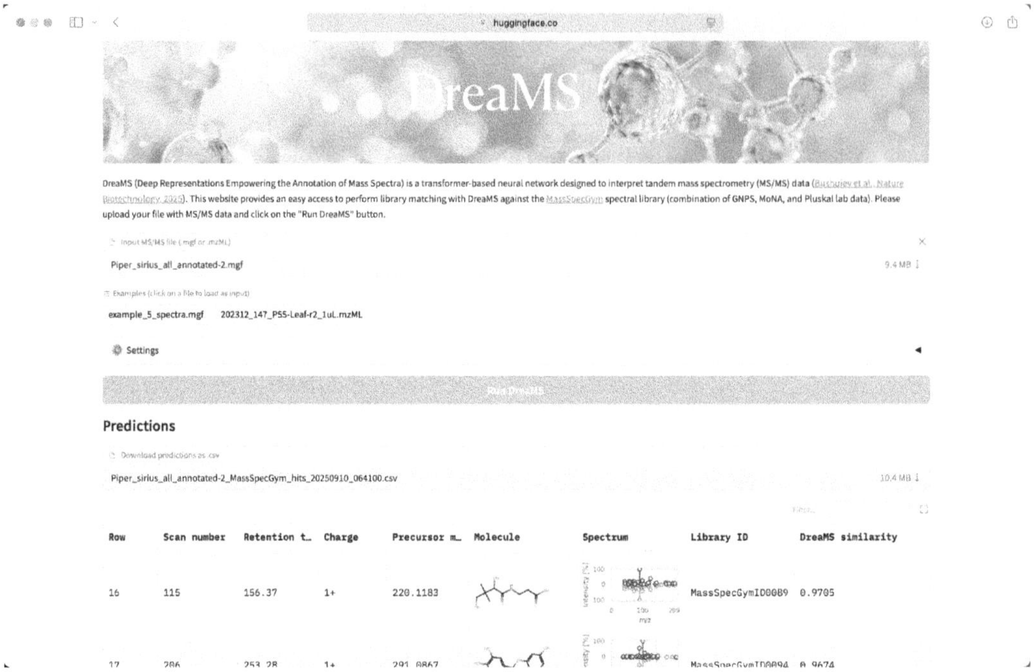

Fig. 10 The DreaMS web interface on HuggingFace Spaces enables users to upload .mgf or .mzML files and perform automated MS/MS annotation via DreaMS-guided library matching against the MassSpecGym reference dataset. The interface computes spectral embeddings, retrieves the top candidate matches, and provides downloadable results directly in the browser

3.5 Molecular Networking and Annotation Overlay

In this tutorial (4_molecular_networking.ipynb), we turn to molecular networking [31]. Molecular networking is a method for organizing and visualizing mass spectrometry data so that molecules with similar chemical structures are grouped together. Each node in the network represents a compound's spectrum, and links between nodes indicate spectral similarities that suggest related structures. By revealing clusters of related molecules, molecular networking helps researchers detect molecular families, discover new analogs of known compounds, and prioritize targets for further study.

DreaMS [9] embeddings lend themselves naturally to this task. Once we compute embeddings, their similarity provides a learned measure of spectral relatedness that we can use to connect nodes in a k-nearest-neighbor graph [32]. DreaMS similarity outperforms common alternatives (modified cosine, spectral entropy, MS2DeepScore [33]) across several criteria that matter for networking: it correlates more strongly with chemical similarity of the underlying compounds (i.e., Tanimoto similarity on Morgan fingerprints), improves library matching and analog search accuracy, and remains robust when spectra are of lower quality. Unlike peak-matching methods that rely on direct alignment of fragment m/z and intensities, DreaMS learns representations of spectra with a neural network, and therefore often

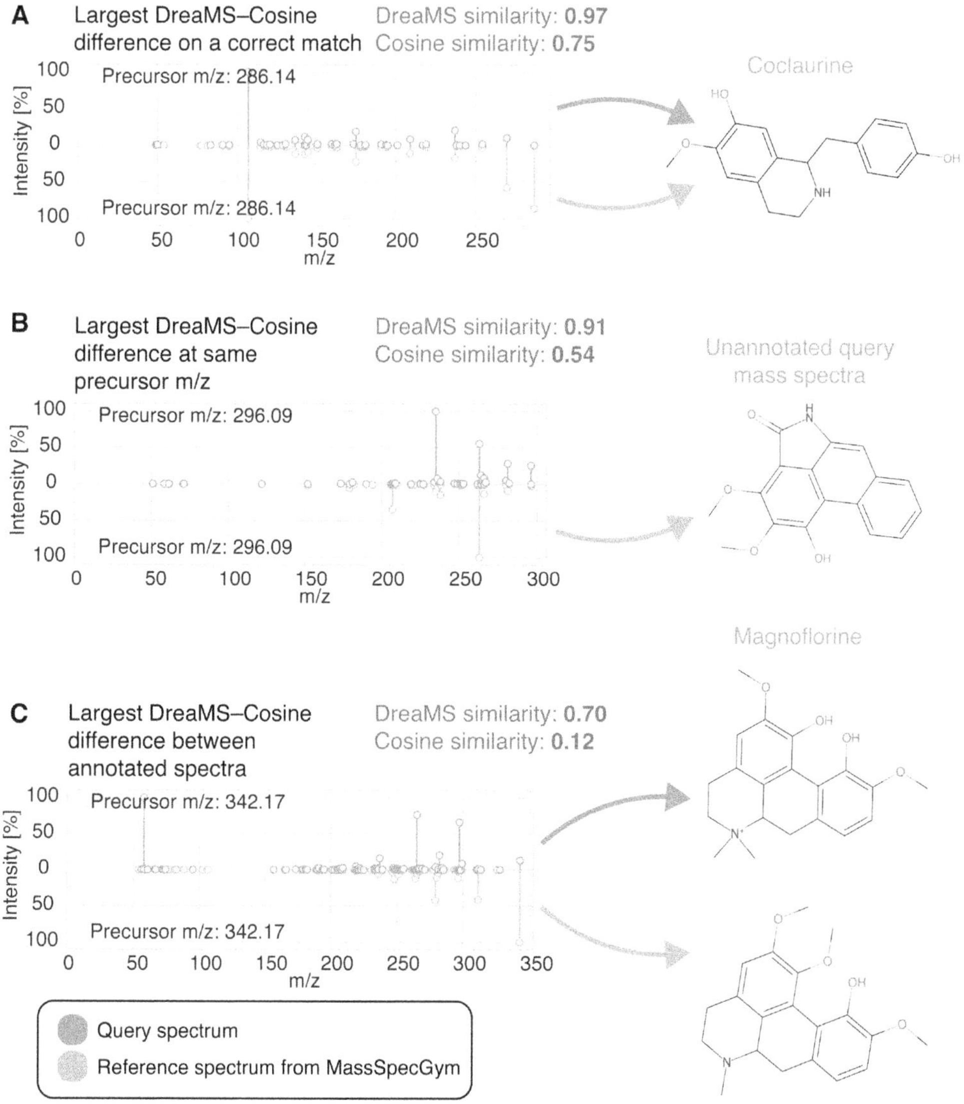

Fig. 11 Examples where DreaMS similarity scoring identifies potential relationships between spectra that would likely remain hidden using modified cosine similarity alone. (**A**) We show the largest difference between DreaMS and cosine similarities for a correctly annotated molecule from the Piper dataset, coclaurine. (**B**) This example highlights one of the largest differences between DreaMS and cosine similarities for spectra derived from the same precursor *m*/*z*. The query spectrum is unannotated, yet we can directly observe similarities between the spectra. The matched library compound is an aporphine alkaloid, which is abundant in the original study. (**C**) We show the largest difference between DreaMS and cosine similarities for two annotated spectra. The query spectrum corresponds to magnoflorine, while the matched library spectrum represents a structural analog. Here, DreaMS preserves a moderate similarity of 0.70, whereas the modified cosine score drops to 0.12, which would otherwise miss the relationship entirely. Together, these examples demonstrate how DreaMS embeddings uncover potential links between spectra and capture chemically meaningful relationships that conventional peak-matching metrics may fail to detect

recovers relationships that conventional metrics miss. Before building a network, we illustrate this effect by inspecting spectrum pairs with high DreaMS similarity but low modified cosine score, pairs that would typically remain unconnected in a standard network (Fig. 11).

```
from sklearn.metrics.pairwise import
cosine_similarity
import numpy as np
import dreams.utils.spectra as su
# Find all pairs with DreaMS similarity > 0.85
sims = cosine_similarity(embs_q)
x, y = np.where(sims > 0.85)
# Define modified cosine similarity function
cos_sim = su.PeakListModifiedCosine(mz_tolerance=
0.05)


# Display hits with low modified cosine similarity
but high DreaMS similarity

max_pairs, i_pairs = 7, 1
for i, (i1, i2) in enumerate(pd.Series(zip(x, y)).
sample(frac=1, random_state=1)):
spec1, spec2 = msdata_q[SPECTRUM][i1],
msdata_q[SPECTRUM][i2]
prec_mz1, prec_mz2 = msdata_q[PRECURSOR_MZ][i1],
msdata_q[PRECURSOR_MZ][i2]
cos_sim_i1_i2 = cos_sim(spec1=spec1, prec_mz1=
prec_mz1, spec2=spec2, prec_mz2=prec_mz2)
su.plot_spectrum(spec=spec1, prec_mz=prec_mz1,
mirror_spec=spec2, mirror_prec_mz=prec_mz2)
```

Having shown that DreaMS finds extra, chemically meaningful links, we now build the network. We start from the preprocessed, deduplicated dataset from the previous tutorials. Within the sample, each query spectrum is connected to its nearest neighbors in the embedding space. We keep only strong links by applying a DreaMS similarity threshold of 0.70 and use a small neighborhood size of 3. As shown in (Fig. 12), coloring nodes by precursor m/z reveals tight within-cluster homogeneity. We export the graph as a GraphML file and open it in Cytoscape [34]. Cytoscape is a free desktop program for viewing and styling networks; you load the GraphML file, color and size nodes by any attribute, and interactively explore clusters without writing code.

To add chemical context without changing the network's shape, we keep the query-only topology and overlay annotation. For each query, we take the top-1 MassSpecGym match and tag the node only if the best DreaMS similarity meets an assignment threshold of 0.80. The exported GraphML includes per-node fields such as has_lib_hit and smiles, which Cytoscape reads directly; with the ChemViz2 [35] extension, we draw 2D structures on nodes and inspect hits interactively (*see* **Notes 17**). For completeness, each edge carries both DreaMS_similarity and modified_cosine_similarity, allowing side-by-side comparison of the learned and peak-based measures.

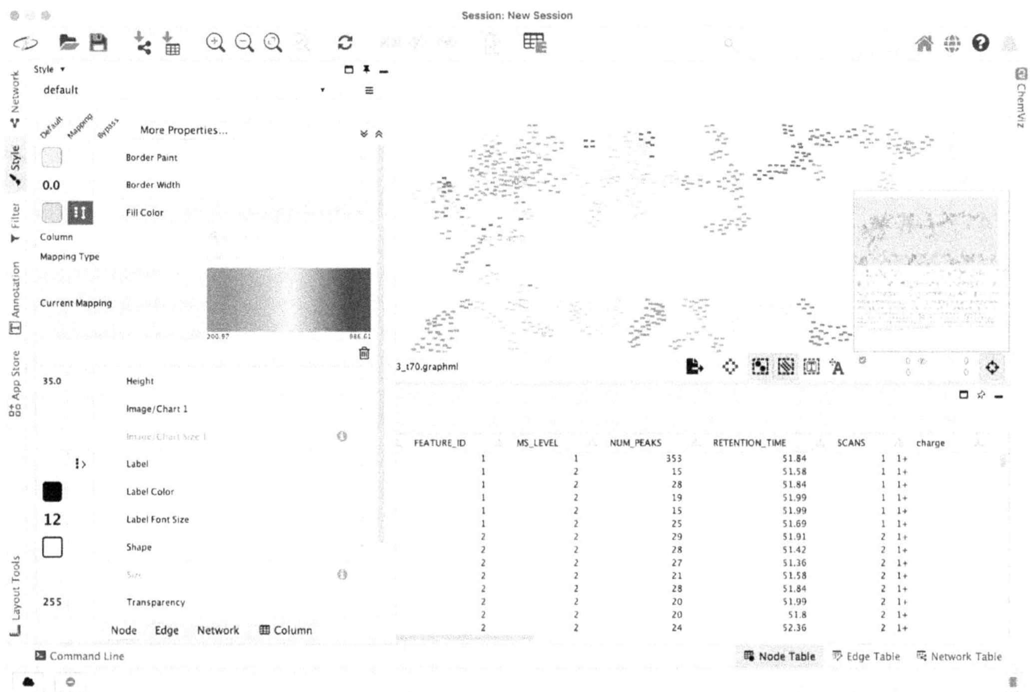

Fig. 12 Visualization of the imported .graphML molecular network in Cytoscape. The nodes are colored by precursor *m/z* values, allowing quick identification of clusters with similar mass ranges. Below the network view, the attribute table displays metadata linked to each node, such as retention time, charge, acquisition method, and other available annotations. The displayed attributes depend on the level of metadata provided in the dataset

In the examined cluster, shown in (Fig. 13), we find a benzyltetrahydroisoquinoline (benzylisoquinoline) family anchored by the correctly annotated higenamine, with DreaMS placing two Piper alkaloids absent from the reference library, *N*-methylhigenamine and norarmepavine, immediately nearby in embedding space. We also observe unannotated query spectra in the same neighborhood for coclaurine and for an *O*-methyl derivative of *N*-methylhigenamine. This topology is chemically coherent: all members share the same scaffold and differ only by small, local edits, single-step N-methylation at the ring nitrogen, and *O*-methylation/demethylation on phenolic positions. These $\pm CH_3$ substitutions leave the benzyltetrahydroisoquinoline core unchanged, so key cleavage sites are conserved, and the spectra remain highly related, both cosine and DreaMS similarities. DreaMS therefore recovers a chemically faithful neighborhood, organizing spectra by scaffold-level similarity and simple methylation ladders, correctly matching the one molecule present in both the reference library and the query set (higenamine) while positioning its close analogs, despite being missing from the library, in the same cluster. This behavior is consistent with the Piper fimbriulatum dataset [23], where benzylisoquinoline alkaloids,

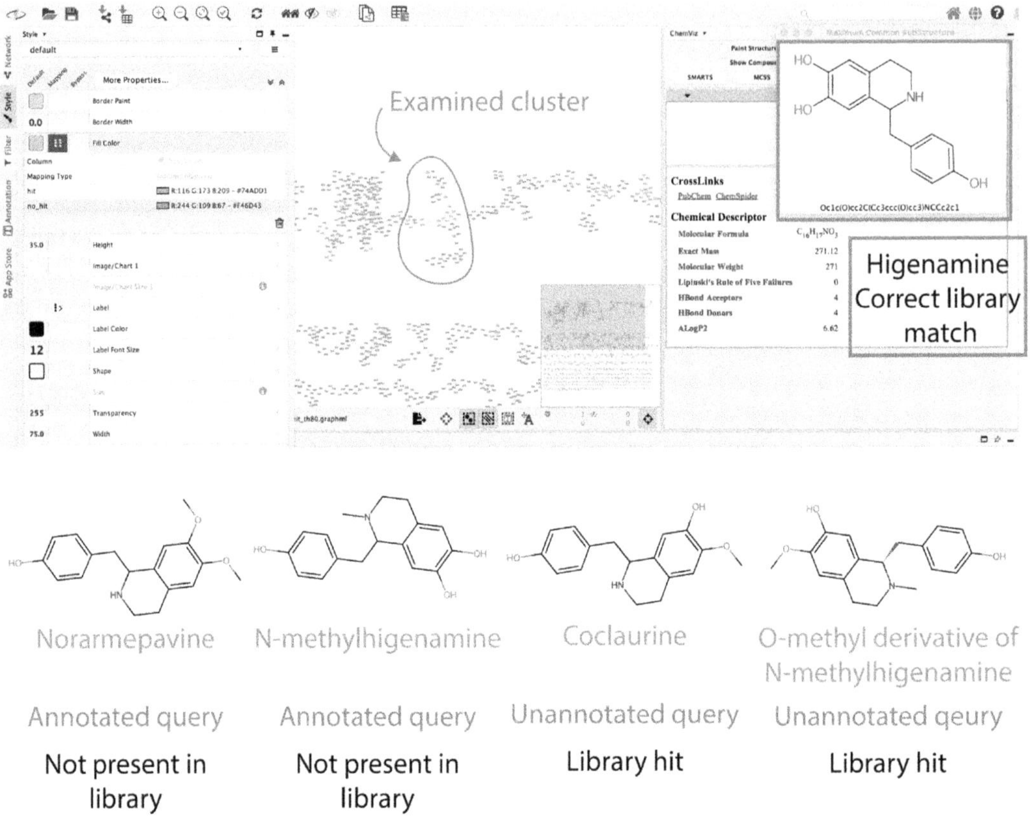

Fig. 13 Visualization of a partially annotated molecular network from the Piper dataset in Cytoscape. Nodes are colored by annotation status: blue nodes represent spectra with a confident match in the MassSpecGym reference library, identified using DreaMS similarity scores above 0.8, while the remaining orange nodes correspond to unannotated spectra. This coloring is configured in Cytoscape by selecting the has_lib_hit attribute under Style → Fill Color. For annotated nodes, SMILES structures are attached and displayed directly on the network using the ChemViz2 extension, enabling rapid inspection of matched molecules. In the examined cluster, higenamine is correctly annotated as the top hit, and we also observe several structural analogs located nearby in the network. Among these are norarmepavine and *N*-methylhigenamine, which are annotated spectra from the Piper dataset [23]. For the remaining unannotated query spectra, DreaMS ranks coclaurine and an *O*-methyl derivative of *N*-methylhigenamine as the top-1 matches, respectively. All five of these molecules share the same benzylisoquinoline scaffold, differing only by small methylation or demethylation changes. This illustrates how DreaMS not only retrieves exact matches when they are present in the reference library but also places structurally related compounds together, enabling the discovery of chemically coherent molecular families

including the armepavine series, form a distinct molecular family anchored by higenamine/coclaurine and connected by predictable N-/O-methylation steps.

Additional examples from the same Piper molecular network further support this pattern. Piperlongumine forms a homogeneous, chemically coherent cluster. The single annotated query spectrum is correctly matched as the top-1 hit to piperlongumine, and the remaining unannotated spectra in the same neighborhood are likewise assigned piperlongumine as their top-1 candidate. A similar pattern is observed for lirididine, present as a MassSpecGym library entry, which also groups into a tight, chemically consistent cluster around its library node. Taken together with the benzylisoquinoline example, these cases demonstrate that DreaMS embeddings organize spectra in a way that respects scaffold-level chemistry, bringing close analogs together and enabling matches even when exact references are absent.

4 Notes

1. DreaMS, and specifically the MSData class, uses predefined column names and requires two main parameters for prediction: spec_col = SPECTRUM and prec_mz_col = PRECURSOR_MZ. If your dataset uses different column names, you must map them correctly when loading the data to ensure proper functioning. To check available column names, you can first load the dataset and run msdata.columns(), then map the correct ones.
2. DreaMS currently supports only positive-ionization, singly charged MS/MS spectra, so filtering for these criteria is an essential step during data preparation.
3. When calling MSData.from_mzml, an .hdf5 file is automatically generated alongside the original.mzML file, using the same base name. For example, loading Path("data/rawfiles/202312_20_P09-Leaf-r1_1uL.mzML") will produce both 202312_20_P09-Leaf-r1_1uL.mzML and 202312_20_P09-Leaf-r1_1uL.hdf5. If the .hdf5 file already exists, MSData automatically detects it and loads the cached version, skipping reconversion.
4. When you run dreams_embeddings for the first time, the model weights are downloaded automatically. This may take some time, as two large files are required: embedding_model.ckpt (~1.2 GB) and ssl_model.ckpt (~1.3 GB). The downloaded models are saved under anaconda3/envs/dreams_mimb/lib/python3.11/site-packages/dreams/models/pretrained and reused in subsequent runs.
5. If a spectrum contains more peaks than the configured limit, only the n most intense peaks are stored in the HDF5 tensor, and the remaining ones are discarded. This limit is controlled via the n_highest_peaks parameter when creating the MSData object. To avoid data loss, set n_highest_peaks to the maximum number of peaks in your dataset. Since zero-padding is efficiently compressed in HDF5, using a high value will not significantly increase file size.
6. Computing DreaMS embeddings for the entire MassSpecGym dataset can take up to an hour, depending on your hardware. We recommend saving the computed embeddings, as we do

in notebook tutorials, so that they can be reused without recalculation.

7. When querying metadata using the matchms library [36], always use lowercase attribute names, even if the dataset uses uppercase columns.
8. If you intend to fine-tune DreaMS, the model weights can be downloaded from https://zenodo.org/records/10997887. The embedding_model.ckpt file is the default model used internally by dreams_embeddings and produces similarity-optimized embeddings. The ssl_model.ckpt file is the original self-supervised model and is intended for customization and fine-tuning, for example, to adapt DreaMS embeddings for specialized tasks such as detecting alkaloids.
9. If you have opened an .hdf5 file in one session and attempt to reuse it in another, the system may block access because the file remains linked to memory. To resolve this, call h5_path.unlink() to release the lock. Alternatively, throughout this tutorial, we use a conditional check before reloading.
10. While .mgf and .mzML are commonly used as input formats, we recommend converting and working with .hdf5 files whenever possible. The HDF5 format preserves all data and metadata, is more efficient for large datasets, and is better suited for machine-learning workflows.
11. If you need to merge multiple datasets into a single HDF5 file using MSData.merge, make sure that all files contain the same columns and define them explicitly with the cols parameter. Otherwise, MSData expects the required set of columns: [SPECTRUM, PRECURSOR_MZ, CHARGE, ADDUCT].
12. MSData automatically stores information about the source file for each spectrum, making it possible to trace spectra back to their original datasets when needed.
13. During quality control, multiple conditions are applied when filtering spectra. As a result, some spectra of interest may be excluded. Refer to the code examples [28] in the tutorial to review these thresholds and adjust them if necessary.
14. Training DreaMS on the MassSpecGym dataset for one epoch typically takes about an hour on a standard workstation.
15. When saving filtered datasets as .hdf5, note that column names may automatically change to lowercase. For example, PRECURSOR_MZ will become precursor_mz. Be aware of this when querying data after saving.
16. File path delimiters differ between operating systems: Linux and macOS use /, while Windows uses \ (or sometimes \\ in Python strings). Ensure that your code handles paths consistently. We recommend using Python's built-in Path from the pathlib module [37] for cross-platform compatibility.

17. When visualizing molecular networks in Cytoscape using ChemViz2, we recommend setting the node attribute name to SMILES (or lowercase smiles) to ensure correct structure mapping and visualization.

Acknowledgments

This chapter was supported by the Czech Science Foundation (GA CR) grant 21-11563M and by the European Union's HORIZON program (ERC TerpenCode, 101170268; ERC FRONTIER, 101097822; and CLARA, 101136607).

Conflict of Interest The authors declare no conflicts of interest that are relevant to the content of this chapter.

References

1. Zhang M, Otsuki K, Li W (2023) Molecular networking as a natural products discovery strategy. Acta Mater Med 2. https://doi.org/10.15212/amm-2023-0007
2. Tian Z, Zhao H, Peter KT et al. (2021) A ubiquitous tire rubber-derived chemical induces acute mortality in coho salmon. Science 371:185–189
3. Qiu S, Cai Y, Yao H et al. (2023) Small molecule metabolites: discovery of biomarkers and therapeutic targets. Signal Transduct Target Ther 8:132
4. de Jonge NF, Mildau K, Meijer D et al. (2022) Good practices and recommendations for using and benchmarking computational metabolomics metabolite annotation tools. Metabolomics 18: 103
5. da Silva RR, Dorrestein PC, Quinn RA (2015) Illuminating the dark matter in metabolomics. Proc Natl Acad Sci USA 112:12549–12550
6. Wang M, Carver JJ, Phelan VV et al. (2016) Sharing and community curation of mass spectrometry data with Global Natural Products Social Molecular Networking. Nat Biotechnol 34:828–837
7. MassBank of North America. https://mona.fiehnlab.ucdavis.edu/. Accessed 12 Aug 2025
8. Mass Spectrometry Data Center, NIST. https://chemdata.nist.gov/. Accessed 12 Aug 2025
9. Bushuiev R, Bushuiev A, Samusevich R, et al. (2026) Self-supervised learning of molecular representations from millions of tandem mass spectra using DreaMS. Nat Biotechnol 44:630–640. https://doi.org/10.1038/s41587-025-02663-3
10. NeurIPS Poster MassSpecGym: a benchmark for the discovery and identification of molecules. https://neurips.cc/virtual/2024/poster/97823. Accessed 10 Dec 2024
11. OpenAI, AchiamJ, Adler S et al (2023) GPT-4 Technical Report. Preprint at arXiv https://doi.org/10.48550/arXiv.2303.08774
12. Horai H, Arita M, Kanaya S et al. (2010) MassBank: a public repository for sharing mass spectral data for life sciences. J Mass Spectrom 45:703–714
13. Brungs C, Schmid R, Heuckeroth S, Mazumdar A, Drexler M, Šácha P, Dorrestein PC, Petras D, Nothias L-F, Veverka V, Nencka R, Kameník Z, Pluskal T (2025) MSnLib: efficient generation of open multi-stage fragmentation mass spectral libraries. Nat Methods. https://doi.org/10.1038/s41592-025-02813-0
14. Kretschmer F, Seipp J, Ludwig M, Klau GW, Böcker S (2025) Coverage bias in small molecule machine learning. Nat Commun 16:554. https://doi.org/10.1038/s41467-024-55462-w
15. Conda Documentation—conda-docs documentation. https://docs.conda.io/en/latest/. Accessed 2 Aug 2025
16. pluskal-lab/DreaMS_MIMB. GitHub. https://github.com/pluskal-lab/DreaMS_MIMB. Accessed 2 Aug 2025
17. roman-bushuiev/MassSpecGym at main. https://huggingface.co/datasets/roman-bushuiev/MassSpecGym/tree/main/data/auxiliary. Accessed 2 Aug 2025
18. Damiani T, Smith J, Hebra T et al (2024) Computational metabolomics reveals overlooked chemodiversity of alkaloid scaffolds

inPiper fimbriulatum. bioRxiv. https://doi.org/10.1101/2024.12.10.627739

19. roman-bushuiev/DreaMS at main. https://huggingface.co/roman-bushuiev/DreaMS/tree/main. Accessed 2 Aug 2025
20. DreaMS - a Hugging Face Space by anton-bushuiev. https://huggingface.co/spaces/anton-bushuiev/DreaMS. Accessed 24 Aug 2025
21. Git Guides—install git. Github. https://github.com/git-guides/install-git
22. Miniconda. In: Anaconda. https://www.anaconda.com/docs/getting-started/miniconda/main. Accessed 8 Aug 2025
23. Damiani T, Smith J, Hebra T et al. (2025) Computational metabolomics reveals overlooked chemodiversity of alkaloid scaffolds in Piper fimbriulatum. Plant J 121:e70086
24. Nothias L-F, Petras D, Schmid R et al. (2020) Feature-based molecular networking in the GNPS analysis environment. Nat Methods 17:905–908
25. API Reference —DreaMS documentation. https://dreams-docs.readthedocs.io/en/latest/modules.html. Accessed 5 Sep 2025
26. Huber F, Verhoeven S, Meijer C et al. (2020) Matchms—processing and similarity evaluation of mass spectrometry data. J Open Source Softw 5:2411
27. Assessing spectral quality—DreaMS documentation. https://dreams-docs.readthedocs.io/en/latest/tutorials/spectral_quality.html. Accessed 2 Sep 2025
28. dreams/utils/dformats.py at 07077222b872095176d16fc1984a4b-9879dcee9d · pluskal-lab/DreaMS. Github. https://github.com/pluskallab/ DreaMS/blob/07077222b872095176d16fc1984a4b-9879dcee9d/dreams/utils/dformats.py
29. Ullman Jeffrey D (2012) Stanford University http://infolab.stanford.edu/~ullman/mmds/book.pdf. Accessed 13 Aug 2025
30. Jafari O, Maurya P, Nagarkar P et al (2021) A survey on Locality Sensitive Hashing algorithms and their applications. Preprint at arXiv https://doi.org/10.48550/arXiv.2102.08942
31. Schmid R, Petras D, Nothias L-F et al. (2021) Ion identity molecular networking for mass spectrometry-based metabolomics in the GNPS environment. Nat Commun 12: 3832
32. Guo G, Wang H, Bell D et al (2003) KNN model-based approach in classification. In: On The Move to Meaningful Internet Systems 2003: CoopIS, DOA, and ODBASE. Springer Berlin Heidelberg, Berlin, Heidelberg, pp 986–996
33. Huber F, van der Burg S, van der Hooft JJJ, Ridder L (2021) MS2DeepScore: a novel deep learning similarity measure to compare tandem mass spectra. J Cheminform 13:84
34. Shannon P, Markiel A, Ozier O et al. (2003) Cytoscape: a software environment for integrated models of biomolecular interaction networks. Genome Res 13:2498–2504
35. chemViz2: Cheminformatics App for Cytoscape. https://www.rbvi.ucsf.edu/cytoscape/chemViz2/. Accessed 16 Aug 2025
36. Welcome to matchms's documentation!—matchms documentation. https://matchms.readthedocs.io/en/latest/. Accessed 3 Sep 2025
37. pathlib—Object-oriented filesystem paths. In: Python documentation. https://docs.python.org/3/library/pathlib.html. Accessed 3 Sep 2025

Check for updates

Chapter 15

ProteoRift Protocols: A Deep Learning Filtering Workflow for Peptide Identification in Mass Spectrometry Data

Bilal Shabbir, Usman Tariq and Fahad Saeed

Abstract

Increasing complexity and volume of mass spectrometry (MS) data have presented new challenges and opportunities for peptide identification in shotgun proteomics. Existing database search algorithms reduce the number of potential candidate peptides by using precursor mass filtering. While useful, relying on a single and simple filter may lead to the exclusion of nonabundant spectra and uncharacterized peptides. In this chapter, we present a machine-learning model, called *ProteoRift*, capable of predicting multiple properties of the peptide (e.g., length, missed cleavages, and modification status) directly from the spectra. Predicting multiple properties helps to apply more targeted filters in the database search and provides up to 12× speedup as compared to mass-only filtering in peptide identification. This chapter describes the general architecture, which is followed by a step-by-step guide for installing and performing database searches using the ProteoRift model. Lastly, we discuss the data preparation and preprocessing steps required for retraining ProteoRift. ProteoRift is an open-source and open-weights model and is available at https://github.com/pcdslab/ProteoRift.

Keywords Proteomics, MS/MS analysis, Peptide identification, Database search, Machine learning, Deep learning, MS data processing

1 Introduction

Mass spectrometry (MS) data are an integral part of modern proteomics that enables large-scale identification and quantification of peptides and proteins across diverse biological contexts. To support the curation, access, and reuse of this data, public MS repositories such as PRIDE [1], ProteomeXchange [2], and MassIVE [3] have been growing over time. The growth of these repositories has given access to high-quality spectral datasets, which supports the adoption of open data-sharing requirements and growing scientific attention to reproducibility and reuse of data [4]. The vast majority of spectra are analyzed through database search algorithms that compare experimental spectra with theoretical peptide candidates. Briefly, the database search algorithms first generate theoretical spectra for given

Frédérique Lisacek (ed.), *Bioinformatics for Mass Spectrometry*, Methods in Molecular Biology, vol. 3071,
https://doi.org/10.1007/978-1-0716-5519-1_15,

peptides in the database and apply mass-based filtering to reduce the candidate peptides. Subsequently, the filtered candidate peptides are scored using heuristic functions to match the experimental spectra with theoretical spectra. These algorithms reduce the search space by filtering candidates with a single property, most often precursor mass. However, this strategy may exclude peptides that are rare, weakly detected, or not already known, because they do not fit neatly within the strict mass filters. In addition, traditional peptide identification tools using heuristic scoring functions with mass filtering may fail to capture complex, nonlinear associations between spectra and peptide sequence [5,6]. This limitation highlights the need for more flexible approaches that go beyond simple mass filtering. In contrast, machine learning (ML) methods, such as SpeCollate [6] and DeepMS [7], can learn complex and nonlinear patterns, which may result in improved peptide identification as more spectral data becomes available.

In proteomics, using peptide mass as a filter to reduce the search space is one of the most fundamental techniques that enables scalable database-search algorithms [8]. Filtering by mass is universally used by database search algorithms since it significantly reduces the search space and improves search accuracy [9–11]. However, current algorithms that rely only on precursor mass and fragment ion masses struggle with the complexity of diverse biological samples, such as in meta-proteomics studies [12–14]. In metaproteomics, the continuous nature of the mass filter can introduce unreasonable search time because of very large databases [15,16]. Moreover, relying only on mass filters leads to many peptides remaining unidentified, even when present in the database [17]. This gap has led to the development of "open-search" and "hybrid-search" strategies [18–21]. Open-search methods increase the mass tolerance window significantly, allowing more candidate peptides to be considered, especially those that might otherwise be missed. MSFragger [21], Crux [22], and SpeCollate [6] have improved peptide identification in proteomics but still rely on filtering based only on the mass of the spectra and corresponding peptides.

Recent advances in machine learning (ML) have enabled the development of deeper and more accurate pipelines for MS data analysis [23–26]. While these ML pipelines can outperform traditional algorithmic or de novo search methods, most ML models have focused narrowly on scoring functions [27]. Incorporating these models into existing workflows remains challenging, as they often identify the same peptide as other algorithm-based approaches. Moreover, many systems biology labs hesitate to adopt these ML solutions due to factors such as the black-box nature of ML models and the perceived uncertainty in their peptide predictions [28–30]. Building complete ML-driven pipelines for MS data search can enable us to fully leverage ML advancements for improving peptide discovery and throughput.

In this chapter, we describe ProteoRift [8], a novel machine learning framework designed to predict peptide properties (e.g., peptide length, missed cleavages, and modification status) directly from tandem mass spectra. In ProteoRift, spectral inputs are used to predict peptide properties before any peptide identification. This approach helps to filter peptides using peptide length, missed cleavages, and modification status in addition to the mass-based filtering used in traditional approaches. ProteoRift achieves 8× to 12× speedup for peptide identifications while maintaining accuracy comparable to conventional approaches in proteomics and other omics (e.g., Metaproteomics and Proteogenomics). The difference between traditional approaches and ProteoRift (*see* **Note 1**) is illustrated in Fig. 1.

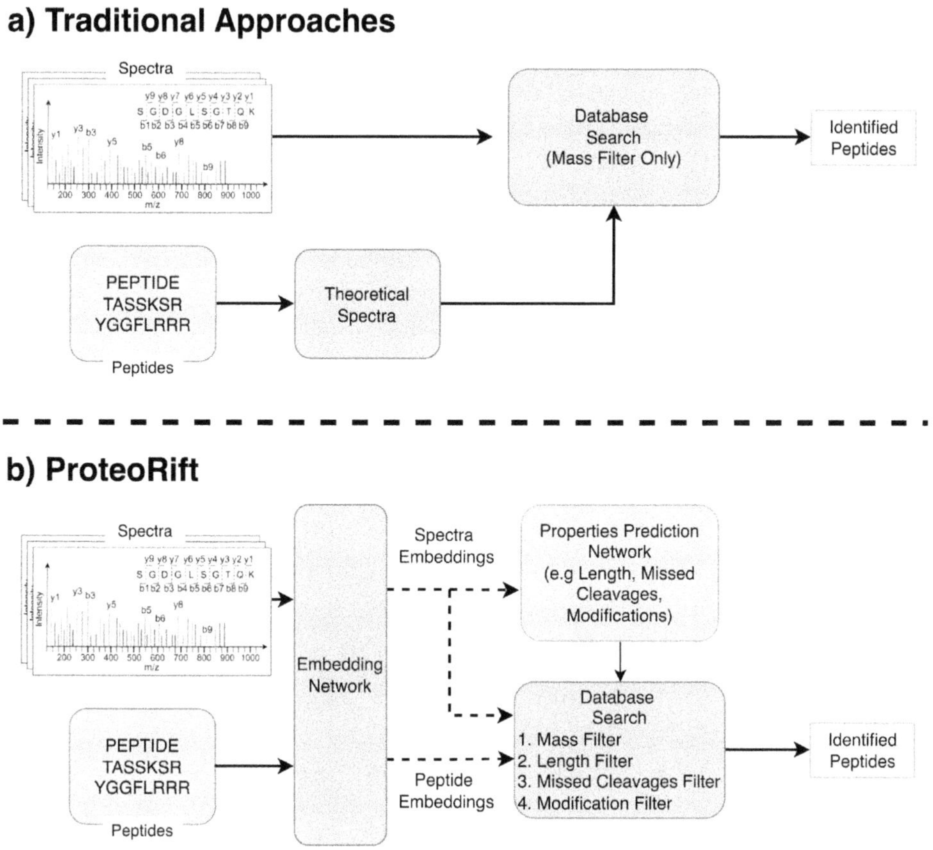

Fig. 1 Workflow of Peptide Identification using traditional approaches and ProteoRift (**a**) Traditional approaches: Theoretical spectra are first generated for peptides and for each experimental spectrum, the search space is reduced by comparing the mass of the theoretical and experimental spectra. (**b**) **ProteoRift:** Experimental spectra are passed through the Spectrum Network (SpeCollate) to predict peptide properties. Peptides are processed through the Peptide Network, and database searches are performed using spectra and peptide embeddings, with mass filtering and additional filters such as peptide length, missed cleavages, and modification status

2 Materials

2.1 Understanding the Data

In proteomics, two fundamental components are required for peptide identification workflows: MS/MS data (hereafter: spectra or spectrum) and a corresponding peptide sequence database. Each spectrum is a series of peaks, with each peak representing a fragment ion's mass-to-charge ratio (m/z) and its intensity. Mass spectra contain fragmentation patterns that encode the amino acid sequence of the original peptide, which can be interpreted computationally. The peptide database contains theoretical peptides against which the spectra are matched. This pairing between experimental data and reference databases is the basis for peptide identification.

2.2 Obtaining the Annotated Data

To retrain ProteoRift, annotated MS/MS data are required. The annotated data would contain the spectra that are matched with their corresponding peptide sequences. These can be acquired using prebuilt spectral libraries or annotated datasets from public repositories such as MassIVE [3] and NIST [9], which offer high-quality annotated spectra across several species. NIST mass spectral library offers curated and annotated spectra in .msp format, covering multiple species and a wide range of experimental conditions. These datasets are readily accessible and can be browsed and downloaded through their online interface (https://chemdata.nist.gov/). MassIVE offers datasets and spectral libraries typically in .mzML or .mzXML format. These datasets include raw or semi-processed spectra, often obtained from peptide identifications. Other repositories such as PRIDE [1] and ProteomeXchange [2] also provide large-scale annotated datasets derived from diverse biological and clinical studies.

2.3 Required Packages

To install and run ProteoRift, ensure the following software dependencies are met:

- Python [31] version >3.8.
- PyTorch [32] version >1.6.
- NumPy [33], Pandas [34], and Scikit-Learn [35] for core data processing.
- CUDA Toolkit [36] version >11.6 for GPU-based acceleration.
- Crux [22] Toolkit, particularly the Percolator module, used for FDR estimation.
- Docker, required to run msconvert [10] for converting raw MS files into MGF format.

2.4 Recommended Hardware

ProteoRift is designed to run on high-performance systems with the following hardware specifications:

- Operating System: Ubuntu 16.04 or newer, or CentOS 8.1 or newer.
- Minimum 12 GB of RAM and 8 CPU cores.
- A CUDA-enabled GPU with at least 12 GB of dedicated memory.

3 Methods

3.1 Data Conversion and Preprocessing

After obtaining the annotated data, the dataset can be converted to standardized mgf format using ProteoWizard tool [10]. After conversion, preprocessing can be applied to the mass spectrometry data. The provided code loads all MGF files from a given directory and starts preprocessing. It first filters spectra by precursor charge. Then, for each spectrum, it keeps only the top N most intense peaks. After that, it restricts peaks to a defined m/z range. It then normalizes intensity values across all spectra to zero mean and unit variance, which makes spectra comparable. The pipeline processes spectra one by one applies these filters in sequence and writes the processed spectra to a new MGF file for later use. This design is flexible and allows integration of additional preprocessing steps when needed.

```
import os
import numpy as np
from pyteomics import mgf
def load_mgf_files(mgf_dir):
 """Collect all MGF files in the directory."""
 mgf_files = [os.path.join(mgf_dir, f) for f in
os.listdir(mgf_dir) if f.endswith(".mgf")]
 return mgf_files
def filter_by_charge(spectra, max_charge=5):
 """Yield spectra with precursor charge <=
max_charge."""
 for spectrum in spectra:
   charge = spectrum['params'].get('charge')
   if charge is None:
      continue # skip spectra with no charge info
 # charge can be '2+' or list ['2+'] - normalize:
 if isinstance(charge, list):
     charge_str = charge[0]
 else:
    charge_str = charge
 try:
     charge_int = int(charge_str.rstrip('+'))
 except Exception:
    continue # skip if charge not parsable
 if charge_int <= max_charge:
    yield spectrum
def filter_top_n_peaks(spectra, n=100):
```

```
 """Yield spectra filtered to top n intensity
peaks."""
  for spectrum in spectra:
     intensities = spectrum.get("intensity array")
    if intensities is None or len(intensities) ==
0:
        continue
   if len(intensities) <= n:
      yield spectrum
      continue
 indices = np.argpartition(intensities, -n)[-n:]
 indices = np.sort(indices)
 # Update arrays
 spectrum["intensity array"] = intensities[indices]
 spectrum["m/z array"] = spectrum["m/z array"]
[indices]
 # charge array might be missing, so check first
 if "charge array" in spectrum:
      spectrum["charge array"] = spectrum["charge
array"][indices]
 yield spectrum
def normalize_intensity_values(spectra):
 """Calculate global intensity mean/std and yield
normalized spectra."""
 all_intensities = []
 # First pass: collect intensities
 spectra_list = list(spectra) # materialize to
iterate twice
 for spectrum in spectra_list:
    intensities = spectrum.get("intensity array",
[])
    all_intensities.extend(intensities)
 if len(all_intensities) == 0:
    raise ValueError("No intensities found for
normalization")
 mean = np.mean(all_intensities)
 std = np.std(all_intensities)
 # Second pass: normalize
 for spectrum in spectra_list:
    intensities = spectrum.get("intensity array",
[])
    norm_intensities = (intensities - mean) / std
    spectrum["intensity array"] = norm_intensities
    yield spectrum
def filter_mz_range(spectra, min_mz=100, max_mz=
7000):
 """Filter peaks in spectra to m/z range [min_mz,
max_mz]."""
  for spectrum in spectra:
    mz = spectrum.get("m/z array")
    intensities = spectrum.get("intensity array")
    if mz is None or intensities is None:
       continue
 mask = (mz >= min_mz) & (mz <= max_mz)
```

```
 if not np.any(mask):
    continue # no peaks in range, skip spectrum
 spectrum["m/z array"] = mz[mask]
 spectrum["intensity array"] = intensities[mask]
 if "charge array" in spectrum:
    spectrum["charge array"] = spectrum["charge
array"][mask]
 yield spectrum
def preprocess_mgf(mgf_dir, output_file
                max_charge=2, top_n_peaks=50
               mz_min=100, mz_max=2000):
   mgf_files = load_mgf_files(mgf_dir)
   # Open spectra generator for all files combined
   spectra = mgf.chain.from_iterable(mgf_files)
   # Step 1: Filter by charge
   spectra = filter_by_charge(spectra, max_charge=
max_charge)
   # Step 2: Filter peaks to top n
   spectra = filter_top_n_peaks(spectra, n=
top_n_peaks)
   # Step 3: Filter peaks by m/z range
   spectra = filter_mz_range(spectra, min_mz=mz_
min, max_mz=mz_max)
   # Step 4: Normalize intensity values
   spectra = normalize_intensity_values(spectra)
   # Write output
   mgf.write(spectra, output_file)
   print(f"Preprocessed spectra written to
{output_file}")
```

3.2 Preprocessing the Spectra into a Fixed-Size Vector

Raw MS spectra often vary in length, resolution, and number of peaks. This irregular nature of spectra makes it difficult to use directly in machine learning models, which usually require inputs of fixed size. To handle the irregularity of the spectra, we can convert the spectra into a fixed-size vector. This can be done by dividing the m/z axis into different bins. This step creates a uniform representation, reduces variability across spectra, and produces inputs that neural networks can process directly. Transforming the spectra into a fixed-size vector also allows the model to compare spectra on the same scale, which improves both learning and prediction. When creating the fixed-size vector, the choice of bin size creates a trade-off between resolution and efficiency. When the bin size is too large, multiple peaks merge into the same bin. It decreases the resolution, and important signals may disappear. When the bin size is too small, the vectors become very long and sparse, which increases both memory usage and computational cost. Such hyper-sparse vectors may also reduce model performance unless model adaptations address the sparsity. Therefore, larger vectors retain more detailed information but require greater resources, whereas smaller vectors reduce cost but risk a loss of precision. The optimal bin size depends on the task and on the computational resources available. Our in-house experimentation has revealed that a bin size of 0.1 Da is a good trade-off and, hence, is used for ProteoRift. It enables us to capture most isotopic peaks and fragment ions with enough precision and preserves important spectral detail. In addition, this parameter avoids the excessive dimensionality that arises from extremely fine bins (e.g., 0.02 Da) and keeps computation and memory demands manageable.

3.3 ProteoRift: Peptide Identification with Deep Learning-Based Filtering

Our novel deep-learning model predicts peptide features (e.g., peptide length, missed cleavages, and modifications) directly from spectra before any database search [8]. Our model uses an embedding network to process the spectrum's mass-to-charge ratio (m/z) and intensities. In ProteoRift, a spectrum is passed through the spectrum network to generate embeddings of the spectrum. This spectrum is then passed through three branches of fully connected layers to predict the peptide length, missed cleavages, and modifications (*see* **Note 2**). For length prediction, two fully connected layers are used with dimensions 256 × 128 and 128 × 1. The length branch predicts peptide lengths ranging from 7 to 30 amino acids. The missed cleavage branch also uses two fully connected layers, 256 × 128 and 128 × 3, as we account for a maximum of two missed cleavages. Finally, the modification branch consists of two fully connected layers with dimensions 256 × 128 and 128 × 2. In the current version of ProteoRift, this branch only determines whether a spectrum is modified or not. Each output branch computes its loss independently, and then three losses are then combined into a weighted sum before performing gradient calculation.

Let LE denote the *length branch*, MC the *missed cleavages branch*, and MO the *modification branch*. The overall weighted loss in ProteoRift is defined as:

$$L = a \cdot L(\mathrm{LE}) + b \cdot L(\mathrm{MC}) + c \cdot L(\mathrm{MO}),$$

where L(.) denotes the loss contribution from a given branch, and *a*, *b*, and *c* are nonnegative real weights that determine the contribution of each branch.

Loss for Length Branch

For the Length branch (LE), the loss is computed using the mean squared error (MSE):

$$L(\mathrm{LE}) = \frac{1}{n} \sum_{i=1}^{n} (y_i - \hat{y}_i)^2,$$

where y_i and $\hat{y}_i$ are true and predicted length.

Loss for Missed Cleavages and Modification Branches

For the Missed Cleavages branch (MC) and Modification branch (MO), the loss is computed using the log-based Softmax (cross-entropy) loss:

$$L = -\sum_{i=1}^{n} y_i \cdot \log(\mathrm{Softmax}(\hat{y}_i)),$$

where y_i and $\hat{y}_i$ are ground truth and prediction, and Softmax is defined as

$$\mathrm{Softmax}(z_j) = \frac{e^{z_j}}{\sum_{k=1}^{K} e^{z_k}}.$$

ProteoRift shares the spectrum features across tasks, and therefore the network captures features that are relevant for all three outputs (length, missed cleavages, and modifications). Each branch then specializes in its own prediction objective, while the weighted combination of their losses ensures balanced training. This design encourages the model to learn a richer representation of the spectrum and improves generalization (*see* **Note 3**). Our results show that ProteoRift achieves up to 92% precision in predicting peptide lengths, 95% for missed cleavages, and 97% for modification status. In addition, our results also indicate that ProteoRift reduces the search space significantly by selecting the candidate peptides within the filters of length, missed cleavages, and modification status, which leads to tremendous speedups while keeping the quality of the peptide deduction comparable to mass-only filters. We further retrained our ML model, SpeCollate [6], by integrating these new filters into the ML workflow. When combined with database-search engines, our filtering models deliver an 8×–12× speedup while maintaining peptide identification accuracy (*see* **Note 4**) comparable to established tools like Crux [22] and MSFragger [21]. The illustrative architecture of ProteoRift is shown in Fig. 2.

3.4 Installation Guidelines for ProteoRift

The following instructions describe how to install and run ProteoRift on a Linux system. The guide covers software prerequisites, hardware requirements, and detailed setup steps, including environment configuration with Miniconda, CUDA installation, and repository setup.

Installing Miniconda (Python Environment Manager)

To create an isolated and reproducible Python environment for ProteoRift, download and run the latest miniconda installer [37] for Linux.

```
#Create a local directory for miniconda
mkdir -p ~/miniconda3
#Download the miniconda from the web using wget
command
wget https://repo.anaconda.com/miniconda/
Miniconda3-latest-Linux-x86_64.sh -O ~/miniconda3/
miniconda.sh
```

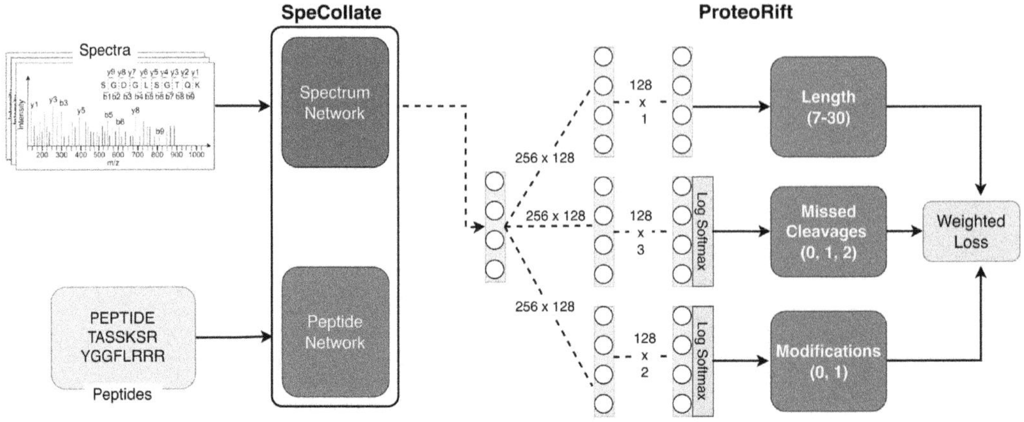

Fig. 2 Architecture of ProteoRift: Spectra and peptides are passed through SpeCollate to generate embeddings. These embeddings are then passed to ProteoRift. ProteoRift uses three separate heads to predict length, missed cleavages, and modification status, and then the loss is computed

```
#Run and Update the installer
bash ~/miniconda3/miniconda.sh -b -u -p ~/
miniconda3
#Execute the Installer
/Miniconda3-[version].sh
```

Installing CUDA Toolkit

The CUDA Toolkit is required for GPU-enabled computation. Install it using the following commands:

```
#Refresh the APT repository
sudo apt-get update
#Install CUDA and GPU support tools
sudo apt-get install cuda
sudo apt-get install nvidia-gds
```

Comprehensive installation instructions can be found in the official NVIDIA documentation [36].

Clone ProteoRift Repository

Clone the ProteoRift repository from our lab github. ProteoRift is provided as an open-source, open-weight ML model available for noncommercial purposes.

```
# Clone the Repo
git clone https://github.com/pcdslab/ProteoRift
#Change the directory to ProteoRift
cd ProteoRift
```

Setup Conda Environment

Install all the dependencies by creating the conda environment from the *yml* file.

```
# Create conda environment
conda env create --file proteorift_env.yml
```

3.5 Perform the Database Search Using ProteoRift Filtering

Once you have configured the environment, you can run an end-to-end peptide identification pipeline that integrates two core models: **SpeCollate** and **ProteoRift**.

1. To begin, prepare the necessary input data:
   - **Spectra Files**: Spectra files need to be converted into mgf format using ProteoWizard [10]. After conversion to the mgf format, put the spectra files in sample_data/spectra.
   - **Peptide Database**: Digest the database using the Crux toolkit [22]. Put the database after digestion in sample_data/peptide_database.
2. Download pretrained weights for SpeCollate and ProteoRift:
   - Download ProteoRift weights from the link: https://github.com/pcdslab/ProteoRift/releases/download/V1.0.0/proteorift_model_weights.pt.
   - Download SpeCollate weights from the link: https://github.com/pcdslab/ProteoRift/releases/download/V1.0.0/specollate_model_weights.pt.
   - Create a subfolder models/ and copy the weights in that subfolder.
3. **Edit Configuration File:** Open the config.ini file and configure the following parameters under the [search] section:
   - model_name: Absolute path to the downloaded ProteoRift model weights (typically named ProteoRift_model_weights.pt).
   - specollate_model_path: Absolute path to the SpeCollate model weights (typically named specollate_model_weights.pt).
   - mgf_dir: Absolute path to the directory containing your spectra files in MGF format.
   - prep_dir: Absolute path to the directory where preprocessed spectra will be stored.
   - pep_dir: Absolute path to the directory containing the peptide database.
   - out_pin_dir: Absolute path to an existing directory where the resulting Percolator.pin files will be written (*see* **Note 5**).

   Set any additional database search parameters according to your requirements.
4. **Preprocess Spectra**: Execute the following command to preprocess the spectra files.
   python read_spectra.py -t u
5. **Run Database Search**: Launch the search using:

python run_search.py

This step generates embeddings for both spectra and peptides, applies filtering, and performs the database search.

6. **Expected Output**

   Upon completion, the pipeline will produce two output files:

   - target.pin: Contains features for target peptide-spectrum matches (PSMs).
   - decoy.pin: Contains features for decoy PSMs, used for downstream false discovery rate (FDR) estimation.

3.6 Retraining or Fine-Tuning ProteoRift Model

ProteoRift was trained on the HCD fragmentation data sets and multiple species that include *Homo sapiens, Vigna mungo, Mus musculus, Methanosarcina mazei, Bacillus, Candidatus endoloripes, Solanum lycopersicum, Saccharomyces cerevisiae*, and *Apis mellifera.* To extend the experiments to additional species or alternative fragmentation methods, ProteoRift might require fine-tuning or retraining. The development of ProteoRift requires MS data preparation for the specific needs of ML-based property prediction. This includes obtaining the annotated data, converting spectra into structured formats compatible with neural networks, standardizing intensities and m/z values. Each of these factors contributes to the generalizability and reproducibility of the model across various proteomics datasets. The complete architectural design and comprehensive evaluation of the *ProteoRift* model have been documented in our recent publication [8].

3.7 Running the ProteoRift Model Training

1. **Edit Configuration File**: Open the config .ini file using any preferred text editor. Set the following parameters accordingly:
   - **mgf_dir**: Specify the absolute path to the directory containing the MGF files.
   - **prep_dir**: Provide the absolute path to the directory where the preprocessed MGF files will be stored.
   - **[ml] Section**: Adjust the model hyperparameters within the [ml] section as needed. This may include settings such as learning_rate, dropout, and others depending on your training requirements.
2. **Configure Weights and Biases (Optional)**: Set up a Weights & Biases(wandb) account and create a new project named *ProteoRift*. The step-by-step guide on how to setup wandb can be found on the following link: https://docs.wandb.ai/quickstart/.
3. **Preprocess Spectra**: Execute the read_spectra.py script in the ProteoRift directory using the command below. This script processes the spectra files, splits them into training, validation,

and test sets, and stores the outputs in the directory specified by prep_dir.

python read_spectra.py -t l

4. **Train the model**: Begin model training by running:

   python run_train.py

4 Notes

1. Many current methods rely only on mass-based filtering, which can limit efficiency, especially in complex searches like proteogenomics or metaproteomics. ProteoRift applies a prune-and-search approach.
2. ProteoRift applies pruning and each step reduces the number of candidate peptides and lowers false positives, but repeated pruning can also decrease accuracy.
3. ProteoRift has three prediction heads for property prediction (e.g., length, missed cleavages, and modification status), and errors from these three prediction heads can accumulate, which can result in fewer identified peptides. This error can occur due to the imbalanced nature of training data, where some peptide features are underrepresented, which leads to less reliable predictions for those cases.
4. ProteoRift relies on the availability of labeled data, and its performance in predicting peptide properties and identifying peptides depends strongly on the quality of labeled data that exists in the training set.
5. This directory must already exist, or the process will terminate with an error.

Acknowledgments

Research reported in this publication was supported by NIGMS of the National Institutes of Health under award number R35GM153434. The content is solely the responsibility of the authors and does not necessarily represent the official views of the National Institutes of Health. This work used the NSF Extreme Science and Engineering Discovery Environment (XSEDE) Supercomputers through allocations: TG-CCR150017 and TG-ASC200004.

Conflict of Interest The authors declare no conflicts of interest that are relevant to the content of this chapter.

References

1. Perez-Riverol Y, Bandla C, Kundu DJ et al. (2025) The PRIDE database at 20 years: 2025 update. Nucleic Acids Res 53(D1):D543–D553. https://doi.org/10.1093/nar/gkae1011
2. Deutsch EW, Bandeira N, Sharma V, et al. (2020) The ProteomeXchange consortium in 2020: enabling 'big data' approaches in proteomics. Nucleic Acids Res 48(D1):D1145–D1152. https://doi.org/10.1093/nar/gkz984
3. MassIVE-KB Peptide Spectral Libraries (2024) https://massive.ucsd.edu/ProteoSAFe/static/massive-kb-libraries.jsp
4. Walzer M, García-Seisdedos D, Prakash A et al. (2022) Implementing the reuse of public DIA proteomics datasets: from the PRIDE database to Expression Atlas. Sci Data 9(1):335. https://doi.org/10.1038/s41597-022-01380-9
5. Burq M, Stepec S, Restrepo J et al (2024) Back to basics: spectrum and peptide sequence are sufficient for top-tier mass spectrometry proteomics identification. Bioinformatics. https://doi.org/10.1101/2024.08.19.606805
6. Tariq MU, Saeed F (2021) SpeCollate: deep cross-modal similarity network for mass spectrometry data based peptide deductions. PLoS One 16(10):e0259349. https://doi.org/10.1371/journal.pone.0259349
7. Wei Q, Li J, Ma J et al. (2025) DeepMS: super-fast peptide identification using end-to-end deep learning method. J Mol Biol 437(17):169237. https://doi.org/10.1016/j.jmb.2025.169237
8. Tariq U, Shabbir B, Saeed F (2026) End-to-end deep attention-based multitask pipeline for predicting uncertainty-quantified peptide properties from mass spectrometry data. Sci Rep 16, 13331. https://doi.org/10.1038/s41598-026-43215-2
9. NIST libraries of peptide tandem mass spectra. https://chemdata.nist.gov/dokuwiki/doku.php?id=peptidew:start
10. Chambers MC, Maclean B, Burke R et al. (2012) A cross-platform toolkit for mass spectrometry and proteomics. Nat Biotechnol 30(10):918–920. https://doi.org/10.1038/nbt.2377
11. Frank A, Tanner S, Bafna V et al. (2005) Peptide sequence tags for fast database search in mass-spectrometry. J Proteome Res 4(4):1287–1295. https://doi.org/10.1021/pr050011x
12. Haseeb M, Saeed F (2021) HiCOPS: high performance computing framework for tera-scale database search of mass spectrometry based omics data, ArXiv Prepr. ArXiv210202286
13. Haseeb M, Saeed F (2019) Efficient shared peak counting in database peptide search using compact data structure for fragment-ion index. In: 2019 IEEE international conference on bioinformatics and biomedicine (BIBM), IEEE, p 275–278
14. Muth T, Renard BY, Martens L (2016) Metaproteomic data analysis at a glance: advances in computational microbial community proteomics. Expert Rev Proteomics 13(8):757–769. https://doi.org/10.1080/14789450.2016.1209418
15. Muth T, Behne A, Heyer R et al. (2015) 'The MetaProteomeAnalyzer: a powerful open-source software suite for metaproteomics data analysis and interpretation'. J Proteome Res 14(3):1557–1565
16. Schiebenhoefer H, Schallert K, Renard BY et al. (2020) A complete and flexible workflow for metaproteomics data analysis based on MetaProteomeAnalyzer and Prophane. Nat Protoc 15(10):3212–3239
17. Eng JK, Searle BC, Clauser KR et al. (2011) A face in the crowd: recognizing peptides through database search. Mol Cell Proteomics 10(11):R111.009522. https://doi.org/10.1074/mcp.R111.009522
18. Nesvizhskii AI (2014) Proteogenomics: concepts, applications and computational strategies. Nat Methods 11(11):1114–1125
19. Chick JM, Kolippakkam D, Nusinow DP et al. (2015) A mass-tolerant database search identifies a large proportion of unassigned spectra in shotgun proteomics as modified peptides. Nat Biotechnol 33(7):743–749
20. Burke MC, Mirokhin YA, Tchekhovskoi DV et al. (2017) The hybrid search: a mass spectral library search method for discovery of modifications in proteomics. J Proteome Res 16(5):1924–1935
21. Kong AT, Leprevost FV, Avtonomov DM et al. (2017) MSFragger: ultrafast and comprehensive peptide identification in mass spectrometry–based proteomics. Nat Methods 14(5):513–520
22. McIlwain S, Tamura K, Kertesz-Farkas A et al. (2014) Crux: rapid open source protein tandem mass spectrometry analysis. J Proteome Res 13(10):4488–4491
23. Altenburg T, Muth T, Renard BY (2021) yhydra: deep learning enables an ultra fast open search by jointly embedding MS/MS spectra and peptides of mass spectrometry-based proteomics. bioRxiv, pp 2021–12
24. Gessulat S, Schmidt T, Zolg DP et al. (2019) Prosit: proteome-wide prediction of peptide

tandem mass spectra by deep learning. Nat Methods 16(6):509–518
25. Tran NH, Zhang X, Xin L et al. (2017) De novo peptide sequencing by deep learning. Proc Natl Acad Sci 114(31):8247–8252
26. Cox J (2023) Prediction of peptide mass spectral libraries with machine learning. Nat Biotechnol 41(1):33–43
27. Yang H, Chi H, Zeng W-F et al. (2019) pNovo 3: precise de novo peptide sequencing using a learning-to-rank framework. Bioinformatics 35(14):i183–i190. https://doi.org/10.1093/bioinformatics/btz366
28. Blundell C, Cornebise J, Kavukcuoglu K et al (20Set the following parameters 15) Weight uncertainty in neural network. In: International conference on machine learning, PMLR, pp 1613–1622
29. Gal Y, Ghahramani Z (2016) Dropout as a Bayesian approximation: representing model uncertainty in deep learning. In: international conference on machine learning, PMLR, pp 1050–1059
30. Lakshminarayanan B, Pritzel A, Blundell C (2017) Simple and scalable predictive uncertainty estimation using deep ensembles. Adv Neural Inf Process Syst 30:6402–6413.
31. Rossum GV, Drake FL (2009) Python 3 Reference Manual. CreateSpace, Scotts Valley, CA 242 https://www.python.org/about/
32. Paszke A, Gross S, Massa F, et al. (2019) PyTorch: An imperative style, high-performance deep learning library. Adv Neural Inf Process Syst 32:8024–8035.
33. Harris CR, Millman KJ, van der Walt SJ, et al. (2020) Array programming with NumPy. Nature 585(7825):357–362. https://doi.org/10.1038/s41586-020-2649-2
34. The pandas development team. (2024). pandas-dev/pandas: Pandas (Version v2.2.2) [Computer software]. Zenodo. https://doi.org/10.5281/zenodo.10957263
35. Pedregosa F, Varoquaux G, Gramfort A, et al. (2011) Scikit-learn: Machine learning in Python. J Mach Learn Res 12:2825–2830.
36. NVIDIA (2025) CUDA Toolkit Documentation. NVIDIA Corporation. https://docs.nvidia.com/cuda/ Accessed 15 Jul 2025
37. Anaconda Inc. (2025) Miniconda. https://docs.anaconda.com/miniconda/ Accessed 15 Jul 2025

Chapter 16

pXg: Comprehensive and Reliable Identification of MHC-I Antigens Using De Novo Peptide Sequencing and Immunotranscriptomic Datasets

Seunghyuk Choi, Shanji Zhang and Eunok Paek

Abstract

De novo peptide sequencing enables the identification of a broad repertoire of major histocompatibility complex class I-associated peptides, providing a valuable source of potential immunotherapeutic targets in cancer. Nevertheless, its practical utility remains limited by the absence of robust methods for confidently determining peptide-spectrum matches and interpreting their proteogenomic information. This chapter presents the pXg workflow, designed to comprehensively and reliably identify peptides derived from de novo sequencing in mass spectrometry-based immunopeptidomics, integrated with matched RNA-seq data. We guide the reader through the use of pXg, covering the rescoring of peptide-spectrum matches and the estimation of a false discovery rate from the popular de novo sequencing tool Casanovo. This ultimately yields rich proteogenomic insights into the identified peptides.

Keywords Tandem mass spectrometry, De novo peptide sequencing, Proteogenomics, Immunopeptidomics, Non-reference peptide, Software

1 Introduction

T cell-mediated immunity plays a pivotal role in the adaptive immune system by eliminating infected or cancerous cells through recognition of specific antigens. This immune response relies on the presentation of major histocompatibility complex (MHC) class I-associated peptides (MAPs) on the cell surface [1]. Leveraging this mechanism, cancer immunotherapy has emerged as a transformative approach that targets tumor-specific MAPs. These MAPs—derived from mutations, gene fusions, aberrant gene expression, abnormal splicing, or unconventional translation from regions previously considered noncoding—can be identified using mass spectrometry (MS)-based immunopeptidomics [2]. The standard approach in immunopeptidomics involves immunoaffinity purification of MHC complexes from detergent-solubilized cell lysates, followed by peptide isolation and

Frédérique Lisacek (ed.), *Bioinformatics for Mass Spectrometry*, Methods in Molecular Biology, vol. 3071,
https://doi.org/10.1007/978-1-0716-5519-1_16,

separation via high-performance liquid chromatography (HPLC). These MAPs are then analyzed with advanced, high-sensitivity mass spectrometry. The resulting MS data are then processed using a computational algorithm to identify peptide sequences [3].

In contrast to standard proteomic searches, immunopeptidomics requires significantly broader search spaces because it considers nonenzymatic and noncanonical peptides. For example, Laumont and colleagues identified 40 tumor antigens through immunopeptidomics profiling of two murine cancer cell lines and seven human primary tumors, with nearly 90% arising from regions previously thought to be noncoding [4]. Pataskar and colleagues showed that interferon-γ-induced codon reassignment (tryptophan to phenylalanine, W > F) could reshape the immunopeptidome, showcasing the potential of tumor antigens capable of eliciting T cell responses [5, 6]. This inherent uncertainty in the search space limits the effectiveness of traditional database searches, which rely on prior genomic annotation, thereby hindering the sensitive detection of tumor antigens.

De novo peptide sequencing, which directly infers peptide sequences from MS/MS spectra without prior knowledge, presents an attractive alternative, especially given that MAPs are typically short (8–12-mers) [7, 8]. Moreover, recent advances in artificial intelligence have significantly improved the accuracy of de novo peptide sequencing, further enhancing its utility for tumor antigen discovery [9–13]. Together with the advancements, several works have recently emerged to obtain confident identification from the de novo peptide sequencing results [8, 14–17]. For example, Spectralis [14] and NovoRank [15] refine peptide-spectrum matches (PSMs) from de novo peptide sequencing using deep learning frameworks. On the other hand, NovoBoard [8], Peptide-PRISM [17], and pXg [16] utilize a target-decoy competition [18] to estimate false discovery rate (FDR), identifying confident PSMs. In contrast to NovoBoard and Peptide-PRISM, pXg requires both a PSM list (.tsv) from any de novo peptide sequencing tools and a matched RNA-seq alignment (.bam) from the same sample, providing more in-depth proteogenomic information for the identified MAPs.

In this chapter, we present the use of pXg to reliably identify comprehensive MAPs from MS-based MHC-I immunopeptidomics and the matched RNA-seq data in MD55A3 melanoma cells with IFNγ treatment [6]. To facilitate this hands-on tutorial, we provide processed datasets including a PSM list (.tsv) from Casanovo [9] and RNA-seq alignment (.bam) from the STAR aligner [19]. It is noteworthy that a PSM list (.tsv or .csv) from any de novo peptide sequencing tools can be used as a pXg input.

2 Materials

2.1 Software

pXg can be downloaded at https://github.com/HanyangBISLab/pXg and requires openjdk 24.0.1 or the latest Java environment. Percolator can be obtained from https://github.com/percolator/percolator.

2.2 Reference Transcriptome and Proteome Datasets

We use the GENCODE V47 basic annotation (.gtf) from https://www.gencodegenes.org as the human transcriptome model. For the protein reference sequences, the UniProt Human reference proteome (.fasta) and common laboratory contaminant sequences (.fasta) are available at https://www.uniprot.org/proteomes/UP000005640 and https://www.thegpm.org/crap, respectively. To minimize the risk of misattributing contaminant-derived peptides, we recommend supplementing the UniProt Human reference proteome with the sequences of common laboratory contaminants before using it as input for pXg.

2.3 Processed Datasets

Together with the reference datasets, pXg also requires a PSM list (.tsv) and RNA-seq alignment (.bam) in the same sample. We deposited the PSM list and RNA-seq alignment in Zenodo (doi.org/10.5281/zenodo.17042918), facilitating a focus on the use of pXg. Briefly, we obtained MS/MS spectra in the ProteomeXchange Consortium from the PRIDE partner repository with the dataset identifier PXD020224 (QEP2_10348_YSA_MD55A3-IFN-1-500-HLA-I_1_311019.raw) and the matched RNA-seq data in the Gene Expression Omnibus (GEO) with accession code GSE142822 (SRR10814106 and SRR10814107) from MD55A3 melanoma cells with IFNγ treatment [6]. The MS/MS spectra and RNA-seq datasets were processed by Casanovo (v5.0.0, https://github.com/Noble-Lab/casanovo) and nf-core/rnaseq pipeline (v3.19.0, https://nf-co.re/rnaseq), respectively [9, 20]. We recommend adding an option "--outSAMunmapped" for the tSTAR aligner [19] when running the nf-core/rnaseq pipeline, considering unmapped reads as a search space for the pXg run. This enables a user to identify peptides derived from fusion genes, virus proteins, and highly mutated regions.

3 Methods

3.1 pXg Workflow

pXg workflow requires four main inputs: a PSM list (.tsv or.csv), RNA-seq alignment files (.bam), a transcriptome model (.gtf), and protein sequences (.fasta, optional) (Fig. 1). It can be executed in two ways: (1) via a command-line interface (CLI), recommended for processing multiple samples and using advanced parameters; (2) via a graphical user interface (GUI), which is ideal for beginners working with standard parameters.

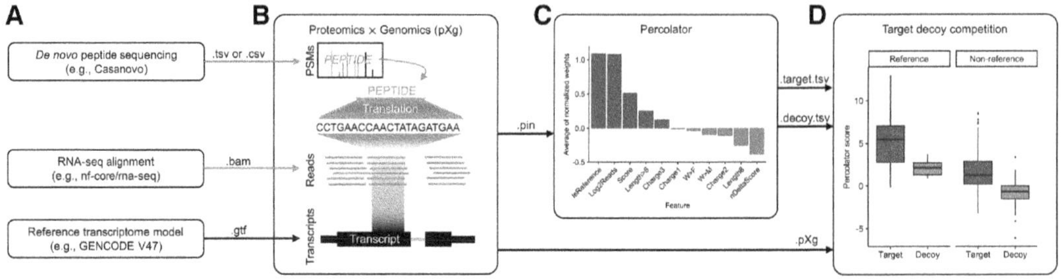

Fig. 1 Overview of workflow for identifying reliable MAPs from de novo peptide sequencing using pXg and Percolator. (**A**) Peptide-spectrum matches (PSMs) obtained by de novo peptide sequencing, RNA-seq alignment files (.bam), and a trancriptome model (.gtf) are provided as input. (**B**) pXg maps the PSMs to RNA-seq reads and annotates their genomic features based on the transcriptome model. (**C**) RNA-supported PSMs are rescored using Percolator. (**D**) High-confidence PSMs are identified at a specified false discovery rate through target-decoy competition.

3.2 Command Line Interface

Run pXg by executing the following command in the command prompt:

pXg Execution with Standard Parameter

```
java
-Xmx32G -jar pXg.v2.4.4.jar \
--psm MD55A3_IFN_1.mztab
\


--peptide_index 2 \
--identifier_index 15 \
--charge_index 12 \
--score_index 9 \
--bam MD55A3_IFN_1.sorted.bam,MD55A3_IFN_2.sorted.
bam \
--gtf gencode.v47.basci.annotation
--fasta UP000005640.crap.fasta \
--aa_variant aa_variant_table.tsv \
--threads 4 \
--output MD55A3_IFN_1
```

Parameter descriptions:

- "-Xmx32G" sets the maximum Java Virtual Machine (JVM) heap size to 32 GB. If pXg runs into memory issues, increase this value (e.g., -Xmx64G for 64 GB).
- "--psm" specifies the path to the PSM list file (.tsv or .csv) from a de novo peptide sequencing tool (e.g., Casanovo in this example).
- "--peptide_index" is the 1-based column index of peptide sequences in the PSM file.
- "--identifier_index" is the 1-based column index of a unique spectrum identifier. To combine multiple columns, provide comma-separated indices without spaces (e.g., --identifier_index 1,2,3).

- "--charge_index" is the 1-based column index for charge state.
- "--score_index" is the 1-based column index for the raw score of the PSM.
- "--bam" specifies the path to the RNA-seq alignment files (.bam). To use multiple bam files (e.g., replicates), provide comma-separated file paths without spaces. The bam files must be sorted by genomic location. To consider MAPs originating from unrecognized regions (e.g., viral proteins and fusion genes), include unmapped reads in the bam files.
- "--gtf" specifies the path to the transcriptome model (.gtf) used in the RNA-seq alignment.
- "--fasta" specifies the path to the protein sequence file (.fasta). It is recommended to append potential laboratory contaminants to avoid misattribution of MAP origins.
- "--aa_variant" specifies the path to the list of amino-acid variants to consider. In this example, "W > F" and "W > M" were considered to identify prevalent amino-acid variants by IFNγ-induced tryptophan depletion [5, 6].
- "--threads" sets the number of threads.
- "--output" specifies the prefix of pXg output files.

pXg produces three output files (Fig. 1B): (1) MD55A3_IFN_1.pXg, which contains detailed genomic annotation for each PSM; (2) MD55A3_IFN_1.pXg.pin, which stores the corresponding proteogenomic features (*see* **Note 1** for more effective feature generation methods); and (3) MD55A3_IFN_1.log, which records the processing log. The.pXg and.pin files serve as input for rescoring and false discovery rate (FDR) estimation, culminating in the generation of the final results.

Rescoring Using Percolator

Run the Percolator by executing the following command in the command prompt:

```
percolator --default-direction Score \
--results-psms MD55A3_IFN_1.target.tsv \
--decoy-results-psms MD55A3_IFN_1.decoy.tsv \
--protein-decoy-pattern XXX_ \
--post-processing-tdc \
--only-psms \
MD55A3_IFN_1.pXg.pin
```

Parameter descriptions:

- "--default-derection" specifies the feature name to be used as the initial search direction.
- "--results-psms" defines the output file path for target PSMs.
- "--decoy-results-psms" defines the output file path for decoy PSMs.

- "--protein-decoy-pattern" indicates the prefix used to specify decoy PSMs.
- "post-processing-tdc" applies Percolator's SVM scoring to remove the lower-scoring PSM (target or decoy) for each spectrum.
- "only-psms" restricts output to PSM-level results.
- The input .pin file must be specified at the end of the command.

Percolator improves PSM scoring by incorporating both standard proteomic features obtained from the de novo peptide sequencing tool and RNA quantification values generated by pXg (Fig. 1C). For each spectrum, only the top-scoring peptide (whether target or decoy) is retained and written to the respective target.tsv or .decy.tsv file.

FDR Estimation

Estimate FDR for each reference and non-reference peptide group by executing the following command in the command prompt:

```
java
-cp pXg.v2.4.4.jar progistar.tdc.TDC \
--input MD55A3_IFN_1.pXg \
--target MD55A_IFN_1.target.tsv \
--decoy MD55A3_IFN_1.decoy.tsv \
--fdr 0.01 \
--output MD55A3_IFN_1
```

Parameter description:

- "--input" specifies the path to the .pXg file.
- "--target" defines the input file path for target PSMs generated by Percolator.
- "--decoy" defines the input file path for decoy PSMs generated by Percolator.
- "--fdr" sets the desired FDR threshold.
- "--output" specifies the output file prefix, generating the final result as [output].fdr.tsv.

The FDRs of reference and non-reference PSMs are estimated separately and reported collectively in the output file (.fdr.tsv) (Fig. 1D). This output includes all target and decoy PSMs, enabling further inspection and downstream analyses.

3.3 Graphical User Interface

Instead of using a command line, run pXg using the following settings in the GUI (Fig. 2). Each file path can be set by clicking the "Choose File" button and selecting the appropriate file from the dialog window. The GUI supports only the basic parameters required to run pXg, configurable by providing values in the corresponding input fields. Finally, the user needs to define the working directory, where all output files will be stored. For clarity, the GUI input fields correspond to the following command-line options:

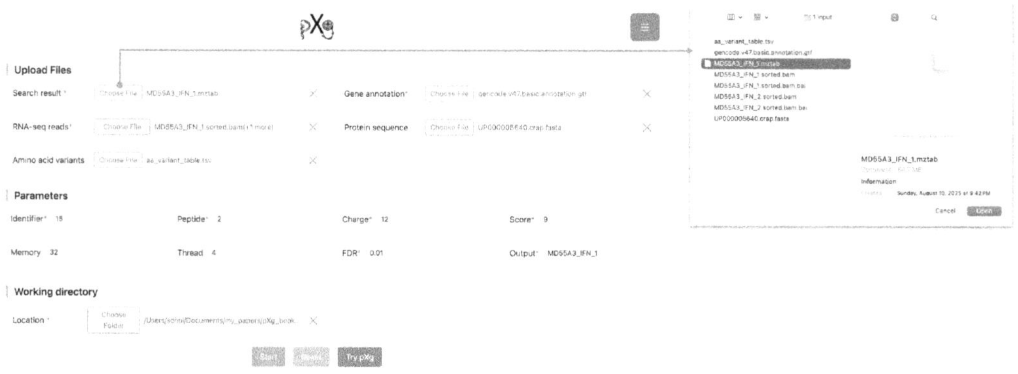

Fig. 2 Screenshot showing the pXg graphical user interface

- "Search result" equals "--psm."
- "Gene annotation" equals "--gtf."
- "RNA-seq reads" equals "--bam." To use multiple bam files (e.g., replicates), select multiple files.
- "Protein sequence" equals "--fasta."
- "Amino acid variants" equals "--aa_variant."
- "Identifier" equals "--identifier_index."
- "Peptide" equals "--peptide_index."
- "Charge" equals "--charge_index."
- "Score" equals "--score_index."
- "Memory" equals "--Xmx."
- "Thread" equals "--threads."
- "Output" equals "--output."

All inputs can be cleared using the "Reset" button. The "Try pXg" button is also provided; it automatically populates the input fields with a toy example to illustrate the appropriate format for each parameter. Once all inputs have been specified, clicking "Start" will sequentially execute pXg, Percolator, and FDR estimation. Upon completion, a summary of results is provided. This summary includes the number of identifications, peptide length distributions, and subclassification of non-reference peptides, helping to understand the overall identification patterns (Fig. 3).

3.4 Description of Output Format

The ".fdr.tsv" file serves as the primary output for downstream analyses. It contains all original output columns from the de novo peptide sequencing tool (e.g., Casanovo) along with 42 additional columns generated by pXg and Percolator. In this section, we describe 22 columns that are most frequently used in downstream analyses. Detailed explanations of the remaining columns are available at https://github.com/HanyangBISLab/pXg.

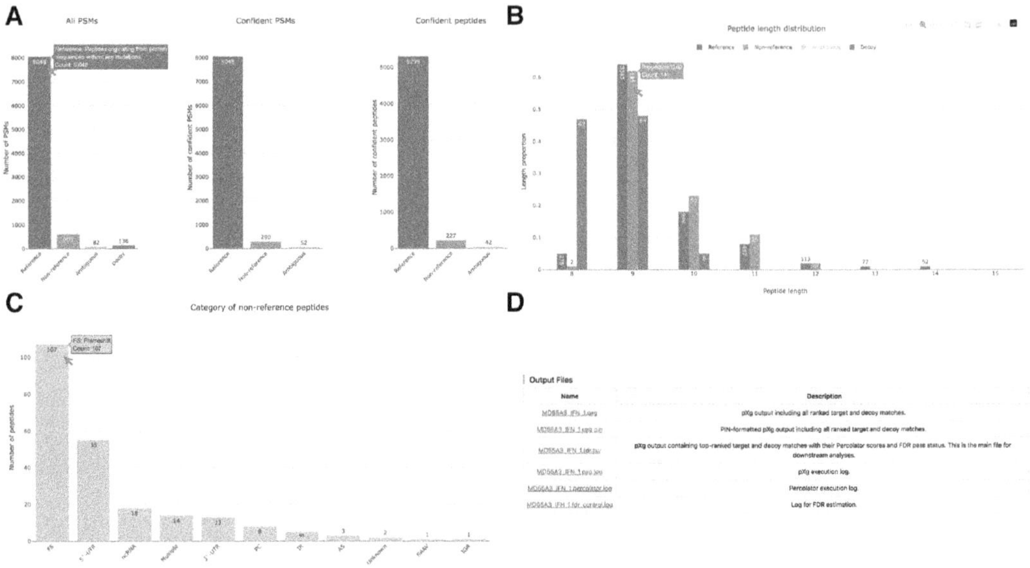

Fig. 3 Summary of results page of the pXg graphical user interface. (**A**) Statistics for all PSMs, confident PSMs, and confident peptides, where identifications passing the user-defined FDR threshold are considered "confident." (**B**) Peptide length distributions of confident and decoy peptides by category (reference, non-reference, ambiguous, and decoy). (**C**) Detailed subclassification of confident non-referencenon-reference peptides. Hovering over graphs (**A–C**) provides detailed information. (**D**) Output files are accessible via hyperlinks

Frequently used columns (22):

- "Label" indicates whether a PSM is a target (1) or decoy (−1).
- "AminoAcidVariant" represents an amino acid substitution in a format [position]:[original amino acid] > [altered amino acid] (e.g., 5:W > F).
- "InferredPeptide" represents a peptide sequence inferred from RNA-seq translation, including post-translational modifications (PTMs).
- "InferredSequence" represents a peptide sequence inferred from RNA-seq translation without PTMs.
- "GenomicLoci" provides genomic coordinates of a peptide, formatted as [chromosome]:[start location]-[end location]. Multiple loci arising from splice junctions are separated by "|" (e.g., chr19:49497583-49497595|chr19:49497917-49497930).
- "Strand" shows the strand orientation of a peptide.
- "Observed/ReferenceLeftFlankNucleotide" reports a left-flanking nucleotide sequence observed in RNA-seq or from the reference genome.
- "Observed/ReferenceNucleotide" reports nucleotide sequence aligned to RNA-seq or reference genome.
- "Observed/ReferenceRightFlankNucleotide" reports a right-flanking nucleotide sequence observed in RNA-seq or from the reference genome.

- "Mutations" specify mutations inferred from RNA-seq. Multiple mutations are separated by "|."
- "MutationStatus" represents whether amino acids are changed due to mutations.
- "TranscriptIDs" list matched transcript identifiers.
- "GeneIDs" list matched gene identifiers.
- "GeneNames" list matched gene names.
- "Events" describe the classification of peptide origin: protein-coding (PC), 5′-/3′-untranslated regions (UTRs), frameshift (FS), noncoding RNA (ncRNA), intron-retention (IR), intergenic region (IGR), alternative splicing (AS), or unknown (when genomic coordinates cannot be fully mapped to the reference genome).
- "FastaIDs" show all matched protein sequences for a peptide.
- "Reads" reports abundance of matched RNA-seq reads, expressed as reads per hundred million (RPHM).
- "IsReference" represents whether a PSM is a reference (true) or a non-reference (false).
- "FDRFilter" indicates whether a PSM passes the specified FDR threshold (Pass) or not (Fail).

Next, we illustrate how nine columns can represent the proteogenomic context of an identified MAP within the Integrative Genomics Viewer (IGV) [21] (Fig. 4). As an example, we selected the peptide *KVAKLVHFL* from the final output file (.fdr.tsv). The values for these nine columns are shown in Fig. 4A, with the corresponding genomic information displayed in IGV after loading the two MD55A3 BAM files (Fig. 4B). Notably, this peptide is encoded on the reverse strand; thus, the peptide sequence should be read in reverse orientation in the IGV "Refseq All" track.

A

Column	Value
InferredSequence	KVAKLVHFL
GenomicLoci	chrx:152767291-152767317
Strand	-
ObservedLeftFlankNucleotide	TGAGGAG
ObservedNucleotide	CAGAAAATGAACCAACTTGGCCACCTT
ObservedRightFlankNucleotide	CCTACTGAGT
GeneName	MAGEA6
Events	PC
IsReference	true

B

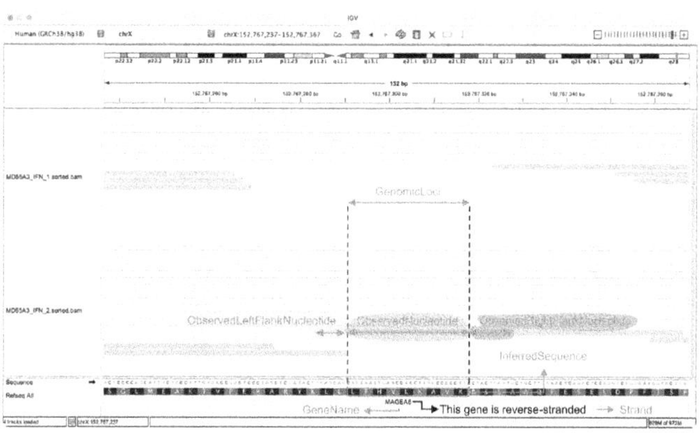

Fig. 4 Example of an identified MAP in MAGEA6. (**A**) Selected columns corresponding to the peptide *KVAKLVHFL* in the pXg format. (**B**) Column details visualized using the Integrative Genomics Viewer (IGV)

3.5 Cautions for Using the .fdr.tsv File in Downstream Analyses

Since the.fdr.tsv file contains both target and decoy PSMs, only confident target PSMs should be retained by filtering out decoy and low-confidence PSMs (Fig. 5A, B). Subsequently, non-reference peptides can be examined by selecting "FALSE" in the IsReference column (Fig. 5C). Classification of reference and non-reference peptides is based on a reference transcriptome model (e.g., GENCODE), which may occasionally conflict with a protein sequence database (e.g., UniProt). Such discrepancies can lead to ambiguous annotation, necessitating additional criteria to clarify them. In the absence of a consensus rule, the approach largely depends on the specific goals of the study.

3.6 Prioritization of Antigens from Identified MAPs

After identifying thousands of MAPs through immunopeptidomics, a critical downstream task is the prioritization of MHC-I antigens, which may represent the most biologically relevant step. Direct assessment of MAP expression in benign tissues by immunopeptidomics would, in principle, be ideal; however, the challenges of detecting low-abundance proteins and the impracticality of obtaining all normal tissues (e.g., brain, heart) from a single patient render this approach infeasible. As an alternative, RNA-seq expression can serve as a proxy for protein expression [22–24]. For well-characterized protein-coding genes, tumor specificity can be evaluated relatively straightforwardly. For instance, RNA expression of MAGEA6 (encoding the peptide *KVAKLVHFL*) can be rapidly examined across 19,788 samples from 54 tissues using GTExPortal [25] (Fig. 6). A similar type of analysis can also be performed using the Human Protein Atlas [26]. However, this gene-centric approach is restricted to peptides with well-annotated origins. As an alternative, peptide-encoding RNA expression can be estimated by directly matching peptides to RNA-seq reads [22]. This method enables a more comprehensive

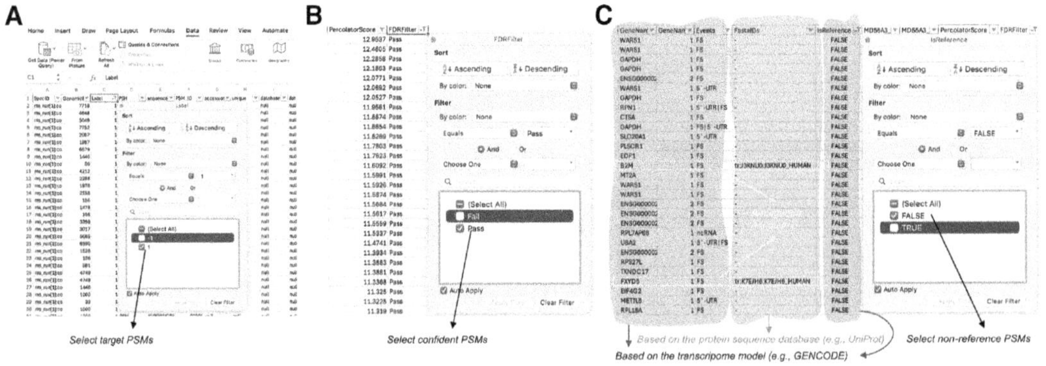

Fig. 5 Selection of confident PSMs from the.fdr.tsv file using Excel and considerations for interpreting non-reference peptides. (**A**) Select target PSMs by marking "1" in the Label column. (**B**) Select confident PSMs passing the user-defined FDR threshold by marking "Pass" in the FDRFilter column. (**C**) View non-reference peptides by marking "FALSE" in the IsReference column. Note that some peptides may map to the provided protein sequence database, resulting in ambiguous annotation

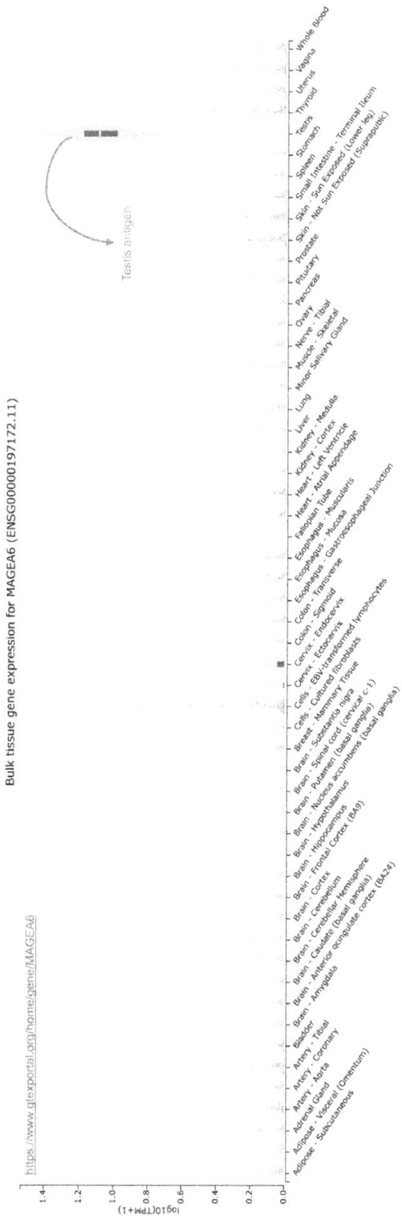

Fig. 6 Example of prioritizing a potential tumor antigen, MAGEA6, using GTExPortal

evaluation of RNA expression, covering peptides derived from novel isoform, but it is computationally intensive and time-consuming when applied across thousands of RNA-seq datasets.

4 Notes

1. We recommend incorporating deep-learning-based feature augmentation into the ".pXg.pin" file using tools such as MSBooster [27], DeepRescore [28], or MS2Rescore [29, 30] to improve both sensitivity and precision of peptide identification.

Acknowledgments

This work was supported by Research Program of the National Research Foundation (NRF) funded by the Korean government (MSIT) (Nos. 2022M3H9A2096188 and 2022-0-00964, National Program for Excellence in Software) and by Institute of Information & communications Technology Planning & Evaluation (IITP) grant funded by the Korea government (MSIT) [No. 2020-0-01373, Artificial Intelligence Graduate School Program (Hanyang University), No. RS-2025-02219317, AI Star Fellowship (Kookmin University)].

Conflict of Interest The authors declare no conflicts of interest that are relevant to the content of this chapter.

References

1. Pishesha N, Harmand TJ, Ploegh HL (2022) A guide to antigen processing and presentation. Nat Rev Immunol 22:751–764
2. Chong C, Coukos G, Bassani-Sternberg M (2022) Identification of tumor antigens with immunopeptidomics. Nat Biotechnol 40:175–188
3. Zhang B, Bassani-Sternberg M (2023) Current perspectives on mass spectrometry-based immunopeptidomics: the computational angle to tumor antigen discovery. J Immunother Cancer 11:e007073
4. Laumont CM, Vincent K, Hesnard L et al. (2018) Noncoding regions are the main source of targetable tumor-specific antigens. Sci Transl Med 10:eaau5516
5. Pataskar A, Champagne J, Nagel R et al. (2022) Tryptophan depletion results in tryptophan-to-phenylalanine substitutants. Nature 603:721–727
6. Bartok O, Pataskar A, Nagel R et al. (2021) Anti-tumour immunity induces aberrant peptide presentation in melanoma. Nature 590:332–337
7. Beslic D, Tscheuschner G, Renard BY et al. (2023) Comprehensive evaluation of peptide de novo sequencing tools for monoclonal antibody assembly. Brief Bioinform 24:bbac542
8. Tran NH, Qiao R, Mao Z et al. (2024) NovoBoard: a comprehensive framework for evaluating the false discovery rate and accuracy of de novo peptide sequencing. Mol Cell Proteomics 23:100849
9. Yilmaz M, Fondrie WE, Bittremieux W et al. (2024) Sequence-to-sequence translation from mass spectra to peptides with a transformer model. Nat Commun 15: 6427
10. Eloff K, Kalogeropoulos K, Mabona A et al. (2025) InstaNovo enables diffusion-powered de novo peptide sequencing in large-scale

proteomics experiments. Nat Mach Intell 7:565–579
11. Tran NH, Zhang X, Xin L et al. (2017) De novo peptide sequencing by deep learning. Proc Natl Acad Sci USA 114:8247–8252
12. Mao Z, Zhang R, Xin L, Li M (2023) Mitigating the missing-fragmentation problem in de novo peptide sequencing with a two-stage graph-based deep learning model. Nat Mach Intell 5:1250–1260
13. Lee S, Kim H (2024) Bidirectional de novo peptide sequencing using a transformer model. PLoS Comput Biol 20:e1011892
14. Klaproth-Andrade D, Hingerl J, Bruns Y et al. (2024) Deep learning-driven fragment ion series classification enables highly precise and sensitive de novo peptide sequencing. Nat Commun 15: 151
15. Seo J, Choi S, Paek E (2025) NovoRank: refinement for de novo peptide sequencing based on spectral clustering and deep learning. J Proteome Res 24:903–910
16. Choi S, Paek E (2024) pXg: Comprehensive identification of noncanonical MHC-I-associated peptides from de novo peptide sequencing using RNA-Seq reads. Mol Cell Proteomics 23:100743
17. Erhard F, Dölken L, Schilling B, Schlosser A (2020) Identification of the cryptic HLA-I immunopeptidome. Cancer Immunol Res 8:1018–1026
18. Elias JE, Gygi SP (2007) Target-decoy search strategy for increased confidence in large-scale protein identifications by mass spectrometry. Nat Methods 4:207–214
19. Dobin A, Davis CA, Schlesinger F et al. (2013) STAR: ultrafast universal RNA-seq aligner. Bioinformatics 29:15–21
20. Ewels PA, Peltzer A, Fillinger S et al. (2020) The nf-core framework for community-curated bioinformatics pipelines. Nat Biotechnol 38:276–278
21. Robinson JT, Thorvaldsdóttir H, Winckler W et al. (2011) Integrative genomics viewer. Nat Biotechnol 29:24–26
22. Cuevas MVR, Hardy M-P, Larouche J-D et al. (2023) BamQuery: a proteogenomic tool to explore the immunopeptidome and prioritize actionable tumor antigens. Genome Biol 24: 188
23. Yi X, Zhao H, Hu S et al. (2024) Tumor-associated antigen prediction using a single-sample gene expression state inference algorithm. Cell Rep Methods 4:100906
24. Kina E, Laverdure J-P, Durette C et al (2024) Breast cancer immunopeptidomes contain numerous shared tumor antigens. J Clin Invest 134. https://doi.org/10.1172/JCI166740
25. GTEx Consortium (2013) The Genotype-Tissue Expression (GTEx) project. Nat Genet 45:580–585
26. Uhlén M, Fagerberg L, Hallström BM et al. (2015) Proteomics. Tissue-based map of the human proteome. Science 347:1260419
27. Yang KL, Yu F, Teo GC et al. (2023) MSBooster: improving peptide identification rates using deep learning-based features. Nat Commun 14: 4539
28. Li K, Jain A, Malovannaya A et al. (2020) DeepRescore: leveraging deep learning to improve peptide identification in immunopeptidomics. Proteomics 20:e1900334
29. Declercq A, Bouwmeester R, Hirschler A et al. (2022) MS2Rescore: Data-driven rescoring dramatically boosts immunopeptide identification rates. Mol Cell Proteomics 21:100266
30. Buur LM, Declercq A, Strobl M et al. (2024) MS2Rescore 3.0 is a modular, flexible, and user-friendly platform to boost peptide identifications, as showcased with MS Amanda 3.0. J Proteome Res 23:3200–3207

SUBJECT INDEX

Frédérique Lisacek (ed.), *Bioinformatics for Mass Spectrometry*, Methods in Molecular Biology, vol. 3071,
https://doi.org/10.1007/978-1-0716-5519-1,

K

L

M

N

O

P

Q

R

S

T

U

V

W

Z

www.ingramcontent.com/pod-product-compliance
Ingram Content Group UK Ltd.
Pitfield, Milton Keynes, MK11 3LW, UK
UKHW050632231026
476911UK00005B/118
* 9 7 8 1 0 7 1 6 5 5 1 8 4 *